# FISH & ENVIRONMENT

# FISH & ENVIRONMENT

*Editor*

**Tobin Danielson**

scitus
academics

**Fish & Environment**

Edited by **Tobin Danielson**

This Book is Printed on Acid-Free Paper. ♾



ISBN: 9781681179476

Library of Congress Control Number: 2017949089



SCITUS Academics LLC,

616, Corporate Way, Suite 2, 4766,

Valley Cottage, NY 10989

www.scitusacademics.com

**Notice**



Printed in United States of America.

# Preface

The first impact aquaculture has is on the surrounding environment the specific operation exists in, with a primary concern being pollution. A successful fish farm, for example, requires fish meal to sustain its stock. As with any other agricultural practices, the consumption of food by the stock results in waste products. Because the population of a species in a fish farm is significantly more dense than what would occur naturally, the waste products generated are more concentrated as well. For a fish farm, if it were left unchecked, the living conditions for the stock would soon become too toxic for survival. And if the possibility exists for those water products to leave the pond or culture tank, surrounding life forms will be negatively impacted too. For cages or nets at sea, those waste products are able to contaminate the surrounding marine areas. is the extension of fishing, resulted from the fact that harvests of wild sources of fish and other aquatic species cannot keep up with the increased demand of a growing human population. Expansion of aquaculture can result with less care for the environment. The first pre-requisite to sustainable aquaculture is clean wate, but bad management of aquatic species production can alter or even destroy existing wild habitat, increase local pollution levels or negatively impact local species. Poorly managed fish farms can also be a haven for disease, such as fish lice. While the lice are detrimental to the health and quality of the fish, other diseases can actually have an impact on the fish farmers and local inhabitants as well. With an expanding population, changing environmental conditions and advancing fishing technologies, there is an increasing challenge to ensure our fish resources continue to be harvested sustainably. This involves taking an integrated approach taking into account a wide range of social, economic and environmental factors in determining how we manage our fish resources.

This volume 'Fish & Environment' covers original studies on the ecology, life history, epigenetics, behavior, physiology, morphology, systematics and evolution of marine and freshwater fishes that deal with the relationship between fishes and their external and internal environment, whether natural or unnatural. The book presents recent research results on the interaction between aquaculture and environment, and includes several case studies all over the world with the aim of improving and performing sustainable aquaculture. A wide variety of students, scientists, researchers and other will benefit from this book.

# Table of Contents

# CHAPTER 1

# Environmental Factors Influencing Fish Species Distribution in Irrigation Channels around Ariake Sea, Kyushu, Japan

**Greshishchev V*, Onikura N and Iyooka H**

*Fishery Research Laboratory, Kyushu University, Tsuyazaki, Fukutsu, Fukuoka, Japan*

## ABSTRACT

**Freshwater ecosystems** have suffered a long history of **anthropogenic** disturbances that are responsible for declining fish populations. Fish populations in irrigation channels, in particular, still remain unprotected and relatively unstudied. This study looked to investigate what environmental variables and anthropogenic disturbances had the greatest influence on fish species distribution in irrigation channels. Fish sampling and environmental data collection was conducted throughout six municipalities, spanning 43 sites, in southern Fukuoka, Kyushu, Japan. A total of 37 fish species and 17 environmental variables were statistically analyzed using detrended correspondence analysis (DCA) and receiver operating characteristic (ROC) curve analysis. DCA results showed that channel width, channel substrate and submerged plants had a statistical correlation with fish species distribution. We believe proper management of **irrigation channel** substrate conditions and macrophyte cover will go a long way in supporting fish species biodiversity and reducing pressure from invasive species. ROC curve analysis revealed a disparity in the environmental preferences of endangered and invasive species. This result suggests that the distribution of endangered species in irrigation channels may be heavily influenced by invasive species. Further research must be done before drafting an effective management solution

**Keywords**: Paddy field; Threatened fish species; Water quality; Freshwater fish; Irrigation ditch

## INTRODUCTION

The start of Japanese **wet-rice cultivation** corresponds to the beginning of the Yayoi period 2300-2800 years ago [1]. Archaeological evidence suggests that some of the oldest rice paddies in the Japanese archipelago are those that have been found in northern Kyushu [2]. During this period land was plentiful and cultivation took place along natural flood plains. There was little need for large scale human induced **irrigation** of **paddy fields**. Naturally, as the population increased the demand for more land and cultivation increased along with it. Wet-rice cultivation would need to take place away from the natural flood plains in order to meet the demands of the rising population. It was not until the Heian period (794-1159) that natural flood plains were dug out as permanent channels and the rerouting of streams and rivers for the irrigation of paddy fields took place [3]. It was of little concern at that time how the rerouting and construction of **irrigation channels** would impact the local freshwater fauna.

Japanese rivers differ from most continental rivers because they are short (max. length: 367 km), steep (average slope: 0.44%), and exhibit very flashy flow regimes [4]. Japanese rivers and streams are host to a rich freshwater fish fauna. There are a total of 321 sub-species, 18 orders, 53 families, and 145 genera [5]. Of these species 88 are endemic to Japan and only 11% of the fish are introduced species [6]. In Kyushu the average number of species per major catchment is 43 [7]. However, the number of those species found in irrigation channels is much lower due to the harsher environmental conditions and lentic nature of the water.

Freshwater ecosystems have suffered a long history of **anthropogenic** disturbances that are responsible for declining fish populations [8]. Some of the prevailing anthropogenic pressures that threaten **biodiversity** include habitat destruction and fragmentation, eutrophication and chemical pollution, invasive alien species, overexploitation and climate change [9]. Up until the 1970's environmental issues were mostly ignored in Japan. Most people used their rivers for waste disposal due to their steep nature. Fortunately, during the past decade water quality has significantly improved, fostered by the enforcement of the Environmental Pollution Prevention Act in 1970 [7]. However, irrigation channels have not encountered the same pressure for rehabilitation. Irrigation channels are fed into by industrial wastewater, agricultural runoff, surface runoff and human sewage that often undergo limited to no treatment. These compounded anthropogenic pressures have drastic effects on the water quality of the irrigation channels.

Due to previous studies largely focusing on fish populations in **riverine environment**s, irrigation channel fish populations have remained relatively unstudied. To increase our understanding of how natural and anthropogenic disturbances shape aquatic communities and individual performance, it is crucial to identify the most vulnerable species and the cause of decline [10]. Scientific identification of the factors threatening biodiversity and understanding of the relative importance of such factors are prerequisite to drafting effective plans [11]. In this study we looked to identify what environmental variables and anthropogenic disturbances had the greatest influence on fish species distribution in irrigation channels and assessed what species may be at risk.

# MATERIALS AND METHODS

## Study area

The study was conducted in the irrigation channels of southwestern Fukuoka Prefecture, Kyushu, Japan (**Figure 1**). The water in these channels originates from the mountains in the east. As water flows downstream it is artificially split into multiple channels and is dispersed throughout the city. These channels serve to transport water and to irrigate paddy fields. Water is pumped up from the channel and is used to irrigate the nearby fields. The unused water is then returned to the channel, subsequently altering the chemical composition of the water. As this process is repeated several times traveling downstream, the **chemical composition** of the water downstream becomes drastically different from that of upstream. The irrigation channels are also fed into by industrial wastewater, agricultural runoff, surface runoff and human sewage that undergo varying stages of treatment depending on the municipality. A total of 43 sites spanning six cities were selected. The Study areas were chosen due to the varying levels of sewage treatment that takes place across the six municipalities (**Figure 1**). The cities do not have access to large **sewage** treatment plants and must rely on on-site treatment units. Roughly seven survey sites per city were chosen. Potential survey sites were selected based on size, connectivity, location and estimated species abundance (high and low). Survey site and sewage treatment maps (**Figure 1**) were created using QGIS ver.2.4.0 [12].

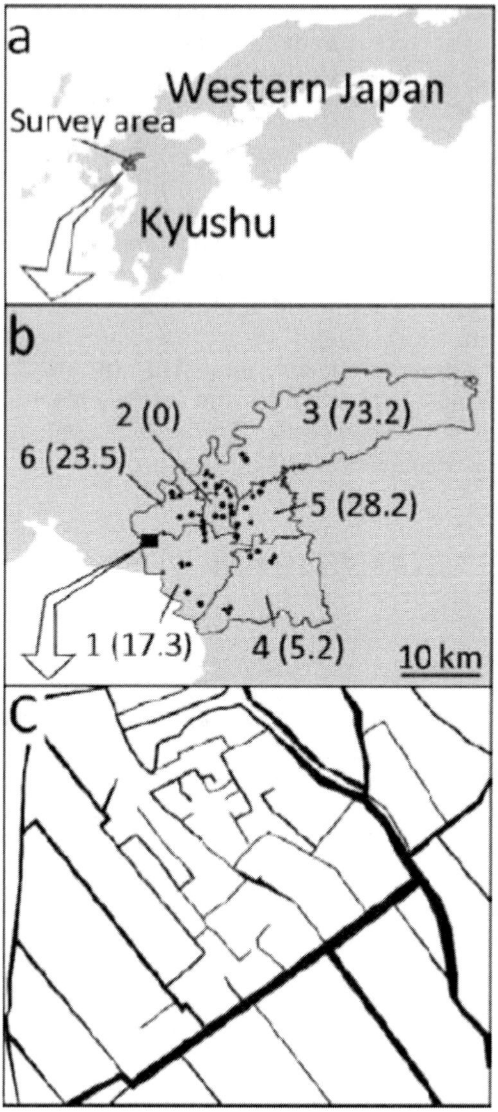


**Figure 1:** Map of survery area located in northern Kyushu Island in western Japan (a), a total of 43 sampling sites across six municipalities (1: Yanagawa City; 2: Oki City; 3: Kurume City; 4: Miyama City; 5: Chikugo City; 6: Okawa City) with the sewage treatment percentage of each municipality writen in parenthesis (b). Aerial view of Irrigation channel network in select area (c).

## Environmental data collection

On site measurement of pH, electric current (EC), current velocity and dissolved oxygen (DO) was conducted throughout multiple days in the months of June and November, 2014. The measurement of pH was done using a pHTestr 30 (Eutech/Oakton instruments, Vernon Hills, IL USA). EC was measured using an ECTestr (Eutech/Oakton instruments, Vernon Hills, IL USA). Current velocity was measured using a VR-301 propeller-type current meter (KENEK Corporation, Tokyo, Japan). DO was measured using a DO-5509 **dissolved oxygen** meter (Lutron Electronic Enterprise, Taipei, Taiwan).

Channel depth and width were measured on-site using a 2 m measuring rod. Each site was divided into ten evenly spaced segments where the depth and width of each segment was measured and recorded.The average depth and width across the ten segments was then used to represent the site as a whole. Concrete revetment and aquatic plant coverage was calculated in a similar fashion. Each site was once again divided into ten evenly spaced segments. In each segment the presence of concrete in the bed and bank was noted. Through the summation of all segments, we were able to calculate the total concrete revetment for that specific site. For aquatic plant coverage, the presence of emergent, floating and submerged plants was recorded for each segment. Again, by totalling each category we were able to calculate the percentage of **aquatic plant** coverage for each site (**Figure 2**).

Water and soil samples were collected at each site and stored for future analysis. The water samples were tested for suspended solids (SS), chemical oxygen demand (COD), total nitrogen (TN) and total phosphate (TP) in accordance to the manuals of the Japan Society for Analytical Chemistry [13]. The soil samples were analyzed through loss on ignition testing in accordance to the standards of the Japanese Geotechnical Society [14].

## Fish sampling and data analysis

Fish sampling was done concurrently with environmental data collection. Sampling was carried out by a team of four researchers equipped with casting nets and hand nets. A minimum of 30-60 minutes was allotted for sample collection and species identification at each site. As irrigation channels differed greatly in size, the allotted time for each site was properly adjusted. Fish fauna identification was accomplished in accordance to Nakabo [15]. After identification, a picture of the fauna was taken and presence/absence data was recorded. All fish captured were released back into the channel.

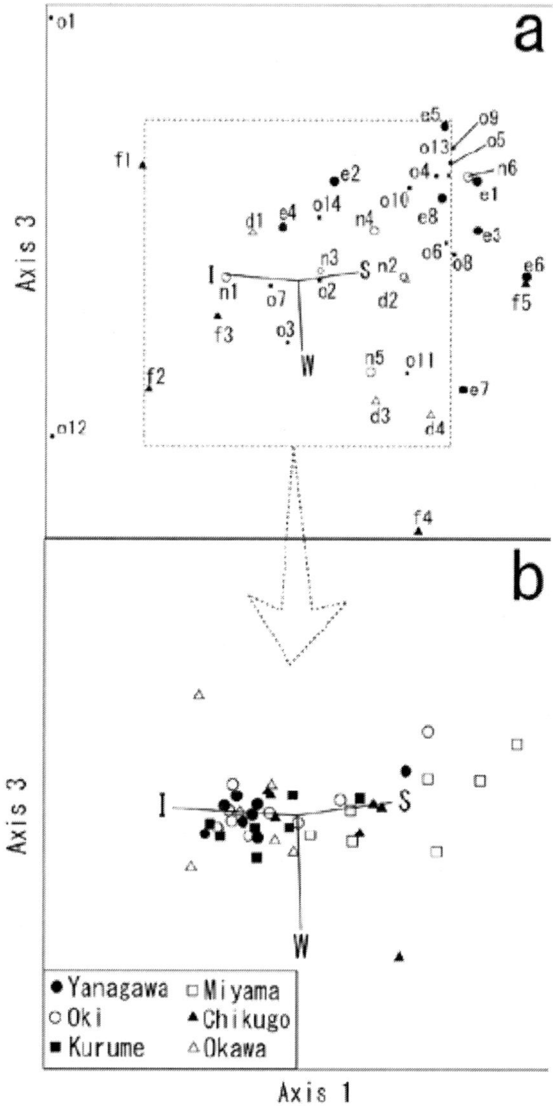


**Figure 2:** Detrended correspondence analysis ordination between axes 1 and 3. (a) Plot of species distributions where solid and open circles represent endangered and near threatened species respectively, solid and open triangles represent foreign and domestic alien species, small solid squares indicate other species. Species IDs are shown near each plot and are referenced in Table 2. (b) Survey site distribution. Environmental variables (I: ignition loss; S: submerged plants; W: width) correlated strongly with axis 1 and/or axis 3 and are indicated as straight lines.

We used detrended correspondence analysis (DCA) to analyze the relationship between 17 environmental variables and species presence/absence distribution of 37 species across 43 sites. The Box-Cox power transformation [16] was applied to improve distribution normality prior to DCA. Statistical analysis software PC-ORD ver. 5.31 (MjM software, Gleneden Beach, Oregon, U.S.A.) was used to perform DCA. A receiver operating characteristic (ROC) curve was used to further analyze the relationship between prominent **environmental variable**s and the presence/absence of each species by using EXCEL add-in software (EXCEL-TOUKEI 2012, SSRI, Shinjuku, Tokyo, Japan). The presence/absence was set to 1/0 in ROC analysis for confirming positive correlation with each environmental variable and 0/1 for confirming negative correlation. When an area under the ROC curve (AUC) was >0.7 it was evaluated as a high or moderate correlation [17]. In addition, ROC curves were also used to determine the optimal cutoff point (COP) for each environmental variable [17].

# RESULTS

## Environmental variables

The average, standard deviation, minimum and maximum values of all measured environmental variables across 43 sites were calculated and recorded in **Table 1**. Nearly all environmental variables, aside from pH, had high variability. This is evident from the large standard deviation values which were often greater or equal to the average. Although pH only had a standard deviation of 0.78, the logarithmic nature of the measurement exacerbates this result. However, the minmax range of pH only varied by a factor of 3 and never exceeded the critical level for fish survival. The wide range between min and max of the variables shows that environmental conditions varied drastically among channels.

## Fish species

The average, min and max species richness across the 43 sites was recorded in **Table 1**. The highest species richness was 17, observed in Miyama City. The lowest species richness was 3, observed in Chikugo City. The average number of species observed per site was 9.67.A total of 37 different species were observed throughout the 43 sites (**Table 2**). Of the 37 species observed eight species were listed on the local red data book as endangered

**Table 1:** The average (Avg), standard deviation (SD), minimum (Min) and maximum (Max) of all measured environmental variables.

| Variable | Avg ± SD | Min–Max |
|---|---|---|
| pH | 7.77 ± 0.78 | 6.58-9.60 |
| EC (μS/cm) | 561.21 ± 236.68 | 3.07-1254 |
| Velocity | 5.38 ± 13.55 | 0-61 |
| Oxygen | 6.92 ± 3.02 | 2.7-22.9 |
| SS (mg/L) | 31.52 ± 32.85 | 1.29-157.00 |
| COD-Mn (mg/l) | 6.90 ± 2.46 | 2.08-10.93 |
| TN (mg/) | 2.66 ± 0.95 | 1.03-6.45 |
| TP (mg-P/l) | 0.31 ± 0.30- | 0.03-1.58 |
| Ignition Loss | 9.41 ± 4.31 | 2.23-20.18 |
| Depth (cm) | 90.40 ± 42.98 | 18.5-250 |
| Width (cm) | 443.50 ± 298.96 | 26.70-1325 |
| Emergent (%) | 23.00 ± 29.01 | 0-100 |
| Floating (%) | 11.52 ± 19.93 | 0-80 |
| Submerged (%) | 7.91 ± 17.12 | 0-70 |
| Conc. Bank (%) | 80.42 ± 26.92 | 0-100 |
| Conc. Bed (%) | 21.63 ± 30 | 0-100 |
| Sp. Richness | 9.67 ± 2.95 | 3-17 |

species [18], four species were domestic aliens and four were foreign invaders. The eight **endangered** and six near threatened freshwater fish species were classified as follows: Critically Endangered: *Acheilognathus tabira* ssp. 2 (Species ID: e1), *Hemigrammocypris rasborella* (e2); Endangered: *Cobitis kaibarai* (e3), *Rhodeus ocellatus kurumeus* (e4), *Rhodeus atremius atremius* (e5); Vulnerable: *Misgurnus anguillicaudatus* (e6), *Sarcocheilichthys variegatus variegatus* (e7), *Tanakia lanceolata* (e8); and Near Threatened: *Abbottina rivularis* (n1), *Acheilognathus rhombeus* (n2), *Biwia zezera* (n3), *Nipponocypris sieboldii* (n4), *Oryzias latipes* (n5), *Tanakia limbata* (n6). The domestic alien species were *Carassius cuvieri* (d1), *Gnathopogon elongatus* (d2), *Opsariichthys uncirostris* (d3) and *Squalidus chankaensis tsuchigae* (d4). The foreign invader species were *Channa argus* (f1), *Gambusiaaffinis* (f2), *Lepomis macrochirus* (f3), *Micropterus salmoides* (f4) and *Paramisgurnus dabryanus* (f5). The remaining species were designated as other and were not listed on the red list. These species include freshwater fish (o1: Carassius auratus beurgeri, o2: C. a. *langsdorfi*, 03: *Cyprinus carpio*, o4: *Hemibarbus barbus*, o5: *Odontobutis obscura*, o6: *Pseudogobio esocinus*, o7: *Pseudorasboraparva*, o8: *Pungtungia herzi*, o9: *Silurus asotus*, o10: *Squalidus gracilis gracilis*, o11: *Zacco platypus*), brackish water fish o12: *Mugil cephalus cephalus*, o13: *Tridentiger brevispinis* and diadromous fish o14: *Rhinogobius* sp.

**Table 2:** Total number of sightings and sighting rankings for all observed species across 43 sites.

| Species ID | Scientific Name | Num. of Sightings | Total sighting rank |
|---|---|---|---|
| **Critically Endangered** | | | |
| e1 | *Acheilognathus tabira* ssp. | 4 | 22 |
| e2 | *Hemigrammocypris rasborella* | 2 | 29 |
| **Endangered** | | | |
| e3 | *Cobitis kaibarai* | 2 | 24 |
| e4 | *Rhodeus ocellatus kurumeus* | 37 | 4 |
| e5 | *Rhodeus atremius atremius* | 1 | 34 |
| **Vulnerable** | | | |
| e6 | *Misgurnus anguillicaudatus* | 1 | 30 |
| e7 | *Sarcocheilichthys variegatus variegatus* | 2 | 25 |
| e8 | *Tanakia lanceolata* | 7 | 16 |
| **Near Threatened** | | | |
| n1 | *Abbottina rivularis* | 25 | 5 |
| n2 | *Acheilognathus rhombeus* | 1 | 35 |
| n3 | *Biwia zezera* | 14 | 13 |
| n4 | *Nipponocypris sieboldii* | 10 | 14 |
| n5 | *Oryzias latipes* | 20 | 9 |
| n6 | *Tanakia limbata* | 6 | 17 |
| **Domestic Alien** | | | |
| d1 | Carassius *cuvieri* | 38 | 3 |
| d2 | *Gnathopogon elongatus* | 1 | 36 |
| d3 | *Opsariichthys uncirostris* | 17 | 11 |

| | | | |
|---|---|---|---|
| d4 | *Squalidus chankaensis tsuchigae* | 6 | 20 |
| **Foreign Invader** | | | |
| f1 | *Channa argus* | 3 | 23 |
| f2 | *Gambusia affinis* | 21 | 8 |
| f3 | *Lepomis macrochirus* | 24 | 6 |
| f4 | *Micropterus salmoides* | 1 | 37 |
| f5 | *Paramisgurnus dabryanus* | 1 | 31 |
| **Other (Native Species)** | | | |
| o1 | *Carassius auratus beurgeri* | 1 | 33 |
| o2 | *Carassius auratus langsdorfi* | 41 | 1 |
| o3 | *Cyprinus carpio* | 18 | 10 |
| o4 | *Hemibarbus barbus* | 2 | 27 |
| o5 | *Odontobutis obscura* | 6 | 18 |
| o6 | *Pseudogobio esocinus* | 6 | 19 |
| o7 | *Pseudorasbora parva* | 41 | 2 |
| o8 | *Pungtungia herzi* | 2 | 26 |
| o9 | *Silurus asotus* | 5 | 21 |
| o10 | *Squalidus gracilis gracilis* | 9 | 15 |
| o11 | *Zacco platypus* | 15 | 12 |
| o12 | *Mugil cephalus cephalus* | 2 | 28 |
| o13 | *Tridentiger brevispinis* | 1 | 32 |
| o14 | *Rhinogobius* sp. | 22 | 7 |

The total number of sightings for each individual species is recorded in **Table 2**. The two most frequently observed species were *C. a. langsdorfi* and *P. parva*. Both were found in 41 out of 43 sites. There were eight species that were observed only once throughout the 43 sites. These species were A. rhombeus, *C. a. beurgeri, G. elongates, M. anguillicaudatus, M. salmoides, P. dabryanus, R. a. attremius* and *T. brevispinis.*

## Relationship between fish fauna and environmental variables

The DCA created 3 axes that explained approximately 60% of the ordination of **fish fauna** in our survey area. The first axis correlated positively with submerged plants and negatively with ignition loss while the third axis correlated negatively with width of channel. Most threatened and near threatened species were plotted on right side, suggesting that these species prefer sites with high submerged plants and low ignition loss. Whilst three foreign invaders were plotted on the left side, suggesting that they are able to adapt to poor sediment quality. Although most sites were plotted around the center of this ordination, all sites of Miyama City were plotted on the right side.

## Relationship between fish distribution and correlated environmental variables

Of the 37 species analyzed, ten had a positive and four had a negative correlation with submerged plants. Six species had positive while another six had negative correlations with width of channel. Fourteen species had a positive while twelve had a negative correlation with ignition loss (**Table 3**). Whilst six endangered species had a positive correlation with submerged plants, three domestic and foreign invader species had a negative correlation with submerged plants. Three endangered species had a negative correlation with channel width where as two invader species had a positive correlation. Five endangered species had a negative correlation with ignition loss while four invader species were positively correlated with ignition loss.

**Table 3:** Receiver operating characteristic (ROC) curve analysis results comparing the three prominent environmental variables with species presence/absence and absence/presence. Positive and negative correlations are noted as P and N respectively. Area under the curve (AUC) values below 0.7 were determined as having a weak correlation and consequently have had their cut-off point (COP) values omitted. Species ID and scientific names are shown in Table 2.

| Species ID | Submerged | | | Width | | | Ignition Loss | | |
|---|---|---|---|---|---|---|---|---|---|
| | P/N | AUC | COP | P/N | AUC | COP | P/N | AUC | COP |
| e1 | P | 0.79 | 10 | | 0.36 | N/A | N | 0.92 | 6.08 |
| e2 | P | 0.93 | 20 | | 0.48 | N/A | | 0.43 | N/A |
| e3 | P | 0.97 | 40 | N | 0.79 | 204.5 | N | 0.85 | 6.08 |

| | | | | | | | | | |
|---|---|---|---|---|---|---|---|---|---|
| e4 | | 0.34 | N/A | N | 0.71 | 569 | | 0.68 | N/A |
| e5 | P | 0.80 | 10 | | 0.33 | N/A | | 0.31 | N/A |
| e6 | P | 0.92 | 40 | N | 0.81 | 148 | N | 0.76 | 6.08 |
| e7 | P | 0.95 | 40 | | 0.55 | N/A | N | 0.91 | 4.67 |
| e8 | | 0.69 | N/A | | 0.34 | N/A | N | 0.74 | 8.62 |
| n1 | | 0.46 | N/A | | 0.33 | N/A | | 0.56 | N/A |
| n2 | | 0.38 | N/A | P | 0.90 | 753 | N | 0.90 | 3.25 |
| n3 | | 0.42 | N/A | N | 0.70 | 468 | | 0.35 | N/A |
| n4 | | 0.62 | N/A | | 0.51 | N/A | | 0.32 | N/A |
| n5 | | 0.61 | N/A | | 0.53 | N/A | | 0.46 | N/A |
| n6 | P | 0.84 | 10 | | 0.33 | N/A | N | 0.90 | 8.62 |
| d1 | N | 0.73 | 20 | | 0.39 | N/A | P | 0.79 | 7.52 |
| d2 | | 0.38 | N/A | P | 0.90 | 753 | N | 0.90 | 3.25 |
| d3 | | 0.60 | N/A | | 0.51 | N/A | N | 0.76 | 9.77 |
| d4 | | 0.62 | N/A | | 0.63 | N/A | N | 0.84 | 8.62 |
| f1 | | 0.58 | N/A | | 0.58 | N/A | P | 0.87 | 11.61 |
| f2 | N | 0.73 | 0 | | 0.45 | N/A | P | 0.65 | N/A |
| f3 | N | 0.72 | 0 | | 0.45 | N/A | P | 0.57 | N/A |
| f4 | | 0.38 | N/A | P | 1 | 1325 | | 0 | N/A |
| f5 | P | 0.92 | 40 | N | 0.81 | 148 | N | 0.76 | 6.08 |
| o1 | | 0.38 | N/A | | 0.62 | N/A | P | 0.90 | 13.81 |
| o2 | | 0.62 | N/A | P | 0.87 | 244 | | 0.49 | N/A |
| o3 | | 0.58 | N/A | | 0.50 | N/A | | 0.65 | N/A |
| o4 | | 0.59 | N/A | P | 0.71 | 434.5 | N | 0.94 | 3.25 |
| o5 | | 0.64 | N/A | | 0.33 | N/A | N | 0.73 | 8.62 |
| o6 | | 0.66 | N/A | | 0.51 | N/A | N | 0.96 | 4.67 |
| o7 | N | 0.95 | 30 | | 0.51 | N/A | | 0.88 | 7.52 |
| o8 | P | 0.87 | 10 | | 0.40 | N/A | N | 0.90 | 4.67 |
| o9 | | 0.58 | N/A | N | 0.82 | 365 | | 0.36 | N/A |
| o10 | | 0.64 | N/A | | 0.39 | N/A | N | 0.72 | 8.62 |
| o11 | | 0.63 | N/A | | 0.54 | N/A | N | 0.91 | 8.62 |
| o12 | | 0.38 | N/A | P | 0.93 | 715 | | 0.61 | N/A |
| o13 | P | 0.80 | 10 | | 0.50 | N/A | N | 0.95 | 2.46 |
| o14 | | 0.49 | N/A | | 0.58 | N/A | | 0.48 | N/A |

# DISCUSSION

## Water and soil quality in Japanese irrigation channels

Japanese rivers and lakes are subject to strict water quality guidelines that outline the acceptable levels for both fish and human use [19]. However, Japanese irrigation channels have not faced the same regulation pressures and to date have not received any strict guidelines regarding acceptable water quality levels for fish populations [19]. There are, however, lesser unenforced guidelines for water quality that outline potential goals. The only form of regulation in regards to water quality in irrigation channels deals with the maintenance of non-toxic levels to human health. Results of our environmental variable analysis (**Table 1**) show that there is a very large range of variability in nearly all measured variables. This means that the water and soil quality of an irrigation channel differs drastically depending on the location of the channel itself and where along the channel the measurement was taken.

Loss on **ignition** is a method used to estimate the organic matter composition of soil as well as the substrate of marine and freshwater ecosystems [20]. In this study loss on ignition was used to assess channel substrate and indirectly the quality of water. The lowest recorded pH out of all the sites was 6.58 recorded in Chikugo City. Uncoincidentally, this site was also recorded as having the highest loss on ignition recorded at 20.18%. However, despite the poor water quality and substrate conditions, this site was recorded as having average species richness [9]. This observation hints that there are more factors at play that govern the distribution of fish species in irrigation channels than simply the overall quality of water and substrate. It appears that sewage treatment coverage had little to do with the overall species richness of a municipality. In fact, the top three cities with the greatest sewage treatment coverage had lower species richness than the bottom three treated cities. Some factors that were more important in determining the overall species richness include survey site location and complexity of the irrigation network. Unfortunately, it is difficult to truly gauge the impacts of anthropogenic pressures on fish species as they might be masked or modified by natural factors that govern the performance of individuals and the structure of aquatic communities [21,22].

The quality of channel substrate was directly correlated to current velocity. Sites with a higher current velocity yielded a much lower loss on ignition. It is also important to note that sites with the highest species richness were all lotic environments. This observation can be explained by the proximity of the irrigation channel to the nearest river or stream. Areas of the irrigation channel that are directly fed into by a river or stream possess a current. These areas act as transitional zones between a river and the agricultural channel. Transitional zones have a higher species richness because fish encompassing all lifestyle preferences (both lentic and lotic) must pass through in order to traverse the channel [23]. Another explanation for this observation is that when water in an irrigation channel branches and travels further downstream, the effects of sedimentation become stronger. Areas upstream that possess a current have a mechanism for cycling nutrients and waste disposal while lotic areas further downstream act as deposition sites and subsequently experience lower water and soil quality.

## Relationships among fish fauna/each species and environmental variables

Results of the detrended correspondence analysis produced three statistically correlated variables. These variables were channel width, channel substrate (Loss on Ignition) and submerged plants. Although most sites were plotted around the center of this ordination, all sites of Miyama City were plotted on the right side. This is an interesting observation because Miyama City was also noted as having the highest species richness. The reason for this has to do with the complexity of the irrigation channel network and the quality of water. Sites of cities plotted around the center of the ordination have heavily branched networks. Although these cities also possess sites with moving water, they are heavily dominated by areas of still water. The abundance of lentic waters reduces the overall water and soil quality and consequently species richness of these cities. Miyama City on the other hand possesses a rather simple irrigation channel network. This simplicity means far less branches on the network which leads to a stronger overall current velocity present throughout the channel. This characteristic has allowed Miyama City to maintain higher water and soil quality levels as well as overall higher species richness.

Previous studies [24,25] have documented that species diversity and stream fish assemblages are affected by wetted width. In this study, channel width had a statistical correlation with species distribution. However, this does not mean that increasing the width of a channel will directly increase fish species diversity. This result supports the findings of Yan et al. [24] and suggests that channel width has an influence on fish species assemblages at the species level.

The quality of channel substrate plays a crucial role in the selection of spawning sites for many fish species found in the channel. Five species observed within the channel have been previously found to directly rely on the substrate as a life history trait and for spawning purposes. These species are *Z. platypus, P. esocinus, S. v. variegatus, N. sieboldii* and *A. rivularis* [26]. Six species of bitterling found in the channel have been found to indirectly rely on the quality of substrate. These species are *R. o. kurumeus, T. limbata, A. tabira ssp. 2, T. lanceolata, R. a. atremius* and *A. rhombeus* [26]. Bitterling are parasitoids that lay their eggs into live bivalves where their embryos develop inside the gills of the bivalve [27]. For this reason, bitterling species are highly dependent on the presence of bivalves in their environment. As substrate quality decreases the presence of bivalves decreases as well. Naturally, as bivalve presence declines, the population size of bitterling species in that environment will decline as well [28].

ROC curve analysis results from this study support the findings of Nakajima et al. and Onikura et al. [26-28]. In this study 7 out of the 11 species previously found to directly and indirectly rely on channel substrate had a very strong negative correlation (AUC>0.90) with Loss on Ignition (**Table 3**). The remaining four species (*N. sieboldii, A. rivularis, R. o. kurumeus and R. a. atremius*) did not have a significant statistical correlation. However, the lack of correlation from these species can be justified. The lack of correlation in *R. a. atremius* may be explained by its very small sample size. Although *R. o. kurumeus* did have a significant sample size it was only marginally off (AUC=0.68) from having a significant correlation.On the other hand, both *N. sieboldii* and *A. rivularis*had meaningful sample sizes but showed no statistical correlation. The reason for this is that both species rely on the substratum for spawning purposes. However, sample collection was conducted during feeding and spawning season which is outside of spawning season. During this time they do not rely on the substratum. If sample collection were to be done during spawning season there would likely be a stronger measurable correlation.

The presence or absence of **macrophytes** can drastically alter fish species diversity and community dynamics of freshwater environments. Macrophytes in lakes and rivers provide cover for fish and macro invertebrates, alter substrate composition, produce oxygen and act as food [29]. In this study the presence of submerged macrophytes was one of the three environmental variables that showed a statistical correlation with fish species distribution. Submerged macrophytes improve the structural heterogeneity of microhabitats in aquatic ecosystems, often providing an important habitat for zooplankton [30]. Submerged macrophytes are utilized by fish for the nursing of juveniles, feeding, and predator avoidance [31]. However, excessive development of free-floating macrophytes on the water surface can reduce the biomass of submerged macrophytes and results in a relatively simple habitat structure [30]. Poor water quality arising from floating weed mats is considered to be the main determinant of reduced fish abundance and diversity [32].

In this study, Of the 37 fish species surveyed, ten had a positive correlation and four had a negative correlation with the presence of submerged macrophytes (**Table 3**). An important observation to note is that three of the four negatively correlated species were domestic and foreign aliens. This shows that irrigation channels are still foreign environments for these species and that they generally prefer environmental similar in conditions to that of their native range of rivers and lakes. The other negatively correlated species was *P. parva.* This species has been noted as having both high life history plasticity and wide-scale habitat use [27]. It was noted by Onikura and Nakajima [33] that structural revetment conditions and macrophyte cover were unimportant for this species. However, the results of this study suggests that when presented with multiple environmental conditions, *P. parva* has a strong aversion (AUC=0.95) to channels with submerged macrophyte cover. The possible reason for this discrepancy may have to do with the scale of the survey area where said research was conducted. The survey sites chosen by Onikura and Nakajima [33] were much smaller in size and all possessed heavy macrophyte cover. In this scenario *P. parva* was found to have no preference as its options were limited and it was forced to adapt to environments with heavy macrophyte cover. However, the survey sites chosen in this study encompassed sites with varying degrees of macrophyte cover. It was in this type of scenario that *P. parva* was found to have a strong aversion to submerged plants due to having the option of preference.

## Relationship between native and invasive species

The most commonly observed domestic alien species was *C. cuvieri.* It was the third most observed species overall and was observed at 38 out of 43 sites (**Table 2**). *C. cuvieri* originates from Lake Biwa and has been widely distributed throughout Japan as a game fish. *C. cuvieri* has been shown to be fairly hardy and adept at avoiding predation at the juvenile stage [34]. DCA results show that *C. cuvieri* is not particularly averted to environments with poor soil and water quality conditions. ROC curve analysis (**Table 3**) shows that *C. cuvieri* may in fact have a statistical preference (AUC=0.79) to areas with poor soil conditions. This is a unique characteristic shared only by three other invasive species (**Table 2**). Carassius auratus beurgeri was also noted as having a strong positive correlation, however, the small sample size puts the validity of the result into question. We hypothesize that the natural hardiness of these invasive species has allowed them to succeed and even show an affinity to areas with poor substrate conditions. Native species are naturally averted to such conditions and leave the invasive species with little competition.

The most commonly observed foreign invasive species was L. Macrochirus. It was the sixth most observed species overall and was present at 24 out of 43 sites (**Table 2**). *L. Macrochirus,* commonly known as bluegill, is one of the most commonly introduced fish species in the world [35]. It is characterized by an opportunistic behaviour that can lead to aggressive dominance [36]. Studies have shown that the introduction of blue gill into Japanese waters induces a range of changes at multiple trophic levels [37]. In irrigation channels *L. Macrochirus* has been shown to have an allopatric distribution with *O. latipes.* Although no direct correlation between these two species was found in this study. ROC curve analysis suggests that L. Macrochirus has a negative correlation (AUC=0.72) with submerged plants. Two other invasive species (d1, f2) were also found to have a negative correlation with submerged plants. On the other hand, six out of eight endangered species and one threatened species had positive correlations to submerged plants. This observation shows that there is a large disparity between the environmental preferences of endangered and invasive species in terms of submerged plants. We believe a possible explanation for this is that the ability of invasive species to outcompete the endangered species in areas with little macrophyte cover has caused the endangered species to have a strong preference or simply an illusion of preference for environments with heavy macrophyte cover.

Endangered species showed a preference for irrigation channels with a narrow width (**Table 3**). Whereas invasive species tended to prefer wider channels. We believe this can be attributed to invasive species a) showing a natural preference to environmental conditions similar to their native range of rivers and lakes b) outcompeting endangered species in wider channels causing a decline in their numbers in those environments.

## CONCLUSION

Fish species distribution was found to be statistically correlated with channel width, channel substrate and submerged plants. Nearly all domestic and endagered species had a positive correlation with submerged plants. Although these macrophytes, in the eyes of a layperson, may seem like unwanted debris clogging up the channel, they are infact a vital part of a healthy ecosystem. In order to support fish species biodiversity we advise against the total clearing of macrophyte cover in irrigation channels. All domestic and endagered species had a negative correlation with substrate quality while nearly all domestic and foreign invaders had a positive correlation.This result sugests that domestic and endagered species strongly rely on good substrate conditions while invasive species are more adaptive. We believe the management of irrigation channel substrate conditions will go a long way in supporting fish species biodiversity and reducing pressure from invasive species. However, further research must be done before drafting an effective management solution.

## ACKNOWLEDGEMENTS

We wish to thank all of the members of the Aquatic Field Science lab at Kyushu University for helping with data collection. We would like to thank Jasso, the Rotary Yoneyama Foundation and the Ogori Rotary Club for providing financial support that made this research possible. A special thanks to the anonymous reviewers for their valuable comments and contribution.

## REFERENCES

1. ImamuraK (1996) Prehistoric Japan: new perspectives on insular East Asia. UCL Press,London
2. Ishikawa H(1992) ZukaiNihonnoJinruiiseki. Tokyo-daigakuShyuppan, Tokyo.
3. EnvironmentAgency of Japan (1981) The 2nd National Survey on Natural Environment: Rivers.Marketing Intelligence Corporation, Tokyo
4. Kawanabe H, MizunoN (1989). Freshwater Fishes of Japan. Yama-kei Publishers, Tokyo.
5. Taniguchi Y, Inoue M, Kawaguchi Y (2001) Stream fish habitatscience and management in Japan. Aquatic Ecosystem Health and Management 4: 357-365.
6. YoshimuraC, Omura T, Furumai H, Tockner K (2005) Present state of rivers and streams inJapan. River Res Appl 21: 93-112.
7. DudgeonD, Arthington AH, Gessner MO, Kawabata Z, Knowler DJ, et al. (2006) Freshwaterbiodiversity: importance, threats, status and conservation challenges.Biological Reviews 81: 163-182
8. JapanBiodiversity Outlook Science Committee (JBOSC) (2010) Report of comprehensiveAssessment of Biodiversity in Japan
9. MacedaAV, Green AJ, De Sostoa A (2014) Scaled body-mass index shows how habitatquality influences the condition of four fish taxa in north-eastern Spain andprovides a novel indicator of ecosystem health. Freshwater Biology 59:1145-1160.
10. Yoshioka A,Miyazaki Y,Sekizaki Y, Suda S,Kadoya T, et al.(2013)A "lost biodiversity" approach to revealing major anthropogenicthreats to regional freshwater ecosystems.*Ecological Indicators*36:348-355.
11. QGISDevelopment Team (2015) QGIS Geographic Information System. Open SourceGeospatial Foundation Project.
12. Hokkaido Branch of the Japan Society for Analytical Chemistry (2005) Water analysis (5thedn.) KAGAKUDOJIN, Shimogyo, Kyoto,Japan.
13. TheJapanese Geotechnical Society (2008) Loss on ignition test method meeting thecriteria of soi.l
14. NakaboT (2002) Fishes of Japan with pictorial keys to the species, English Edition.Tokai University Press. Shibuya-ku, Tokyo, Japan.

15. BoxGEP, Cox DR (1964)An analysis of transformations. Journal of the RoyalStatistical Society26: 211-252.
16. AkobengAK (2007) Understanding diagnostic tests 3: receiver operating characteristiccurves. ActaPaediatrica 96: 644-647.
17. FukuokaPrefecture (2014) FUKUOKA RED DATA BOOK 2014. Fukuoka Pref, Hakataku, Fukuoka,Japan.
18. Urase T(2011) Clear explanations of water quality and environmental science. PleiadesPublishing House, Azumino, Nagano, Japan.
19. FrangipaneG, Pistolato M, Molinaroli E, Guerzoni S, Tagliapietra D (2008) Comparison ofloss on ignition and thermal analysis stepwise methods for determination ofsedimentary organic matter. Aquatic Conservation 19: 24-33.
20. Gasith A, Resh V (1999) Streams in Mediterranean climateregions: abiotic in?uences and biotic responses to predictable seasonal events.Annual Review of Ecological Systems 30: 51-81.
21. FerreiraT, Caiola N, Casals F, Cortes R, Economou A, et al. (2007) Ecological traits of?sh assemblages from Mediterranean Europe and their responses to humandisturbances. Fisheries Management and Ecology 14: 473-481.
22. Terra BF, dos Santos ABI, Aurajo FG (2010)Fishassemblage in a dammed tropical river: an analysis along the longitudinal andtemporal gradients from river to reservoir. NeotropIchthyol 8: 599-606.
23. YanYZ, Xiang XY, Chu L, Zhan YJ, Fu CZ (2011) Influences of local habitat andstream spatial position on fish assemblages in a dammed watershed, the QingyiStream, China. Ecol Freshwater Fish 20: 199-208.
24. YanYZ, Wang H, Zhu R, Chu L, Chen YF (2012) Influences of low-head dams on thefish assemblages in the headwater streams of the Qingyi watershed, China. EnvironBiol Fish 96: 495-506.
25. HossainMM, Perhar G, Arhonditsis GB, Matsuishi T, Goto, et al. (2013) Examination ofthe effects of largemouth bass (*Micropterussalmoides*)and bluegill (*Lepomismacrochirus*) onthe ecosystem attributes of lake Kawahara-oike, Nagasaki, Japan. EcologicalInformatics 18: 149-161.
26. OnikuraN, Nakajima J, Eguchi K, Miyake T, Kawamura K, et al. (2008) Present Distributions of Exotic Species in Creeks Around Sea of Ariake and Yatsushiro, Northwestern Kyushu, Japan. Journal of Japan Society on Water Environment 31:395-401.
27. Nakajima J, Shimatani Y, Itsukushima R, Onikura N (2010) Assessingriverine environments for biological integrity on the basis of ecologicalfeatures of fish. Advances in River Engineering 16: 449-454.
28. Onikura N, Nakajima J, Miyake T, Kawamura K, Fukuda S (2011) Predicting distributions of seven bitterling ?shes in northern Kyushu, Japan.Icthyol Res 59: 124-133.
29. FlemingJP and Dibble ED (2014) Ecological mechanisms of invasion success in aquaticmacrophytes. Hydrobiologia 746: 23-27.
30. JYChoi, Jeong KS, La GH, Joo KH (2014a) Effect of removal of free-floatingmacrophytes on zooplankton habitat in shallow wetland. Knowledge and Managementof Aquatic Ecosystems 414: 1-11.
31. VritilekM and Reichard M (2012) An indirect effect of biological invasions: the effectof zebra mussel fouling on parastisation of unionid mussels by bitterling fish.Hydrobiologia 696 : 205-214
32. OnikuraN and Nakajima J (2012) Age, growth and habitat use of the topmouthgudgeon, *Pseudorasboraparva* in irrigation ditcheson northwestern Kyushu Island, Japan. Journal of Applied Ichthyology 29:186-192.
33. Choi JY,Jeong KS, Kim SK, La GH, Chang KH, et al. (2014b) Role of macrophytes asmicrohabitats for zooplankton community in lentic freshwater ecosystems ofSouth Korea. Ecological Informatics 24: 177-185.
34. PernaCN, Cappo M, Pusey BJ, Burrows DW, Pearson RG (2012) Removal of aquatic weedsgreatly enhances fish community richness and diversity: an example from

theBurdekin River floodplain, tropical Australia. River Research and Applications28: 1093-1104.

35. TaguchiT, Miura Y, Krueger D, Sugiura S (2014) Dispersal, migration, and predation ofjuvenile crucian carps after release from rice fields. Environ Biol Fish 98:679-690.

36. Lowe S,Browne M, Boudjelas S (2000) 100 of the World's Worst Invasive Alien Species: ASelection from the Global Invasive Species Database. Invasive SpeciesSpecialist Group, Auckland New Zealand.

37. MarchettiMP, Moyle PB, Levine R (2004) Invasive species profiling? Exploring thecharacteristics of non-native fishes across invasion stages in California. Freshwater Biology 49: 646-661.

# CHAPTER 2

# Evaluating Detectability of Freshwater Fish Assemblages in Tropical Streams: Is Hand-Seining Sufficient?

**Amy E. Deacon, Rajindra Mahabir, Devan Inderlall, Indar W. Ramnarine, Anne E. Magurran**

*Centre for Biological Diversity, School of Biology, University of St Andrews, St Andrews, UK*

*Department of Life Sciences, The University of theWest Indies, St Augustine Campus, St Augustine, Trinidad and Tobago*

## ABSTRACT

Unprecedented threats to natural ecosystems mean that accurate quantification of biodiversity is a priority, particularly in the tropics which are underrepresented in monitoring schemes. Data from a freshwater fish assemblage in Trinidad were used to evaluate the effectiveness of hand-seining as a survey method in tropical streams. We uncovered large differences in species detectability when hand-seining was used alone, in comparison with when hand-seining and electrofishing were used together. The addition of electrofishing increased the number of individuals caught threefold, and increased the biomass fivefold. Some species were never detected using hand-seining, resulting in significant underestimates of species richness; rarefaction curves suggest that even when hand-seining effort increases, species richness is still underestimated. Diversity indices (Shannon and Simpson index) reveal that diversity was also significantly lower for hand-seined samples. Furthermore, the results of multivariate analyses investigating assemblage structure also differed significantly depending on whether they were based on hand-seined data alone, or a combination of hand-seining and electrofishing. Despite the extra equipment and maintenance required, these findings underline the value of including electrofishing when sampling tropical freshwater streams.

**Keywords:** Trinidad, Electrofishing, Alpha diversity, Detectability, Species richness, Sampling methods, Gear bias

# INTRODUCTION

Detectability is an important but often overlooked factor in biodiversity assessment and monitoring (Buckland et al. 2011). Depending on the survey methods and the habitat, certain species are more easily sampled than others (Boulinier et al. 1998). This bias is important, as measures of diversity assume that samples are representative of the community sampled (Magurran 2004). Variation in detectability between species can therefore have important implications for interpreting measures of diversity (see Yoccoz et al. 2001).

Issues of detectability are particularly pertinent to freshwater communities. Despite covering just 1% of the Earth's surface, freshwater habitats are estimated to support 6% of all described species – a fact that is not yet reflected in the corresponding amount of research effort (Abell 2002; Dudgeon et al. 2006). Neotropical freshwaters are also one of the most threatened habitats in the world (Balmford et al. 2002; WWF 2014), yet knowledge of their biodiversity is limited. Understanding and monitoring neotropical freshwaters is necessary for their conservation, and effective sampling is needed to address this research gap. Detectability needs to be considered in these investigations (Gotelli and Colwell 2011), especially as different fish species can have markedly different habits, ranging from those that tend to reside in the water column, including those that form shoals, to benthic species that hide in crevices or under stones on the river bed.

In freshwater streams, the most effective way to get around issues of detectability is to use a piscicide such as rotenone, which kills all fish in the treated stretch without biases related to ecology or size (Głowacki and Penczak 2005), yet in doing so destroys the biodiversity the investigator has set out to quantify.

Both hand-seining and electrofishing are commonly-used, non-lethal methods for surveying fish assemblages in shallow streams and rivers (e.g., Maciolek and Timbol 1980; Willis et al. 2005; Gerstner et al. 2006). There are advantages and disadvantages to each, both in terms of detectability and practicalities. Hand-seining uses simple, easy-to-transport equipment and is considered efficient in terms of cost and labour. However, it is less well-suited to streams characterised by obstructions such as rocks and vegetation which make effective hand-seining harder, both by preventing the smooth passage of the net, and by providing places for benthic fish to seek refuge. Conversely, electrofishing can be effective at drawing fish out of cover (Dauble and Gray 1980; Cote and Perrow 2006) and is thought to be less selective overall than many other methods (see Fièvet et al. 1999; Hendricks et al. 1980). Nonetheless, it is not well suited to deep, turbid pools or very low conductivity conditions. There are also safety considerations for the operator, especially when using a backpack, and if using a bank-side generator then there is heavy equipment to transport (Onorato et al. 1998). Furthermore, there are ethical considerations for the fish; there can be mortality if individuals inadvertently sustain prolonged exposure to electric current, albeit at low levels (Snyder 2003; Cote and Perrow 2006).

Both approaches are potentially subject to gear-related size-biases. Electrofishing tends, on average, to catch larger individuals than hand-seining (Dauble and Gray 1980; Wiley and Tsai 1983). Mechanisms may include the differential vulnerability of different-sized individuals to immobilisation by electric current (Dolan and Miranda 2003) as well as behavioural differences (Zalewski 1983; Bohlin et al. 1989).

Many researchers advocate the use of a combination of different non-lethal sampling methods with the aim of achieving a more representative sample of the community than any one method alone (Colwell and Coddington 1995; Onorato et al. 1998; Ebner and Morgan 2013), for example combining electrofishing and seining (Fago 1998). However, an additional consideration is that, especially in the tropics, expensive electrofishing equipment is not always available. This gear can also be more difficult to maintain in hot, humid conditions. Tropical forest streams are often hard to access, and necessitate transporting heavy equipment through difficult terrain (Ebner et al. 2014). This leads us to question whether electrofishing is worth the considerable additional effort in such environments.

Here we examine detectability in a tropical freshwater fish assemblage using hand-seining alone, and hand-seining in combination with electrofishing. We do this in the context of the assessment of diversity and compare the abundance, diversity and assemblage structure of fishes caught using the two approaches. We focus on the numerous parallel streams of Trinidad's Northern Range that have been widely used as a model system for evolutionary and ecological studies (see Magurran 2005), and where the fish fauna is well documented (Phillip et al. 2013). Our aim is to use this well-replicated system to assess whether surveys of tropical freshwater stream fishes that rely solely on hand-seining provide adequate assessments of biodiversity and are free from bias due to detectability issues.

# METHODS

## Sampling methods

### *Location and site features*

Sixteen sites along the southern slopes of Trinidad's Northern Range (Fig. 1) were surveyed repeatedly (13 times) over a 3-year period as part of a longitudinal biodiversity study. Each site consisted of a variety of habitats, including pools and riffles, and many of the sites included portions of undercut banks. Water conductivity was measured before each survey; almost all readings fell between 100 and 400 µS/cm (mean: 252 µS/cm ± SD 97). No sites were considered 'low' conductivity (>43 µS/cm), which can reduce electrofishing efficiency in some neotropical streams (Allard et al. 2014).

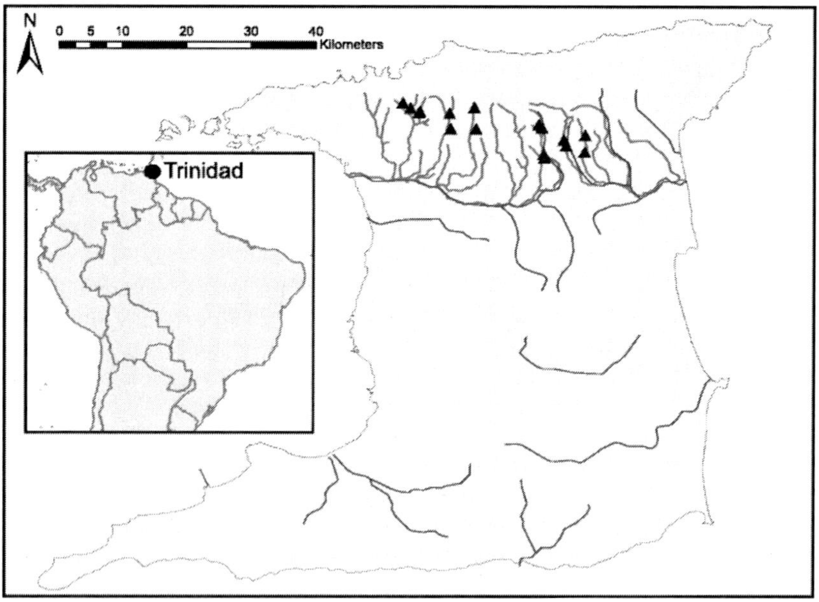


**Figure. 1** Trinidad's position in relation to South America, and the distribution of the 16 sampling sites (*black triangles*) along the streams of the southern slopes of the island's Northern Range mountains. Other major rivers are also shown

Surveys always took place between 08:00 and 11:00 and were spread throughout the year to include both dry and rainy season conditions. Water depth was measured at 1 m intervals along three transverse transects at each site; maximum depth at any site was 1.88 m and the mean depth over all transects at all sites was 24 (±SD 22) cm. Turbidity was measured on a visually assessed scale of 1–5, where one is clear and five is opaque (see Deacon et al. 2015 for details). The median turbidity was 1.28 (with a lower quartile of 1 and an upper quartile of 2); we report the median value here in light of the categorical nature of the scale. Substrata varied within and between sites; on average the stream bed was covered by 36.9% (±SD 9.5) fine substrates (silt, sand, gravel) and 30.5% (±SD 10.4) coarse substrates (cobble, boulders). Flow rate was low to moderate across sites with a mean of 0.31 (±SD 0.15) m/s, and only occasionally exceeded 1 m/s.

Each site comprised a 50 m stretch of stream, blocked at both ends by seine nets. During a survey, each stretch was systematically fished from one end to the other, first using a two-person hand seine (64 mm mesh), and then again with electrofishing equipment.

**Hand-seining**

Different sized hand seines were used depending on the dimensions of the site; all were 1.5 m high and between 2 and 3 m long, with wooden poles, floats along the top and weights along the bottom. The seine was pulled wherever physically possible along the stretch (all parts of the stream with the exception of the shallowest riffles and smallest pools); three passes were made and these constituted one sample. Each pass consisted of the operators starting at one end of the site and systematically fishing from one end to the other.

**Electrofishing**

The electrofishing equipment consisted of a portable generator (Yamaha EF1000is) on the bank, linked to two hand-operated poles with a safety switch. Both operators wore rubber waders, and while one operated the poles, the other stood nearby with a landing net and bucket to scoop up the stunned fishes. A double pass was made along the entire length of the 50 m stretch. The operators were experienced in applying these techniques in these environments and the same two operators were involved in all surveys.

**Fish handling**

Fish were kept in shaded buckets, separated according to the method with which they were caught. Care was taken to separate predators from their prey and to keep densities low. Fishes were identified to species level and weighed individually to the nearest gram using a small plastic container on a portable electronic balance. This represents wet weight and is referred to as 'biomass' from here on. The fish caught by hand-seining were not released until the electrofishing was complete. All fish were released unharmed at the end of each survey.

## Analysis

### Species differences

Abundance data from all 208 surveys (13 surveys at each of the 16 sites) were combined and the percentage of total catch attributed to hand-seining alone, and for both methods together, was calculated for each species. Any species that were represented by fewer than five individuals over all 16 sites were omitted from this analysis, as meaningful percentages could not be calculated from such a small number (i.e. a single individual has too much influence on the overall percentage). These species were, however, included in all subsequent analyses (i.e. computations of $\alpha$ diversity and multivariate analyses).

### Alpha (α) diversity

To test how sampling method influences our assessment of α diversity, we calculated the abundance of each species at each of the 16 sites (number of individuals per species, per site, summed over the 13 repeat surveys) for both fishing methods (hand-seining alone and hand-seining and electrofishing data combined). All species caught were included in these analyses. We also calculated total number of individuals (N) (summed over 13 surveys), and mean fish biomass (mean of 13 surveys) at each site for both methods (hand-seining, and hand-seining and electrofishing combined). These metrics are widely used in the monitoring of freshwater assemblages.

We next calculated three popular measures of α diversity (richness, Shannon and Simpson diversity) for each of the 208 surveys; first using the hand-seining data alone, and again using the data from the hand-seining and electrofishing combined. Species richness (S) is the most intuitive measure of diversity, but most affected by sampling effort. The Shannon index (H′) takes the relative abundance of different species into account while the Simpson index (1/D) places greater weight on species dominance (Magurran 2004). To evaluate differences between sampling methods, species richness and Shannon diversity (log transformed, to meet ANOVA assumptions) were analysed using a two-way ANOVA with method and season as fixed factors. Because Simpson index values did not meet the assumptions of a parametric test, even after transformation, a non-parametric analogue of ANOVA [Scheirer-Hare-Ray test (Dytham 2011)] was employed. Surveys conducted between June and December were considered rainy season surveys, while those conducted between January and May were considered dry season surveys (http://www.metoffice.gov.tt/Climate).

Rarefaction provides researchers with a means of making a fair comparison between sites or studies where there has been a difference in sampling effort (Gotelli and Colwell 2001). However, estimates of richness obtained using rarefaction may be biased if different sampling methods favour different taxa. To ascertain whether this problem arises when the different fishing methods employed in our study are used, we estimated species richness for the entire assemblage (that is, all 16 sites) by computing sample-based rarefaction for hand-seined data alone and data from hand-seining and electrofishing combined. These calculations were carried out using Estimate S v. 9.1.0 (Colwell 2013) to implement eq. 17 in Colwell et al. (2012). For each sampling method a rarefaction curve, with 95% confidence intervals, was constructed.

**Multivariate methods**

Multivariate methods are widely used in ecology to summarise and interpret community data. We applied three multivariate techniques: PERMANOVA (Anderson 2001), cluster analysis (Galili 2015), and NMDS (Oksanen et al. 2008), to shed additional light on the extent to which the two fishing methods (hand-seining and combined hand-seining and electrofishing) shape our perception of biodiversity in these neotropical streams.

PERMANOVA, an informative and appropriate alternative to ANOVA when testing for differences in ecological data (Anderson 2001), is based on distance measures and uses permutation to generate pseudo-F ratios. We employed PERMANOVA to evaluate differences between the two fishing methods (hand-seined data and combined data). This analysis was computed using overall species abundances at each site (with a log10 x + 1 transformation) in R (RCoreTeam 2016), using the adonis function in the vegan package (Oksanen et al. 2008).

Cluster analysis provides a visual summary of the relationship between entities for which there are multivariate data and is often used to represent the relationship between localities. We used this approach to examine the pattern of classification of sites obtained using the different fishing methods. A cluster analysis of sites, one per method, was constructed in R (RCoreTeam 2016) using Euclidian distance and the 'complete' clustering algorithm. Using the dendextend package in R

(Galili 2015), a tanglegram was then constructed to illustrate the concordance in clustering outcomes achieved using the different methods.

Nonmetric Multidimensional Scaling (NMDS) is an ordination method based on distance data and an effective tool for describing ecological patterns. We thus employed NMDS [metaMDS in the vegan package (Oksanen et al. 2008)] to represent the ecological placement of fish species, relative to one another, as deduced when hand-seining is used alone, or when hand-seining data and electrofishing data are combined. All univariate statistical tests were performed using IBM SPSS v.22. Multivariate statistics were performed using R v.3.3.2 (RCoreTeam 2016).

# RESULTS

## Species differences

As is typical of ecological surveys, some species were common, while others were rare (Fig. 2). However, the coverage of a given species, that is its relative abundance in the overall data set, depended on the sampling method. For some species hand-seining accounted for the majority of individuals caught, while other species were completely missed using this method. Indeed, the percentage of total catch using hand-seining varied greatly according to species, spanning a wide range of values from 87% to zero (Fig. 2).

Three species [the swamp eel *Synbranchus marmoratus* (Bloch, 1795), the river goby *Awaous banana* (Valenciennes, 1837) and the knife fish *Gymnotus carapo* (L., 1758)] were never caught in the hand-seined samples even though electrofishing revealed their presence at four sites (on a total of five occasions). A further five species were caught by both methods but detected more often by electrofishing (<25% of individuals caught were caught by hand-seining): the silver catfish *Rhamdia quelen* (Quoy & Gaimard, 1824), the armoured catfishes *Ancistrus maracasae*(Fowler, 1946) and *Hypostomus robinii* (Valenciennes, 1840), the featherfin tetra *Hemigrammus unilineatus* (Gill, 1858), and the wolf-fish *Hoplias malabaricus* (Bloch, 1794). Two species were represented well by the hand-seining (>80% of individuals caught were caught by hand-seining): *Odontostilbe pulchra* (Gill, 1858) and the swordtail sardine *Corynopoma riisei* (Gill, 1858). Around half of the total individuals caught of the blue acara *Andinoacara pulcher* (Gill, 1858), the two-spotted sardine *Astyanax bimaculatus* (L., 1758),the mountain stream sardine *Hemibrycon taeniurus* (Gill, 1858) and the hunchback sardine *Roeboides dientonito* (Schultz, 1944) were caught using hand-seining.

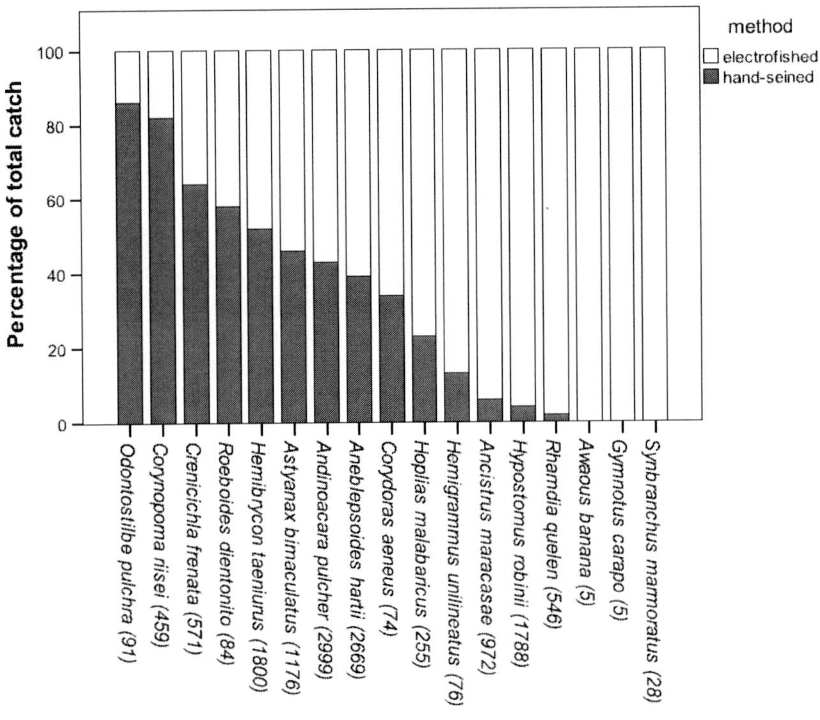


**Figure. 2** Species ranked according to the % caught by hand seining, illustrating considerable between-taxa variation. Total N in parentheses on the x-axis label

The Trinidadian guppy, *Poecilia reticulata*, was not well-sampled by hand-seining or electrofishing; its small size (<3 cm) means that it passes through the mesh of the seine and the landing net. Guppies were present at all sites and sampled separately using dip nets during our surveys.

## Alpha (α) diversity

Conclusions regarding total fish abundance were heavily dependent on fishing method. The number of individuals caught increased threefold, and the biomass recorded per site increased by five times, when data compiled using both methods combined were compared to the results from hand-seined sampling alone (Table 1).

**Table 1** Hand-seining vs combined method. Includes all fish caught over 13 visits to each of the 16 sites

| | Hand-seining only | Both methods combined | *p*-value |
|---|---|---|---|
| Total N | 4958 | 14,015 | n/a |
| Total biomass (g) | 25,523 | 128,393 | n/a |
| Mean biomass (g) per visit | 122 (SE ± 32) | 617 (SE ± 98) | n/a |
| Mean S per site | 7.31 (SE ± 0.74) | 9.75 (±SE0.93) | <0.01** |
| Total S | 17 | 21 | n/a |
| Mean Shannon (H) per site | 0.50 (SE ± 0.10) | 0.92 (SE ± 0.13) | <0.001*** |
| Mean Simpson (1/D) per site | 1.40 (SE ± 0.14) | 2.02 (SE ± 0.24) | <0.001*** |

Inferences about assemblage diversity were also shaped by methodological approach. Higher estimates of diversity were obtained when the combined data were used: species richness (ANOVA $F_{1,60} = 9.94$ $p < 0.001$; Table 1), Shannon index ($F_{1,60} = 14.33$; $p < 0.001$; Table 1) and Simpson index ($SS/MS_{total} = 13.69$; $p < 0.001$; Table 1). There was no effect of season (all tests: $p > 0.100$), nor significant interactions between method and season (all tests: $p > 0.100$) for any of the diversity metrics tested (species richness, Shannon index or Simpson index).

Our analyses further show that rarefaction cannot compensate for the methodological differences. Rarefaction curves illustrate the difference in estimates of species richness between hand-seining and combined fishing methods, and also show both curves reaching an asymptote at a similar point along the x-axis (Fig. 3).

## Multivariate measures of diversity

Multivariate approaches add additional support to the conclusion that fishing method profoundly influences perception of diversity, as well as helping to visualise these differences. The PERMANOVA revealed significant differences in assemblage structure between hand-seined samples and the combined hand-seined and electrofished samples ($F_{1,30} = 5.128$; $p < 0.002$; Fig. 4). Differences in the placement of sites in a cluster analysis is evident from the tanglegram (Fig. 5) and contrasts in the apparent ecological relationship of species apparent from the NMDS plots (Fig. 4). Taken together these results provide compelling evidence that conclusions about biodiversity and the structure of these fish assemblages are shaped by the sampling methods used to obtain the data (Fig. 5).

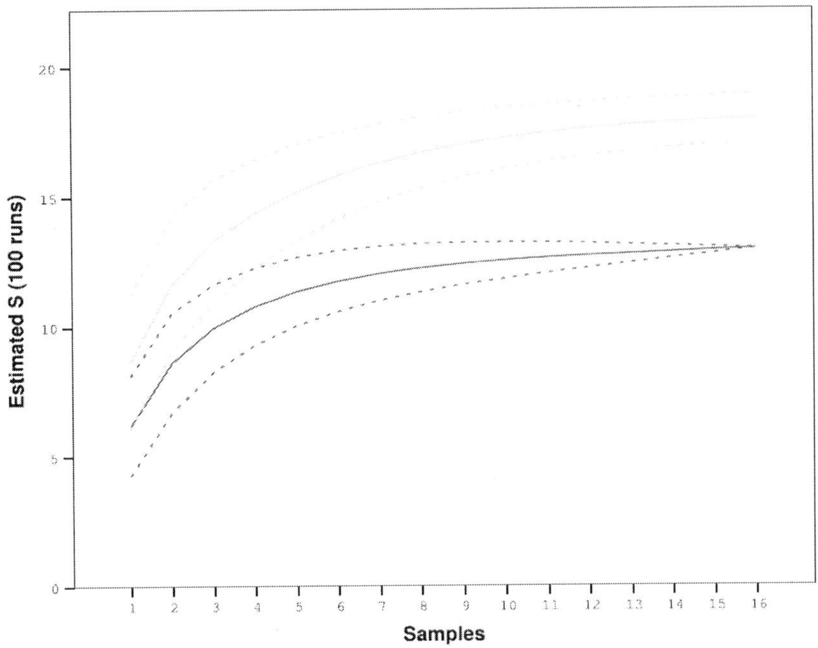


**Figure. 3** Rarefaction curves for hand-seined (dark grey) and combined (light grey) fish species richness data (S(est)), including 95%confidence intervals. Calculated using Estimate S rarefaction (100 runs)

# DISCUSSION

A striking finding of this study is that sampling method has a far-reaching influence on the conclusions drawn about these tropical fish assemblages. Hand-seining was better at capturing some species than others, and tended to be more successful at sampling open water than benthic species. Had hand-seining been the only sampling technique used here, each of the assemblage properties measured (species richness, Simpson index, Shannon index and multivariate properties) would have been considerably different, and the subsequent conclusions would have been potentially misleading.

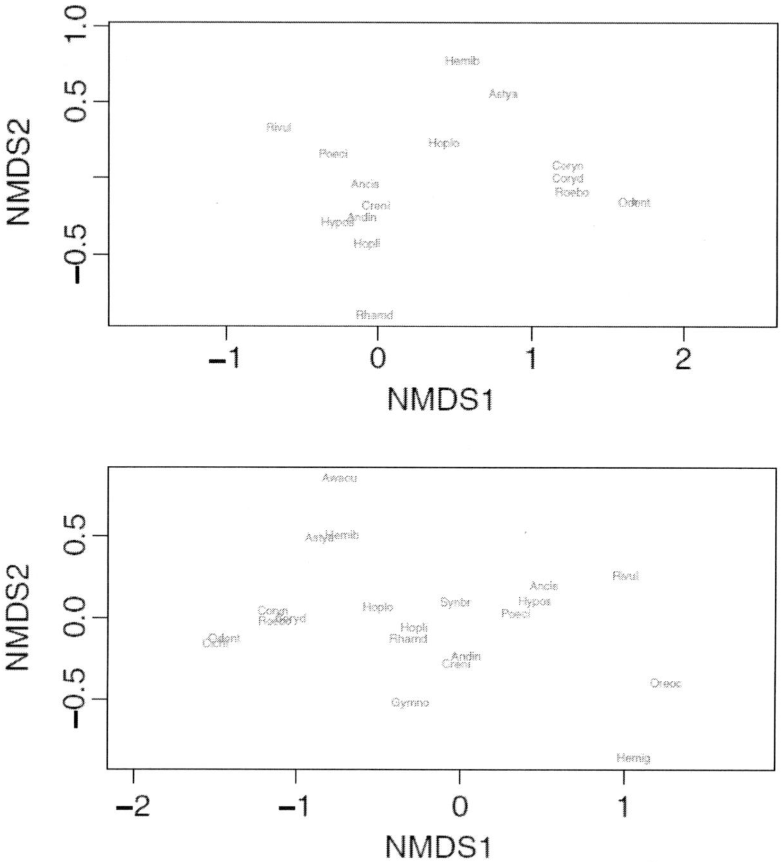


**Figure. 4** NMDS plots for hand-seined (*above*) and combined (*below*) fish assemblage data. Stress values were 0.13 for the hand-seining data, and 0.08 for the combined data (values under >0.2 are generally considered a 'good' representation of the data). Key to species: Ancis = *Ancistrus maracasae;* Andin = *Andinoacara pulcher;* Astya = *Astyanax bimaculatus;* Awaou = *Awaous banana;* Cichl = *Cichlasoma taenia;* Coryd = *Corydoras aeneus;* Coryn = *Corynopoma riseii;* Creni = *Crenicichla frenata;* Gymno = *Gymnotus carapo;* Hemib = *Hemibrycon taeniurus;* Hemig = *Hemigrammus bilineatus;*Hopli = *Hoplias malabaricus* Hoplo = *Hoplosternum littorale;* Hypos = *Hypostomus robinii;* Odont = *Odontostilbe pulcra;* Oreoc = *Oreochromis mossambicus;*Poeci = *Poecilia reticulata;* Roebo = *Roeboides dientonito;* Rhamd = *Rhamdia quelen;*Rivul = *Anablepsoides hartii;* Synbr = *Synbranchus marmoratus*

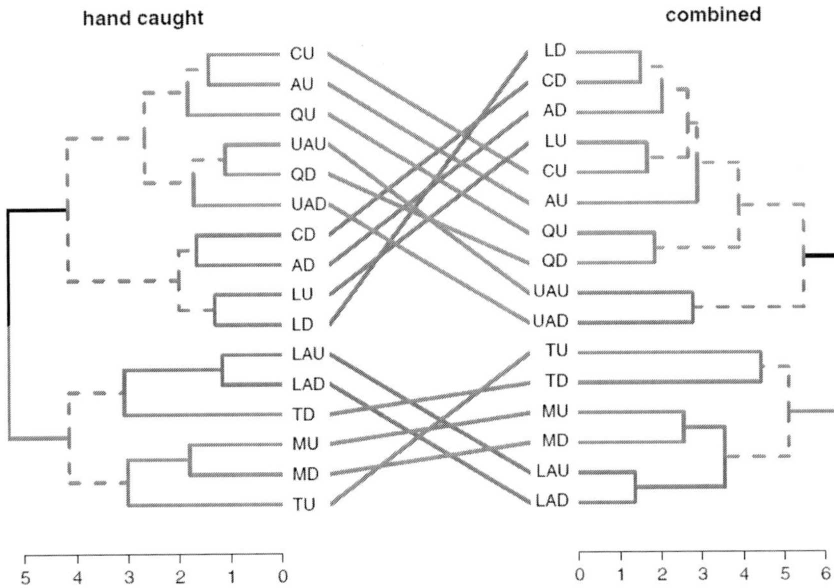


**Figure. 5** Tanglegram showing hierarchical clustering of sites according to similarities in fish assemblage, for hand-seined data (*left*) and combined data (*right*). Key to sites: AD = Acono Down; AU = Acono Up; CD = Caura Down; CU = Caura Up; LAD = Lower Aripo Down; LAU = Lower Aripo Up; LD = Lopinot Down; LU = Lopinot Up; MD = Maracas Down; MU = Maracas Up; QD = Quare Down; QU = Quare Up; TD = Turure Down; TU = Turure Up; UAD = Upper Aripo Down; UAU = Upper Aripo Up

The species that were best-represented by the hand-seining tended to be the open water, shoaling characids. These species form large pelagic shoals, often in mixed species groups (Lowe-McConnell 1975). Although the order effects of this study prevented a direct comparison between the two methods, it appears that seining is best-suited to capturing shoals of fish, and this is primarily what it seine nets are designed to do (Lapointe et al. 2006; Mercado-Silva and Escandón-Sandoval 2008).

In contrast, the three species that were never caught by hand-seining (despite being detected by the electrofishing) were all solitary benthic fishes that tend to be found hiding in vegetation or on the river bottom where they can easily evade a seine net (Angermeier and Karr 1983). Similarly, although occasionally caught in the hand seine, the benthic catfishes were overwhelmingly better-represented in the electrofishing samples. This finding is supported by Mazzoni et al. (2000), who also found electrofishing to be effective for benthic species with cryptic habits, such as *Rhamdia* spp. and *Synbranchus* spp. The result of this bias in the current study is that several benthic species would have been severely underrepresented - or even assumed absent – had electrofishing not been used.

Not only does the hand-seined rarefaction curve plateau at a lower species richness, but the narrow 95% confidence intervals it yields could lead to the false conclusion that all species in the habitat have been detected, and thus provide a dramatic underestimate of species richness. In fact, the difference in these curves can be attributed to the several species that are never, or very rarely, detected by hand-seining. This suggests that the difference in species richness is independent of effort, and that the estimate would never converge on the combined methods richness total, even given further hand-seining passes. This is an important observation as the premise of rarefaction is that a fair comparison is possible once sampling effort is controlled. What the analysis does highlight is that sampling effort implies not just the number of samples, but also the mode of sampling – a distinction easily overlooked. Of course, it is not possible to be sure that the combined methods detected every single species in the habitat, but it is clear that the hand-seining alone did not.

Our findings suggest that it is not simply the number of species that may be wrongly estimated in the absence of electrofishing. Had hand-seining alone been used to survey these sites, the proportional measures of diversity (Shannon and Simpson index) and multivariate analyses would have each pointed towards different interpretations of the structure and composition of these assemblages than those based on the results from the fuller dataset (with electrofishing). Such differences are not trivial, as they could lead to different decisions or recommendations regarding the management and conservation of freshwater habitats. Practitioners routinely use these types of indices and analyses to get a more nuanced picture of the structure of a community or assemblage. For example, multivariate analyses can reveal differences in β diversity, concerning similarities between sites in the composition and structure of assemblages.

The comparison of α diversity indices is especially common in biodiversity monitoring programmes, therefore it is important to be aware that choice of sampling gear can have a significant effect on these. In a similar study of tropical Mexican streams, Mercado-Silva and Escandón-Sandoval (2008) also found that hand-seining significantly underestimated these types of metrics (in this case species richness, Shannon index and biomass). Such differences are not solely attributable to missing the rare, solitary species when hand-seining. For example, the benthic catfish, *Hypostomus robinii,* was one of the most abundant species across our sites, yet would have been considered a rare species if the hand-seine data alone had been examined. Our findings emphasise the importance of taking detectability, and how this might be affected by fishing methods, into account when interpreting measures of assemblage diversity.

Although in our study, all three α diversity measures were simultaneously affected by gear type, this is not always the case, For example, Onorato et al. (1998) found that despite the proportions of different taxa differing between electrofished and hand-seined samples in streams in Alabama, species richness remained the same. These findings support the suggestion that electrofishing is better and more consistent in efficacy than seining (Wiley and Tsai 1983; Mercado-Silva and Escandón-Sandoval 2008) and that it is the most effective method for small to medium streams (Mazzoni et al. 2000). Results from electrofishing in French Guiana (Allard et al. 2014) and South Africa (Weyl et al. 2013) compare well with those obtained using destructive sampling with rotenone. Despite this, it is important to be aware of other stream properties that can affect the efficacy of either method. Conductivity is a particularly important one for electrofishing in neotropical streams, and although it was not an issue at our sites, low water conductivity (>43 μS/cm) can reduce the efficacy of electrofishing (Allard et el. 2014), and in these cases careful use of rotenone may need to be considered.

## Recommendations

Given that electrofishing improved the detectability of fish assemblages at our sites, we recommend that it is employed alongside hand-seining in surveys of tropical streams, both in rainy and dry seasons. We are not the first to advocate the use of a combination of methods: Mercado-Silva and Escandón-Sandoval (2008) compared the two methods directly in their Mexican stream surveys, and found that at least one species was consistently underrepresented when electrofishing was conducted alone. Similarly, Pusey et al. (1998) found that combined electrofishing and hand-seining in streams of Queensland, Australia produced better estimates of total fish abundance, among other metrics, and Fago (1998) recommended a combination of different methods (including hand-seining and electrofishing) to get an accurate idea of species composition, based on Wisconsin stream surveys. The strength of our study is that it supports these findings using data from over 14,000 fish caught during a total of 208 surveys at 16 sites over 3 years, as well as demonstrating that multiple measures of biodiversity may be affected if electrofishing is not employed alongside hand-seining.

# CONCLUSIONS

Despite the potential challenges of acquiring, maintaining, transporting and operating electrofishing equipment in the tropics, our study suggests that the additional effort will be rewarded with greater detection, resulting in more comprehensive and representative samples. This finding has important implications for the conservation and management of biodiversity. Rapid anthropogenic changes to freshwater habitats in many tropical regions, including to the very streams surveyed here (Deacon et al. 2015), mean that monitoring the effects of disturbance and land use change on biodiversity is increasingly important (Dudgeon et al. 2006). Neotropical assemblages in general are under growing threat (WWF 2014), while the ability of policy makers to respond appropriately is hampered by sparse data (Abell 2002). An understanding of the strengths and limitations of survey methods (depending on the habitat and the questions being addressed), is vital if we are to collect high quality data on threatened fish assemblages.

# ACKNOWLEDGEMENTS

This work was funded by a European Research Council grant (BIOTIME 250189). AEM also acknowledges the Royal Society. The authors are grateful to K. Deonarinesingh and A. Deonarinesingh for fieldwork assistance, F. Jones for constructing Fig. 1, and B. Ebner and two anonymous reviewers for comments on an earlier draft of this manuscript.

## Compliance with Ethical Standards

Research was conducted in line with the ethical guidelines of the European Research Council, the University of St Andrews' Animal Welfare and Ethics Committee, and The University of the West Indies' Policy and Procedures on Research Ethics.

# REFERENCES

1. Abell R (2002) Conservation biology for the biodiversity crisis: a freshwater follow-up. Conserv Biol 16:1435–1437
2. Allard L, Grenouillet G, Khazraie K, Tudesque L, Vigouroux R, Brosse S (2014) Electrofishing efficiency in low conductivity neotropical streams: towards a non-destructive fish sampling method. Fish Manag Ecol 21:234–243

3. Anderson MJ (2001) A new method for non-parametric multivariate analysis of variance. Austral Ecol 26:32–46
4. Angermeier PL, Karr JR (1983) Fish communities along environmental gradients in a system of tropical streams. Environ Biol Fish 9:117–135. doi:10.1007/BF00690857
5. Balmford A, Bruner A, Cooper P, Costanza R, Farber S, Green RE, Jenkins M, Jefferiss P, Jessamy V, Madden J, Munro K (2002) Economic reasons for conserving wild nature. Science 297:950–953. doi:10.1126/science.1073947
6. Bohlin T, Hamrin S, Heggberget T, Rasmussen G, Saltveit S (1989) Electrofishing — theory and practice with special emphasis on salmonids. Hydrobiologia 173:9–43. doi:10.1007/bf00008596
7. Boulinier T, Nichols JD, Sauer JR, Hines JE, Pollock KH (1998) Estimating species richness: the importance of heterogeneity in species detectability. Ecology 79:1018–1028. doi:10.1890/0012-9658(1998)079[1018:esrtio]2.0.co;2
8. Buckland ST, Studeny AC, Magurran AE, Newson SE (2011) Biodiversity monitoring: the relevance of detectability. In: Magurran AE, McGill BJ (eds) Biological diversity: Frontiers in measurement and assessment. Oxford University Press, Oxford, pp 25–36
9. Colwell RK (2013) EstimateS: statistical estimation of species richness and shared species from samples. Version 9 and earlier. User's Guide and application Published at: http://purloclcorg/estimates
10. Colwell RK, Coddington JA (1995) Estimating terrestrial biodiversity. In: Hawksworth DL (ed) Biodiversity: measurement and estimation. The Royal Society/Chapman & Hall, London
11. Colwell RK, Chao A, Gotelli NJ, Lin S-Y, Mao CX, Chazdon RL, Longino JT (2012) Models and estimators linking individual-based and sample-based rarefaction, extrapolation and comparison of assemblages. J Plant Ecol 5:3–21
12. Cote IM, Perrow MR (2006) Fish. In: Sutherland WJ (ed) Ecological census techniques: a handbook. Second Edition edn. Cambridge University Press, Cambridge
13. Dauble DD, Gray RH (1980) Comparison of a small seine and a backpack electroshocker to evaluate nearshore fish populations in rivers. Progress Fish-Cultur 42:93–95
14. Deacon AE, Shimadzu H, Dornelas M, Ramnarine IW, Magurran AE (2015) From species to communities: the signature of recreational use on a tropical river ecosystem. Ecol Evol 5:5561–5572. doi:10.1002/ece3.1800Central
15. Dolan CR, Miranda LE (2003) Immobilization thresholds of electrofishing relative to fish size transactions of the. Am Fisheries Soc 132:969–976. doi:10.1577/t02-055
16. Dudgeon D, Arthington AH, Gessner MO, Kawabata Z-I, Knowler DJ, Lévêque C, Naiman RJ (2006) Freshwater biodiversity: importance, threats, status and conservation challenges. Biol Rev 81:163–182. doi:10.1017/s1464793105006950
17. Dytham C (2011) Choosing and using statistics: A biologist's guide. Chichester, West Sussex, UK: Wiley-Blackwell
18. Ebner BC, Morgan D (2013) Using remote underwater video to estimate freshwater fish species richness. J Fish Biol 82:1592–1612
19. Ebner BC, Fulton CJ, Cousins S, Donaldson JA, Kennard MJ, Meynecke J-O, Schaffer J (2014) Filming and snorkelling as visual techniques to survey fauna in difficult to access tropical rainforest streams. Mar Freshwater Res 66:120–126
20. Fago D (1998) Comparison of littoral fish assemblages sampled with a mini-fyke net or with a combination of electrofishing and small-mesh seine in Wisconsin lakes. N Am J Fish Manag 18:731–738
21. Fièvet É, Bonnet-Arnaud P, Mallet JP (1999) Efficiency and sampling bias of electrofishing for freshwater shrimp and fish in two Caribbean streams, Guadeloupe Island. Fish Res 44:149–166. doi:10.1016/S0165-7836(99)00058-2
22. Galili T (2015) Dendextend: an R package for visualizing, adjusting, and comparing trees of hierarchical clustering. Bioinformatics doi:10.1093/bioinformatics/btv428

23. Gerstner CL, Ortega H, Sanchez H, Graham DL (2006) Effects of the freshwater aquarium trade on wild fish populations in differentially-fished areas of the Peruvian Amazon. J Fish Biol 68:862–875. doi:10.1111/j.0022-1112.2006.00978.x
24. Głowacki Ł, Penczak T (2005) Species richness estimators applied to fish in a small tropical river sampled by conventional methods and rotenone. Aquat Living Resour 18:159–168
25. Gotelli NJ, Colwell RK (2001) Quantifying biodiversity: procedures and pitfalls in the measurement and comparison of species richness. Ecol Lett 4:379–391
26. Gotelli NJ, Colwell RK (2011) Estimating species richness. In: Magurran AE, McGill BJ (eds) Biological diversity: frontiers in measurement and assessment. Oxford University Press, Oxford, pp 39–54
27. Hendricks ML, Hocutt CH, Stauffer JR (1980) Monitoring of fish in lotic habitats. In: Hocutt CH and Stauffer JR (eds) Biological monitoring of fish. DC Heath, Lexington, MA pp 205–232
28. Lapointe NWR, Corkum LD, Mandrak NE (2006) A comparison of methods for sampling fish diversity in shallow offshore waters of large rivers. N Am J Fish Manag 26:503–513. doi:10.1577/m05-091.1
29. Lowe-McConnell RH (1975) Fish communities in tropical freshwaters: their distribution, ecology, and evolution. Longman, London
30. Maciolek JA, Timbol AS (1980) Electroshocking in tropical insular streams. Progress Fish-Cultur 42:57–58. doi:10.1577/1548-8659(1980)42[57:eitis]2.0.co;2
31. Magurran AE (2004) Measuring biological diversity. Blackwell Publishing, Oxford
32. Magurran AE (2005) Evolutionary ecology: the Trinidadian guppy. Oxford series in ecology and evolution. Oxford University Press, Oxford
33. Mazzoni R, Fenerich-Verani N, Caramaschi EP (2000) Electrofishing as a sampling technique for coastal stream fish populations and communities in the southeast of Brazil. Rev Bras Biol 60. doi:10.1590/s0034-71082000000200003
34. Mercado-Silva N, Escandón-Sandoval DS (2008) A comparison of seining and electrofishing for fish community bioassessment in a Mexican Atlantic slope montane river. N Am J Fish Manag 28:1725–1732
35. Oksanen J, Kindt R, Legendre P, O'Hara B, Simpson GL, Solymos P, Stevens MHH, Wagner H (2008) vegan: Community Ecology Package. R package version 1.15–1, URL http://CRAN.R-project.org/package=vegan
36. Onorato DP, Angus RA, Marion KR (1998) Comparison of a small-mesh seine and a backpack electroshocker for evaluating fish populations in a north-central Alabama stream. N Am J Fish Manag 18:361–373
37. Phillip DA, Taphorn DC, Holm E, Gilliam JF, Lamphere BA, López-Fernandez H (2013) Annotated list and key to the stream fishes of Trinidad & Tobago. Zootaxa 3711:1–64
38. Pusey BJ, Kennard MJ, Arthur JM, Arthington AH (1998) Quantitative sampling of stream fish assemblages: single- vs multiple-pass electrofishing. Aust J Ecol 23:365–374. doi:10.1111/j.1442-9993.1998.tb00741.x
39. RCoreTeam (2016) R: A language and environment for statistical computing. http://www.R-project.org. R Foundation for Statistical Computing, Vienna
40. Snyder DE (2003) Invited overview: conclusions from a review of electrofishing and its harmful effects on fish. Rev Fish Biol Fish 13:445–453
41. Weyl O, Ellender B, Woodford D, Jordaan M (2013) Fish distributions in the Rondegat River, cape floristic region, South Africa, and the immediate impact of rotenone treatment in an invaded reach. Afr J Aquat Sci 38:201–209
42. Wiley ML, Tsai C-F (1983) The relative efficiencies of electrofishing vs. seines in piedmont streams of Maryland. N Am J Fish Manag 3:243–253
43. Willis SC, Winemiller KO, Lopez-Fernandez H (2005) Habitat structural complexity and morphological diversity of fish assemblages in a Neotropical floodplain river. Oecologia 142:284–295. doi:10.1007/s00442-004-1723-z

44. WWF (2014) Living planet report 2014: species and spaces, people and places. Gland, Switzerland
45. Yoccoz NG, Nichols JD, Boulinier T (2001) Monitoring of biological diversity in space and time. Trends Ecol Evol 16:446–453
46. Zalewski M (1983) The influence of fish community structure on the efficiency of electrofishing. Aquac Res 14:177–186

# CHAPTER 3

# Influence of Environmental Features at Multiple Scales and Spatial Structure on Stream Fish Communities in a Tropical Agricultural Region

*Camilo A. Roa-Fuentes & Lilian Casatti*

**Zoology and Botanic Department, São Paulo State University, São José do Rio Preto, Brazil**

## ABSTRACT

Land cover change for agricultural expansion exerts strong pressures on tropical streams. We examined the relative contributions of catchment land cover, local environmental features, and spatial patterns on stream fish communities ($n$ = 43). Explained variation in fish community structure was partitioned (p-RDA) between the explained fractions as follows: pure local 10%, pure catchment 1%, pure spatial 3%, and shared environmental and spatial 7%. Among the possible reasons for the lack of effect of catchment-scale variables on stream fish structure we can cite (1) weak associations between catchment-scale and local-scale features that are affecting stream fish communities, (2) major influence of past land cover on the present-day fish community structure compared to current land cover, or (3) the small variability in catchment variables (e.g. homogeneous landscape) is not able to overcome the impact of local-scale variables. The best practice to improve the ecological integrity of streams would be the restoration at the catchment-scale, which is impossible for agricultural lands. In areas with intense land use for agriculture and pasture, our study indicates that restoration of riparian forests and manipulation of local conditions may be an option to reduce the impacts of anthropogenic activities on the stream fish communities.

**Keywords:** Hierarchical classification, stream ecology, riparian cover, riparian width, reach-scale variables, Neotropical stream fish, agricultural landscape

# INTRODUCTION

Every key aspect of river ecosystems, river geomorphic systems, and river chemical systems starts in headwater streams (Freeman et al.2007) Headwater streams provide ecosystem services from local to global, serve as ecological refuges, and supply water for human activities (Lehner et al. 2011) Human activities commonly associated with headwater stream modification include land cover change, road construction, mining, agricultural drainage, and reservoir construction. Land cover change for agricultural expansion exerts strong pressures on freshwater ecosystems through augmented inputs of nutrients, contaminants, sediments, and modification of riparian zones (Sala et al.2008;Laurence et al 2014) Particularly in tropical streams, land cover change is projected to have a paramount effect on biodiversity when compared with other factors, such as climate and biotic exchange (Sala et al. 2008)

The concept of hierarchical classification of stream systems provides a useful framework for integrating the multiple scales over which ecological processes that organize stream communities are supported (Frissell et al.1986Angermeier & Winston1998 feld 2013) An important assumption of this framework is that the structure and dynamics of stream habitat (e.g. pool/riffle morphology and sequence) and biological patterns are determined by the surrounding catchment (Hynes 1975 frissell et al1986) By considering the environmental features in a hierarchical framework, it is expected that catchment-scale variables impact the local-scale variables (i.e. at the reach-scale), which then in turn affect stream fish community structure (Fitzpatrick et al. 2001 wang et al 2003Dembkowski & Miranda 2014)

Although there is a substantial number of studies that address this issue (Lammert & Allan 1999 Diana et al.2006 johnson et al 2007Cianfrani et al. 2012 including in tropical streams (Cruz et al. 2013Casatti et al. 2015Santos et al.2015) the relative importance of catchment- versus local-scale factors for stream biota has remained elusive (Roth et al. 1996Cianfrani et al. 2012) Studies addressing the issue are necessary because the assessment of catchment-stream relationships using a hierarchical classification can aid in conservation, restoration, and management of freshwater systems (Frissell et al. 1986 Cianfrani et al. 2012 feld 2013 Dembkowski & Miranda2014) and provides a more efficient and holistic view of these (Dembkowski & Miranda2014)

This study addresses the relative contributions of catchment land cover, local environmental features, and spatial patterns on stream fish community structure, and discusses possible implications for management and/or restoration of streams located in heavily modified tropical landscapes. We expected that both local and catchment variables will influence the stream fish community structure since catchment variables would have indirect connections with ichthyofauna through their direct influences on local variables (Fitzpatrick et al. 2001 Wang et al.2003 Dembkowski & Miranda2014) We also hypothesized that spatial structure should be weak or non-significant because, as for stream organisms assessed within single, small drainage basins, environmental control prevails over spatial constraints (Heino & Mykrä 2008 Heino et al.2015and references therein)

# METHODS

## Study area

The 43 stream reaches were located in the Turvo-Grande and São José dos Dourados basins in the northwest region of São Paulo State, south-eastern Brazil (Figure 1). These two basins belong to the same biogeographical province, where the fish communities have a shared evolutionary history (Géry1969) The study area is located in the Serra Geral geological formation, presenting a relatively flat slope and plains of quaternary fluvial sedimentary nature (IPT 1999) The soil has a high erosive potential since it is composed of unconsolidated sand and clay sediments (Silva et al.2007) The climate is tropical and hot, with two well-defined periods: a dry season with lower rainfall (28 mm mean) and cooler temperatures (21.4 °C mean) between June and September, and a wet season between December and February with higher rainfall (208.9 mm mean) and hotter (25.8 °C mean) temperatures (IPT 1999;CEPAGRI2017)

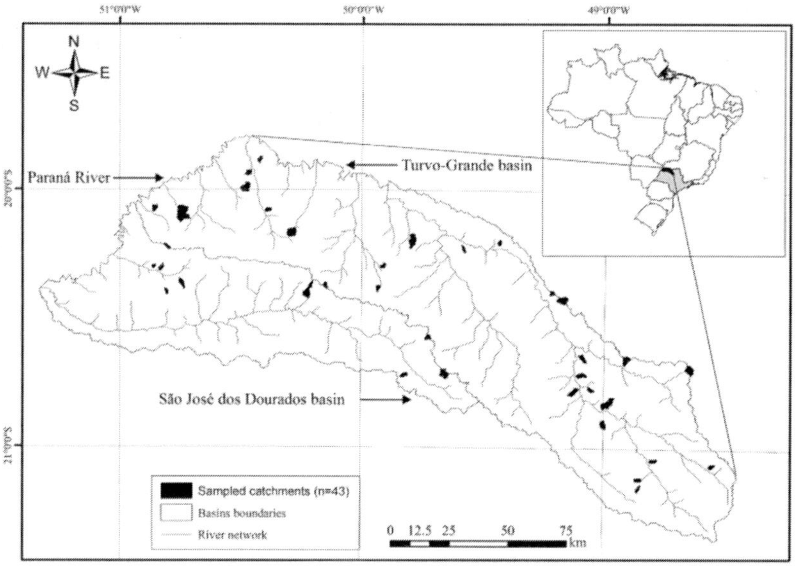


**Figure** 1. Sampling units (43 different catchments = 43 independent reaches) along São José dos Dourados and Turvo-Grande River basins at northwest region of São Paulo State (gray area in the country map), south-eastern Brazil. Some catchments were neighbors therefore it is not possible to differentiate on the map.

Originally, this region was covered by semi-deciduous seasonal forest (Silva et al. 2007) however, the landscape has been fragmented since the beginning of the last century (1900) for the development of coffee crops, followed by livestock grazing (Victor et al.2005) and more recently by sugarcane plantations (Rudorff et al.2010) Currently, the native vegetation is restricted to less than 4% of its original area, distributed in small and unconnected fragments embedded within agricultural matrices (Nalon et al. 2008) or limited to riparian areas (Silva et al.2007) As in other São Paulo state river basins (e.g. Corumbataí basin; Gerhard & Verdade2016) the stream fish fauna of the study area may be considered as homogenized due to species introductions (Rahel2002Casatti et al. 2009) habitat destruction and simplification led by an extensive, dynamic, and long history of land cover change (Victor et al. 2005;silva et al 2007 rudorff et al 2010)

## Site selection and land cover characterization

For site selection, first we mapped the land cover in the São José dos Dourados and Turvo-Grande basins through the digital processing of LANDSAT-5/TM satellite images for 2004, 2005, 2007 and 2011 (221-74, 221-75 e 222-74 scenes; 30 m spatial resolution). Four classes of land cover were defined through visual estimate using Google Earth™: native forest, pasture, sugarcane, and other land cover. Second, we preselected catchments with an area between 400 and 1400 ha (first to third-order streams according to the Strahler system). Using this procedure we identified 128 catchments, from which we selected 43 reaches (from independent catchments) for study based on accessibility and landowner support. Finally, to increase the reliability of the land cover data, we re-classified land cover in each of 43 selected catchments, but this time using orthorectified aerial photographs ('orthophotos') with a 1 m spatial resolution (years 2010/2011). Overall, eight classes of land cover were defined using orthophotos (Table A1), but we grouped them into four broad classes: native forest, pasture, sugarcane, and other land cover.

**Table A1**. Description of land cover classes.

| Land cover class | Description |
|---|---|
| Native forest | Areas occupied by the different native forest found in the study area (i.e. savanna and semi-deciduous seasonal forest). |
| Herbaceous and shrub vegetation | Areas occupied by herbaceous and shrub vegetation, which also included species of cattail (*Typha* spp.) |
| Pasture | Areas used for intensive and extensive livestock, also including areas covered by native grasses but to a lesser extent. |
| Sugarcane | Areas covered with sugarcane culture (*Saccharum* spp.) or that showed evidences of being used for this culture. |
| Perennial culture | Areas covered with perennial cultures (e.g. *Citrus* spp., *Hevea* spp., *Coffea* spp.) |
| Reforestation | Included planted forests and restored areas with *Eucalyptus* spp. or *Pinus* spp. |
| Urban area | Include towns and villages. |
| Other land cover | It groups rural installations, temporary cultures, highways and exposed soil. |

For the digital preparation, processing, and classification of LANDSAT-5/TM satellite images and orthophotos, we used ERDAS IMAGINE 9.2 and ArcGis 9.3 software. LANDSAT-5/TM satellite images were provided by the Instituto Nacional de Pesquisas Espaciais (INPE). Orthophotos were supplied by Empresa Paulista de Planejamento Metropolitano SA – EMPLASA (CLU N° 060/14). For the sugarcane land cover class, the CANASAT project (sugarcane crop monitoring in Brazil; Rudorff et al. 2010) provided data about area location and culture of sugarcane in São Paulo State for 2012.

## Fish sampling

In each stream, a 75 m-long reach was blocked using 5-mm-mesh stop nets according to standardized methods of fish collection for the region (see Casatti et al.2009 for details) The stream reaches were sampled using a Smith Root Model LR-24 backpack electrofisher (pulsed DC, 50–990 V, 1–

120 Hz, 40 A peak max, 400 W) with settings adjusted based on ambient conditions (i.e. with the quick set-up feature activated, which automatically sets output voltage, frequency, and duty cycle) and on observations of fish behavior and recovery times. A two-pass electrofishing technique was conducted for a total of 45 minutes of standardized time in each reach. Each electrofishing pass was conducted from downstream to upstream, covering from bank to bank in order to sample all available habitats. Captured specimens were fixed in 10% formalin solution and transferred to a 70% EtOH solution. Fishes were identified to species and counted. Voucher specimens were deposited at the fish collection of the Departamento de Zoologia e Botânica da Universidade Estadual Paulista (DZSJRP 19264–19326), São José do Rio Preto, São Paulo, Brazil. Fish sampling was carried out in the dry season between July and September 2013.

## Predictor variables

### Local-scale features

Ten local-scale descriptors related to in-stream features, riparian physical structures, water physicochemical properties, stream morphology, habitat composition, and heterogeneity were estimated at each reach (Table A2). We measured these descriptors following standardized protocols for the study area. These descriptors are commonly used in studies of stream fish community structure in this region (see Casatti et al. 2009)

### Catchment-scale features

For each catchment, the proportions of native forest, pasture, sugarcane, and other land cover in the whole catchment area, as defined by the orthophotos processing (years 2010/2011), were grouped in three different sets comprising a total of 12 catchment descriptors (Figure 2and Table A2). The definition of the sets was based on Strayer et al. (2003) as follows: catchment, which includes the land cover in the entire catchment area from each sampling site; drainage network, which comprises land cover within a 30 m buffer zone around the river network (minimum width established by the current Brazilian Forest Code); and local radius, which includes the land cover within a circle (150-m radius) centered on the sample site (adapted from Strayer et al.2003 figure2)

**Table A2.** Predictor variables measured at the 43 sampling sites, Turvo-Grande and São José dos Dourados River basins at the northwest region of São Paulo State, south-eastern Brazil. Mean and standard deviation (SD) of each variable are also presented.

| Category | Variable | Unit | Transformation | Code | Mean | ±SD |
|---|---|---|---|---|---|---|
| | *Local-scale variables* | | | | | |
| Stream physical habitat condition | Physical Habitat Index | Absolute value | Square root | phi | 87.56 | 33.00 |
| Mesohabitat | Pools | Proportion | Logit | pool | 0.20 | 0.13 |
| Physicochemical descriptors | Water temperature | °C | Square root | tem | 19.91 | 3.57 |
| Substratum | Unconsolidated substrate: clay, silt, and sand | Proportion | Logit | usu | 0.68 | 0.21 |
| Stream morphology | Mean width | m | Square root | wid | 1.74 | 0.86 |
| | Standard deviation of width | – | – | widSD | 0.59 | 0.25 |
| | Mean depth | m | Square root | dep | 0.20 | 0.08 |
| | Mean water velocity | m/s | Square root | vel | 0.14 | 0.06 |
| | Standard deviation of water velocity | – | – | velSD | 0.07 | 0.05 |
| Marginal vegetation | Grasses (mostly *Brachiaria* spp.) | Proportion | Logit | gra | 0.38 | 0.27 |
| | *Catchment-scale variables* | | | | | |
| Entire catchment | Native forest | Proportion | Logit | Cfor | 0.09 | 0.08 |
| | Pasture | Proportion | Logit | Cpas | 0.35 | 0.30 |
| | Sugarcane | Proportion | Logit | Csug | 0.38 | 0.34 |
| | Other land cover | Proportion | Logit | Coth | 0.18 | 0.12 |
| Drainage network | Native forest | Proportion | Logit | Nfor | 0.28 | 0.23 |
| | Pasture | Proportion | Logit | Npas | 0.11 | 0.08 |
| | Sugarcane | Proportion | Logit | Nsug | 0.01 | 0.02 |
| | Other land cover | Proportion | Logit | Noth | 0.59 | 0.25 |
| Local land cover | Native forest | Proportion | Logit | Lfor | 0.25 | 0.24 |
| | Pasture | Proportion | Logit | Lpas | 0.28 | 0.26 |
| | Sugarcane | Proportion | Logit | Lsug | 0.18 | 0.23 |
| | Other land cover | Proportion | Logit | Loth | 0.29 | 0.23 |
| | *Spatial variables* | | | | | |
| Network distance | PCNM eigenvectors with positive spatial correlation | – | – | $V_i$ | - | - |

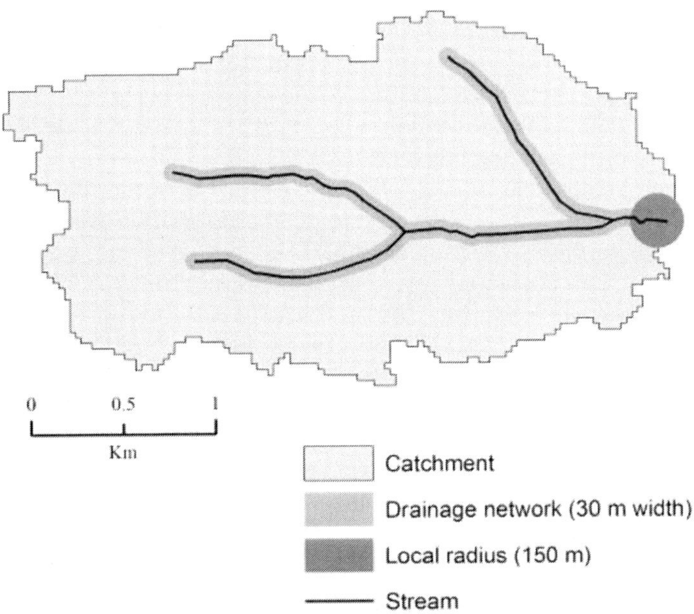


**Figure** 2. Schema illustrating a hypothetical catchment representing the three different sets at which land cover was assessed.

**Spatial variables**

We considered distance between sites as network distance (i.e. the distance between sites concerned with the dispersal pathways dictated by the riverine dendritic network, Brown & Swan 2010;altermatt 2013) The network distance between sites was calculated using Hawth's Analysis Tool (Beyer2004)forArcGIS 9.3.

From the network distance matrix, principal coordinates of neighborhood matrix (PCNM) was calculated to generate spatial variables. We retained only PCNM eigenvectors with positive spatial correlation since we were mainly concerned with patterns produced by spatially contagious processes (Borcard et al.2011) For this, we used the 'PCNM' function from the PCNM package (Legendre et al. 2013) The threshold value used in the PCNM analysis was the minimum distance giving connected network. All analyses were carried out in the R environment (R Development Core Team 2014)

## Analytical methods

First, we conducted a principal components analysis (PCA) to describe the main environmental gradients. The PCA was performed separately for each scale, namely, one for the local-scale and another for the catchment-scale. For PCA, environmental variables were standardized to zero mean and unit variance. Subsequently, exploratory data analysis (box-plot and quantile–quantile plots) was used to assess predictor variable normality and the presence of outliers. Since we used a linear model based on multiple linear regression (see explanation below), data transformations were applied to (1) linearize the relationships, (2) make the frequencies of distributions as symmetric as possible, and (3) reduce the effect of outlier data (Legendre & Legendre 2012) Consequently, the logit transformation was applied to the proportion variables (Warton & Hui 2011) and for variables in other units, square root transformation was used (Table A2). Since collinearity among predictor variables should be reduced prior to RDA (Legendre & Legendre 2012) we checked for strong linear dependencies, assessed separately within each scale, using a Spearman's correlation $\rho \geq 0.7$ as a threshold (Dormann et al. 2013) Forest in the catchment and network, pasture in the catchment and local radius, and sugarcane in the catchment and local radius displayed $\rho \geq 0.7$. We discarded forest in the catchment, and pasture and sugarcane in the local radius for the subsequent statistical analyses (Table A3). Afterward, predictor variables were standardized to zero mean and unit variance to be used in the successive analyses.

**Table A3.** Spearman's correlation (rho) for catchment-scale features (see Table A2 for features codes). In bold, features with rho ≥ 0.7.

| | Cfor | Cpas | Csug | Ceth | Nfor | Npas | Nsug | Neth | Lfor | Lpas | Lsug |
|---|---|---|---|---|---|---|---|---|---|---|---|
| Cpas | −0.2 | | | | | | | | | | |
| Csug | 0.1 | −0.9 | | | | | | | | | |
| Ceth | 0.0 | 0.2 | −0.5 | | | | | | | | |
| Nfor | **0.8** | −0.4 | 0.4 | −0.2 | | | | | | | |
| Npas | 0.0 | 0.6 | −0.3 | 0.1 | 0.0 | | | | | | |
| Nsug | 0.3 | −0.0 | 0.4 | −0.0 | 0.3 | 0.1 | | | | | |
| Neth | −0.7 | 0.2 | −0.2 | 0.1 | −0.5 | −0.3 | −0.4 | | | | |
| Lfor | 0.5 | −0.2 | 0.2 | −0.2 | 0.6 | 0.0 | 0.0 | −0.6 | | | |
| Lpas | −0.2 | **0.7** | −0.6 | 0.2 | −0.4 | 0.5 | −0.1 | 0.2 | −0.4 | | |
| Lsug | 0.1 | −0.7 | **0.8** | −0.3 | 0.3 | −0.3 | 0.3 | −0.2 | 0.1 | −0.6 | |
| Leth | −0.2 | 0.1 | −0.2 | 0.3 | −0.5 | −0.1 | −0.1 | 0.4 | −0.6 | −0.1 | −0.3 |

To reduce the initial number of predictor variables we performed a forward selection procedure (Blanchet et al. 2008 tableA5) To prevent the inflation of Type I error in a forward selection procedure, a global test needs to be done prior to any variable selection (Blanchet et al. 2008) Therefore, we carried out three global tests (10,000 permutations each): one with the local-scale variables, another with the catchment-scale variables, and last one with the PCNM eigenvectors with positive spatial correlation. We also used two stopping criteria in the forward selection procedure: (1) the corrected alpha level of rejection of null hypothesis (i.e. only if the global test was significant, the corresponding set was used to continue with forward selection procedure) and (2) the adjusted $R^2$ statistic ($R^2$adj) of the global test (Blanchet et al.2008)

Finally, to identify the influence of local-scale, catchment-scale, and spatial features on stream fish community structure we performed a partial redundancy analysis (p-RDA). The p-RDA allowed the variance partitioning of the response variable (i.e. community structure assessed through the abundance of species) into fractions explained by each set of explanatory variables (i.e. local-scale, catchment-scale, and spatial features) (Legendre et al.2011Legendre & Legendre2012) The Hellinger transformation was applied to the fish data prior to the p-RDA (Legendre & Gallagher2001) however, before applying the Hellinger transformation the abundance data were square-root-transformed in order to reduce the asymmetry of the species distributions as recommended by Legendre & Legendre (2012) We used permutation procedures (10,000 permutations; $p < 0.05$) to test the partial contribution of each predictor variable set and to test the significance of environmental variables assessed individually. For forward selection and p-RDA analyses, we used the 'forward.sel' and 'rda' functions of 'packfor' (Blanchet et al.2008) and 'vegan' (Oksanen et al. 2015) packages, respectively, in the R environment (R Development Core Team2014)

# RESULTS

A total of 3859 individuals belonging to 51 fish species, 17 families, and 6 orders were collected in the 43 stream reaches (Table A4). The first two PCA axes accounted for 46% (local-scale variables) and 59% (catchment-scale variables) of the variation in environmental features (Figure 3). Regarding the local-scale variables, PC1 represented a gradient from streams with more pools and better physical habitat conditions to streams with more grasses and unconsolidated substrate; PC2 represented streams with faster waters to wider channels (Figure 3(A)). For catchment-scale variables, the PC1 revealed the gradient from streams with more sugarcane in the catchment and forest in the network contrasting to those ones with more pasture in the catchment; PC2 separated streams with more forest and pasture in the network contrasting to those ones with other land uses (Figure 3(B)).

**Table A4**. Fish species collected in the 43 stream reaches.

| Order and families, species and authors | Codes |
|---|---|
| Characiformes | |
| Parodontidae | |
| *Apareiodon affinis* (Steindachner 1879) | Apaaff |
| *Apareiodon piracicabae* (Eigenmann 1907) | Apapir |
| *Parodon nasus* Kner 1859 | Parnas |
| Curimatidae | |
| *Cyphocharax vanderi* (Britski 1980) | Cypvan |
| Crenuchidae | |
| *Characidium gomesi* Travassos 1956 | Chagom |
| *Characidium zebra* Eigenmann 1909 | Chazeb |
| Characidae | |
| *Astyanax altiparanae* Garutti & Britski 2000 | Astalt |
| *Astyanax fasciatus* (Cuvier 1819) | Astfas |
| *Astyanax paranae* Eigenmann 1914 | Astpar |
| *Hasemania* sp. | Hassp |
| *Hemigrammus marginatus* Ellis 1911 | Hemmar |
| *Hyphessobrycon eques* (Steindachner 1882) | Hypequ |
| *Knodus moenkhausii* (Eigenmann & Kennedy 1903) | Knomoe |
| *Moenkhausia sanctaefilomenae* (Steindachner 1907) | Moesan |
| *Oligosarcus pintoi* Amaral Campos 1945 | Olipin |
| *Piabina argentea* Reinhardt 1867 | Piaarg |
| *Planaltina britskii* Menezes, Weitzman & Burns 2003 | Plabri |
| *Serrapinnus heterodon* (Eigenmann 1915) | Serhet |
| *Serrapinnus notomelas* (Eigenmann 1915) | Sernot |
| Acestrorhynchidae | |
| *Acestrorhynchus lacustris* (Lütken 1875) | Acelac |
| Erythrinidae | |
| *Erythrinus erythrinus* (Bloch & Schneider 1801) | Eryery |
| *Hoplias malabaricus* (Bloch 1794) | Hopmal |
| Lebiasinidae | |
| *Pyrrhulina australis* Eigenmann & Kennedy 1903 | Pyraus |
| Siluriformes | |
| Callichthyidae | |
| *Aspidoras fuscoguttatus* Nijssen & Isbrücker 1976 | Aspfus |
| *Callichthys callichthys* (Linnaeus 1758) | Calcal |
| *Corydoras aeneus* (Gill 1858) | Coraen |
| *Hoplosternum littorale* (Hancock 1828) | Hoplit |
| *Lepthoplosternum pectorale* (Boulenger 1895) | Leppec |
| *Megalechis thoracata* (Valenciennes 1840) | Megtho |
| Loricariidae | |
| *Hisonotus francirochai* (Ihering 1928) | Hisfra |
| *Hypostomus ancistroides* (Ihering 1911) | Hypanc |
| *Hypostomus* cf. *nigromaculatus* (Schubart 1964) | Hypnig |
| *Pterygoplichthys ambrosettii* (Holmberg 1893) | Pteamb |
| Pseudopimelodidae | |

| | |
|---|---|
| *Pseudopimelodus mangurus* (Valenciennes 1835) | Psemon |
| Heptapteridae | |
| *Imparfinis* cf. *schubarti* (Gomes 1956) | Impsch |
| *Pimelodella* cf. *avanhandavae* Eigenmann 1917 | Pimava |
| *Rhamdia quelen* (Quoy & Gaimard 1824) | Rhaque |
| Gymnotiformes | |
| Gymnotidae | |
| *Gymnotus paraguensis* Albert & Crampton 2003 | Gympar |
| *Gymnotus* cf. *sylvius* Albert & Fernandes-Matioli 1999 | Gymsyl |
| Sternopygidae | |
| *Eigenmannia trilineata* Lopez & Castello 1966 | Eigtri |
| Cyprinodontiformes | |
| Rivulidae | |
| *Melanorivulus pictus* (Costa 1989) | Melpic |
| Poeciliidae | |
| *Phalloceros harpagos* Lucinda 2008 | Phahar |
| *Poecilia reticulata* Peters 1859 | Poeret |
| Synbranchiformes | |
| Synbranchidae | |
| *Synbranchus marmoratus* Bloch 1795 | Synmar |
| Perciformes | |
| Cichlidae | |
| *Cichlasoma paranaense* Kullander 1983 | Cicpar |
| *Crenicichla britskii* Kullander 1982 | Crebri |
| *Crenicichla haroldoi* Luengo & Britski 1974 | Crehar |
| *Geophagus brasiliensis* (Quoy & Gaimard 1824) | Geobra |
| *Laetacara araguaiae* Ottoni & Costa 2009 | Laeara |
| *Oreochromis niloticus* (Linnaeus 1758) | Orenil |
| *Satanoperca pappaterra* (Heckel 1840) | Satpap |

**Table A5.** Results of forward selection procedure for stream fish community structure and each set of predictor variables (i.e. local, catchment, spatial). For variable codes, see Table A2. In bold $p < 0.05$.

| Predictor variables | Global model | | Selected variables |
|---|---|---|---|
| | $R^2_{Adj}$ | $p$ | |
| Local | 0.19 | **<0.001** | dep, gra, poo, tem, wid, uns |
| Catchment | 0.05 | **0.035** | Csug |
| Spatial (PCNM) | 0.09 | **0.001** | PCNM3, PCNM1, PCNM2 |

*Poecilia reticulata*, *Knodus moenkhausii*, *Astyanax altiparanae*, and *Aspidoras fuscoguttatus* were the most abundant species, accounting for 65% of total abundance. Overall, a small proportion of variability in fish abundance data was accounted for by the predictor variables ($R^2_{Adj.}$ = 0.20, $p$ = 0.001). Among the predictor variables, environmental variables were more influential than spatial variables. Explained variation in fish community structure was partitioned between the fractions as follows: pure local 10% ($p$ = 0.001), pure catchment 1% ($p$ = 0.166), pure spatial 3% ($p$ = 0.045), and shared environmental and spatial location 7% (fraction not testable).

The environmental variables significantly related to community structure were mean depth, proportion of grasses (mostly *Brachiaria* spp.), proportion of pools, water temperature, mean width, and proportion of unconsolidated substrate (include clay, silt, and sand; Table 1). Sugarcane proportion in the entire catchment, the only catchment-scale variable retained by the forward selection procedure, was not significant (Table 1). Only one PCNM eigenvector with positive spatial correlation influenced fish community structure but its effect was weak (Table 1).

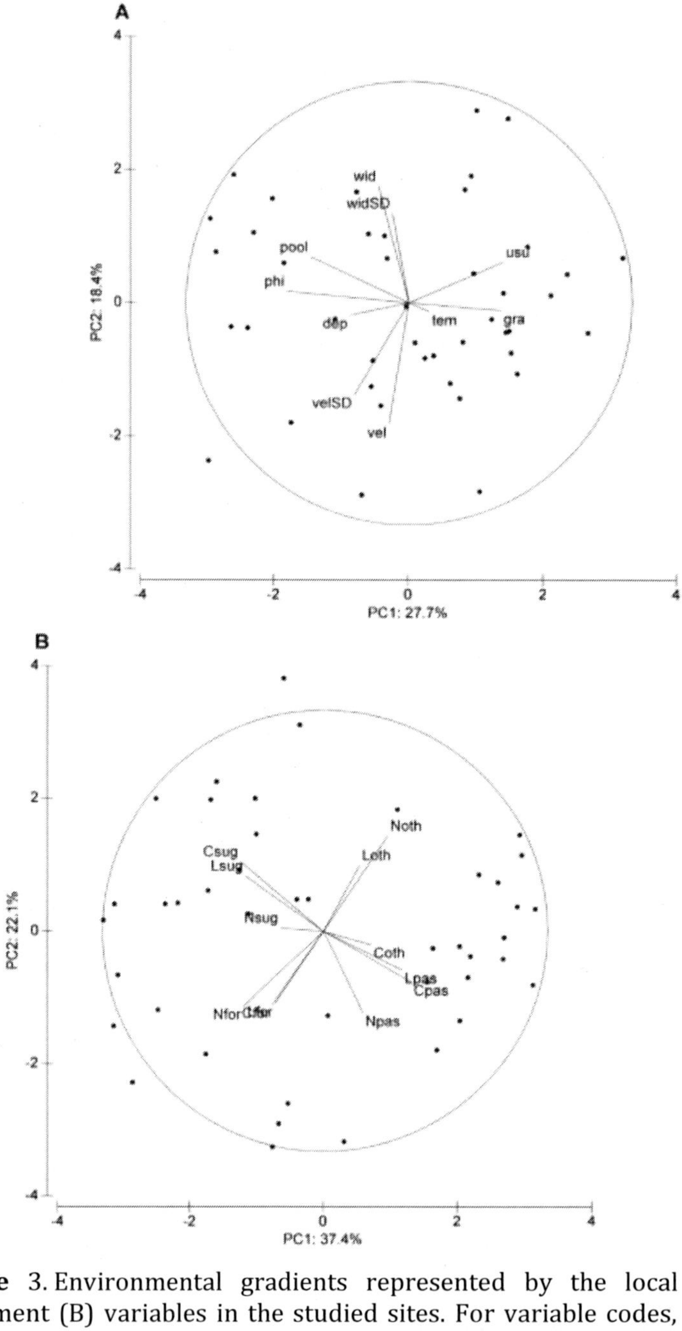


**Figure** 3. Environmental gradients represented by the local (A) and catchment (B) variables in the studied sites. For variable codes, see Table A2.

**Table** 1. Predictor variables significantly related to fish community structure. Test based on 10,000 permutations. In bold $p < 0.05$.

| | Variance | $F$ | $p$ |
|---|---|---|---|
| Mean depth | 0.04 | 3.91 | **0.001** |
| Proportion of grasses | 0.03 | 2.46 | **0.002** |
| Proportion of pools | 0.03 | 2.48 | **0.002** |
| Water temperature | 0.03 | 2.36 | **0.004** |
| Mean width | 0.02 | 2.07 | **0.008** |
| Proportion of unconsolidated substrate | 0.02 | 1.80 | **0.027** |
| Proportion of sugarcane in the catchment | 0.02 | 1.42 | 0.108 |
| PCNM eigenvector 1 | 0.02 | 2.04 | **0.010** |
| PCNM eigenvector 3 | 0.01 | 1.25 | 0.215 |
| PCNM eigenvector 2 | 0.01 | 0.85 | 0.656 |

The biplot of p-RDA depicting the pure local-scale fraction effect (i.e. considering the local variables that were significantly related to community structure; Table 1) indicated that *Poecilia reticulata* and *Hypostomus* cf. *nigromaculatus* were mainly associated with wide and shallow sites, while *Imparfinis* cf. *schubarti*, *Rhamdia quelen*, and *Moenkhausia sanctaefilomenae*displayed preference for deep and narrow environments (Figure 4). Species such as *Gymnotus*cf. *sylvius* and *Aspidoras fuscoguttatus* were associated with sites displaying high proportion of grasses, whereas the proportion of pools influenced the abundance of *Astyanax altiparanae* and *H.* cf. *nigromaculatus*, among others species (Figure 4).

## DISCUSSION

The objective of this study was to identify and compare local-scale, catchment-scale, and spatial determinants of stream fish communities in a highly modified tropical region. We found that local-scale variables contributed most to the total explained variability of fish community structure whereas catchment-scale and spatial features displayed non-significant or weak relationships with stream fish diversity. These results do not support our initial hypotheses that both local and catchment environmental features will explain variation in the stream fish community structure. Influences of local-scale environmental factors on the structure of stream fish communities have been recognized widely. For instance,

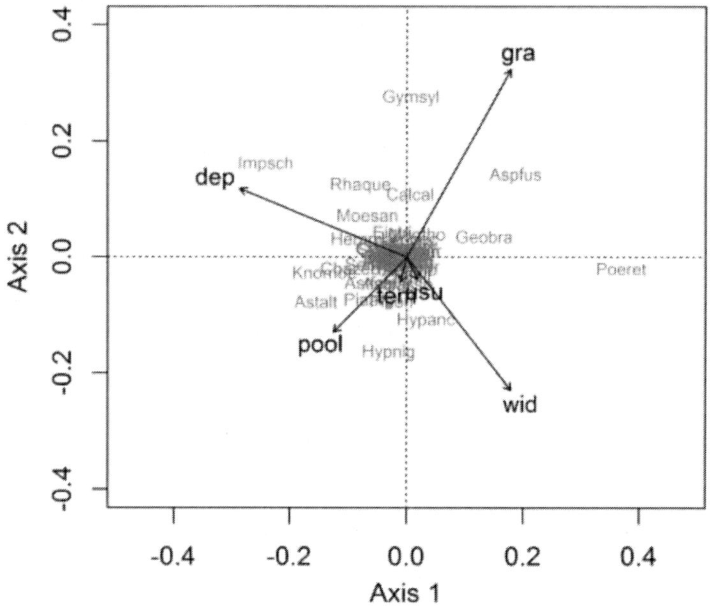


**Figure** 4. Biplot of p-RDA depicting the pure local-scale fraction effect on stream fish community structure. Note that the biplot (scaling = 2) shows only pure effect associated with local-scale variables, excluding the effects of catchment-scale and space location. For species codes, see Table A4.

variation in fish community structure has been associated with the addition of deeper areas (Sheldon 1968 Schlosser 1982) widespread propagation of marginal vegetation as grasses (mostly *Brachiaria* spp.; Casatti et al. 2009) increased extent and volume of pool habitats (Schlosser 1982) and stream width (Angermeier & Karr 1983Lammert & Allan1999) These local-scale variables are related with the vertical habitat used by fish (Sheldon1968) availability of food resources (Angermeier & Karr1983) and structural complexity of stream habitat (Casatti et al. 2009Montaña & Winemiller 2010 Carvalho & Tejerina-Garro 2015) These results suggest that species' niche differences, in terms of environmental characteristics, contribute to the variability in stream fish community structure at the spatial scale evaluated by us (Heino & Mykrä 2008) This outcome is in accordance with species sorting, the typical mechanism structuring stream communities (Heino & Mykrä 2008; Carvalho & Tejerina-Garro2015)

On the other hand, previous studies have identified that catchment-scale variables, together with local-scale features, are also important determinants of stream fish fauna. For instance, land cover variables, as proportion of agriculture, measured at entire catchment, drainage network and local radius, influence stream fish community structure in the temperate streams (Fitzpatrick et al.2001 Strayer et al.2003) In tropical streams, although studies about scale-related effects are still scarce, the findings are similar to those of temperate environments. For example, in tropical agricultural landscapes both catchment- and local-scale features affected both functional and taxonomic diversity of stream fish communities (Cruz et al.2013; Bordignon et al.2015Santos et al.2015but see Casatti et al.2015) However, it should be mentioned that although catchment variables are important (Roth et al. 1996), in most cases local-scale variables accounted for the most variation in fish community structure (Lammert & Allan 1999; Wang et al.2003Diana et al.2006Johnson et al.2007)

In our study, among the possible reasons for the lack of effect of catchment-scale variables on stream fish structure could be (1) weak associations between catchment-scale and local-scale features that are affecting stream fish communities (Heino et al.2008) (2) greater influence of past land cover on the present-day fish community structure than current land cover (Harding et al.1998) due to stream fish communities crossing the threshold of response to catchment modification, suggesting that post-threshold communities display a weak response to additional agricultural intensification (Balmford 1996 Fitzpatrick et al.2001) or (3) the low variability in catchment variables (e.g. homogeneous landscape) is not able to overcome the impact of local-scale variables (Heino et al.2007; Casatti et al.2015)

Despite the significant effect of the environment on the stream fish community structure, its contribution was relatively low (< 20% of explained variation). This could be related to the occurrence of a considerable number of less abundant and rare fish species in the data set. According to Heino and Mykrä (2008) the more prevalent these species are, the greater the amount of community structure that will not well-explained by environmental features. In addition, in communities with a high ratio of habitat generalist species, as in our study area (Casatti et al.2009) environmental variations might not necessarily result in strong variations in local community structure (Heino 2013) Stream environments are characterized by recurrent disturbances in terms of flash floods (Heino & Mykrä 2008 Marques et al.2013) and, in agricultural landscapes, by sporadic events of high pesticides concentration, increased water temperatures, and low oxygen concentrations (Hawkins et al. 2015) These random and spatially variable disturbances could also restrain the association between stream fishes and the environment, leading to important amounts of variation that are not accounted by environmental factors (Heino & Mykrä 2008Heino et al.2015)

Although spatially structured environment component could hamper the explanation of the relative importance of environmental and spatial control of community structure (Heino & Mykrä 2008) we found a weak effect of the pure spatial fraction (i.e. network distance between sites) on the stream fish community structure. This supports our hypothesis that spatial structure should be weak, as dispersal limitation will not be important at this spatial extent (i.e. within a drainage basin; Heino & Mykrä 2008 Heino et al.2015) Weak spatial structure coupled with the absence of strong environmental influence on the community structure could suggest that the fish communities at studied sites were already homogenized. This would be related to the fact that the most sensitive species to changes in environmental features had already been extinct within the system (Brook et al.2003) However, it should be noted that dispersal is an important determinant of stream fish population and community structure (Skalski & Gilliam 2000) and even though it is expected that it would not influence the community structure at the spatial extension evaluated by us, data about dispersal rates and distances for majority of freshwater organisms are lacking almost completely (Bohonak & Jenkins 2003; Heino & Mykrä2008) This obstructs the use of more appropriate divisions according with different dispersal abilities of species (Heino & Mykrä 2008) In addition, the influence of barriers built massively in agricultural landscapes (e.g. dams used by livestock; *personal observation*) on spatial structure is, as far as we know, still unknown. Future studies on dispersal in Neotropical stream fish will be fundamental to better understand the processes mediated by space, as studies for temperate stream fish suggest that variability in phenotypic traits is related to variability in movement behavior (Skalski & Gilliam 2000) and may explain, for example, cases of species invasions (Shigesada et al. 1995Skalski & Gilliam 2000)

## Ecological applications

The assessment of environmental features from a hierarchical perspective offers a suitable framework for integrating the multiple scales over which ecological processes organize stream fish communities (Angermeier & Winston1998) allows identification of appropriate spatial scale(s) at which environmental features impact the freshwater systems (Feld 2013) and gives essential information to direct funds in stream and catchment restoration (Sheldon et al. 2012;feld 2013) Results of this study reveal, in an unexpected way, that stream fish community structure in a tropical agricultural landscape was affected mainly by features operating at a single spatial scale (i.e. local scale). Therefore, local-scale variables, such as depth, proportion of grasses, proportion of pools, water temperature, width, and proportion of unconsolidated substrate could be manipulated for stream restoration purposes.

Despite of these findings, stream restoration is a major challenge for ecologists. For example, if the intention of restoration is the ecological integrity recovery, the enhancement of physical stream characteristics alone (i.e. channel reconfiguration and/or improving in-stream structural complexity) should not be the main guide for restoration efforts (Palmer et al. 2010). It is important to consider that other factors control stream biodiversity (e.g. water quality, disturbance regime, regional species pools, differences in species dispersal capacity, degraded hydrological regimes; Palmer et al.2010and references therein). Obviously, restoration at the catchment-scale is more effective in improving the ecological integrity of streams (Lake et al.2007 Palmer et al.2010) and therefore should be the first action to be considered. However, in areas like those studied herein, it is unlikely that agricultural lands would be replaced by forests in order to improve ecological integrity. Therefore, among other strategies, restoration of riparian forests plus manipulation of some local features may mitigate the impacts of anthropogenic activities in streams (Saunders et al.2002 Palmer et al.2010 Sweeney & Newbold 2014)

## ACKNOWLEDGMENTS

We thank to J.O. Zeni, M.C. Molina, E. Manzano and our colleagues from the Ichthyology Laboratory for their help during laboratory and fieldwork; IBILCE-UNESP for facilities; IBAMA for the collecting license (5580-1/11435); landowners for permission to conduct research on their properties; Frederico T.S. Miranda and Márcia S. Morinaga for help with land cover data; and Francisco Langeani and Fernando R. Carvalho for help with fish identification. We also appreciate the critical reading and suggestions from the two anonymous reviewers.

## REFERENCES

1. Altermatt F. 2013. Diversity in riverine metacommunities: a network perspective. Aquatic Ecol. 47:365–377
2. Angermeier PL, Karr JR. 1983. Fish communities along environmental gradients in a system of tropical streams. Environ Biol Fishes. 9:117–135
3. Angermeier PL, Winston MR. 1998. Local vs. regional influences on local diversity in stream fish communities of Virginia. Ecology. 79:911–927
4. Balmford A. 1996. Extinction filters and current resilience: the significance of past selection pressures for conservation biology. Trends Ecol Evol. 11:193–196
5. Beyer HL. 2004. Hawth's analysis tools for ArcGIS. Available from: http://www.spatialecology.com/htools
6. Blanchet FG, Legendre P, Borcard D. 2008. Forward selection of explanatory variables. Ecology. 89:2623–2632

7. Bohonak AJ, Jenkins DG. 2003. Ecological and evolutionary significance of dispersal by freshwater invertebrates. Ecol Lett. 6:783–796
8. Borcard D, Gillet F, Legendre P. 2011. Numerical ecology with R. New York (NY): Springer-Verlag
9. Bordignon CR, Casatti L, Pérez-Mayorga MA, Teresa FB, Brejão GL. 2015. Fish complementarity is associated to forests in Amazonian streams. Neotrop Ichthyol. 13:579–590
10. Brook BW, Sodhi NS, Ng PKL. 2003. Catastrophic extinctions follow deforestation in Singapore. Nature. 424:420–423
11. Brown BL, Swan CM. 2010. Dendritic network structure constrains metacommunity properties in riverine ecosystems. J Anim Ecol. 79:571–580
12. Carvalho RA, Tejerina-Garro FL. 2015. Environmental and spatial processes: what controls the functional structure of fish assemblages in tropical rivers and headwater streams? Ecol Freshw Fish. 24:317–328.
13. Casatti L, Ferreira CP, Carvalho FR. 2009. Grass-dominated stream sites exhibit low fish species diversity and dominance by guppies: an assessment of two tropical pasture river basins. Hydrobiologia. 632:273–283
14. Casatti L, Teresa FB, Zeni JO, Ribeiro MD, Brejão GL, Ceneviva-Bastos M. 2015. More of the same: high functional redundancy in stream fish assemblages from tropical agroecosystems. Environ Manage. 55:1300–1314
15. [CEPAGRI] Centro de Pesquisas Meteorológicas e Climáticas Aplicadas à Agricultura. 2017. Clima dos municípios paulistas: São José do Rio Preto [Climate of the municipalities of the State of São Paulo: São José do Rio Preto]. Available from: http://www.cpa.unicamp.br/outras-informacoes/clima_muni_559.html
16. Cianfrani CM, Sullivan SM, Hession WC, Watzin MC. 2012. A multitaxonomic approach to understanding local- versus watershed-scale influences on stream biota in the Lake Champlain basin, Vermont, USA. River Res Appl. 28:973–988
17. Cruz BB, Miranda LE, Cetra M. 2013. Links between riparian landcover, instream environment and fish assemblages in headwater streams of south-eastern Brazil. Ecol Freshw Fish. 22:607–616
18. Diana M, Allan JD, Infante D. 2006. The influence of physical habitat and land use on stream fish assemblages in southeastern Michigan. Am Fish Soc Symp. 48:359—374
19. Dembkowski DJ, Miranda LE. 2014. Environmental variables measured at multiple spatial scales exert uneven influence on fish assemblages of floodplain lakes. Hydrobiologia. 721:129–144
20. Dormann CF, Elith J, Bacher S, Buchmann C, Carl G, Carré G, García Marquéz JR, Gruber B, Lafourcade B, Leitão P, et al. 2013. Collinearity: a review of methods to deal with it and a simulation study evaluating their performance. Ecography. 36:27–46
21. Feld CK. 2013. Response of three lotic assemblages to riparian and catchment-scale land use: implications for designing catchment monitoring programmes. Freshw Biol. 58:715–729
22. Fitzpatrick FA, Scudder BC, Lenz BN, Sullivan DJ. 2001. Effects of multi-scale environmental characteristics on agricultural stream biota in Eastern Wisconsin. J Am Water Resour Assoc. 37:1489–1507
23. Freeman MC, Pringle CM, Jackson CR. 2007. Hydrologic connectivity and the contributions of stream headwaters to ecological integrity at regional scales. J Am Water Resour Assoc. 43:5–14
24. Frissell CA, Liss WJ, Warren CE, Hurley MD. 1986. A hierarchical framework for stream habitat classification: viewing streams in a watershed context. Environ Manage. 10:199–214
25. Gerhard P, Verdade LM. 2016. Stream fish diversity in an agricultural landscape of southeastern Brazil. In: Gheler-Costa C, Lyra-Jorge MC, Verdade LM, editors. Biodiversity in agricultural landscapes of southeastern Brazil. Berlin: De Gruyter Open; p. 206–224

26. Géry J. 1969. The fresh-water fishes of South America. In: Fitkau EJ, editor. Biogeography and ecology in South America. The Hague: Dr. W. Junk; p. 828–848
27. Harding JS, Benfield EF, Bolstad PV, Helfman GS, Jones EBD. 1998. Stream biodiversity: the ghost of land use past. Proc Natl Acad Sci USA. 95:14843–14847
28. Heino J. 2013. The importance of metacommunity ecology for environmental assessment research in the freshwater realm. Biol Rev Camb Philos Soc. 88:166–178.
29. Heino J, Mykrä H. 2008. Control of stream insect assemblages: roles of spatial configuration and local environmental factors. Ecol Entomol. 33:614–622
30. Heino J, Mykrä H, Kotanen J, Muotka T. 2007. Ecological filters and variability in stream macroinvertebrate communities: do taxonomic and functional structure follow the same path?Ecography. 30:217–230
31. Heino J, Mykrä H, Kotanen J. 2008. Weak relationships between landscape characteristics and multiple facets of stream macroinvertebrate biodiversity in a boreal drainage basin. Landsc Ecol. 23:417–426
32. Heino J, Melo AS, Siqueira T, Soininen J, Valanko S, Bini LM. 2015. Metacommunity organisation, spatial extent and dispersal in aquatic systems: patterns, processes and prospects. Freshw Biol. 60:845–869
33. [IPT] Instituto de Pesquisas Tecnológicas. 1999. Diagnóstico da situação atual dos recursos hídricos e estabelecimento de diretrizes técnicas para a elaboração do Plano da Bacia Hidrográfica do São José dos Dourados—minuta [Diagnosis of the current situation of water resources and establishment of technical guidelines for the elaboration of the São José dos Dourados River Basin Plan]. Available from: http://www.sigrh.sp.gov.br/public/uploads/documents/7001/volume-1.pdf
34. Hawkins CP, Mykrä H, Oksanen J, Vander Laan JJ. 2015. Environmental disturbance can increase beta diversity of stream macroinvertebrate assemblages. Glob Ecol Biogeogr. 24:483–494
35. Hynes HBN. 1975. The stream and its valley. Verh Int Ver Theor Ang Limnol. 19:1–15.
36. Johnson RK, Furse MT, Hering D, Sandin L. 2007. Ecological relationships between stream communities and spatial scale: implications for designing catchment-level monitoring programmes. Freshw Biol. 52:939–958
37. Lake PS, Bond N, Reich P. 2007. Linking ecological theory with stream restoration. Freshw Biol. 52:597–615
38. Lammert M, Allan JD. 1999. Assessing biotic integrity of streams: effects of scale in measuring the influence of land use/cover and habitat structure on fish and macroinvertebrates. Environ Manage. 23:257–270
39. Laurance WF, Sayer J, Cassman KG. 2014. Agricultural expansion and its impacts on tropical nature. Trends Ecol Evol. 29:107–116
40. Legendre P, Gallagher ED. 2001. Ecologically meaningful transformations for ordination of species data. Oecologia. 129:271–280.
41. Legendre P, Legendre LFJ. 2012. Numerical ecology. 3rd ed. Amsterdam: Elsevier
42. Legendre P, Borcard D, Blanchet FG, Dray S. 2013. PCNM: MEM spatial eigenfunction and principal coordinate analyses. R package version 2.1-2/r109. Available from: http://R-Forge.R-project.org/projects/sedar/
43. Legendre P, Oksanen J, ter Braak CJF. 2011. Testing the significance of canonical axes in redundancy analysis. Methods Ecol Evol. 2:269–277
44. Lehner B, Liermann CR, Revenga C, Vörösmarty C, Fekete B, Crouzet P, Döll P, Endeja n M, Frenken K, Magome J, et al. 2011. High-resolution mapping of the world's reservoirs and dams for sustainable river-flow management. Methods Ecol Evol. 9:494–502
45. Marques LC, Ceneviva-Bastos M, Casatti L. 2013. Progressive recovery of a tropical deforested stream community after a flash flood. Acta Limnol Bras. 25:111–123
46. Montaña CG, Winemiller KO. 2010. Local-scale habitat influences morphological diversity of species assemblages of cichlid fishes in a tropical floodplain river. Ecol Freshw Fish. 19:216–227

47. Nalon MA, Matto ISA, Franco GADC. 2008. Diretrizes para a conservação e restauração da biodiversidade no Estado de São Paulo [Guidelines for the conservation and restoration of biodiversity in the State of São Paulo]. São Paulo: Instituto de Botânica. Chapter 2, Meio físico e aspectos da fragmentação da vegetação [Physical environment and aspects of fragmentation of vegetation]; p. 14–21
48. Oksanen J, Blanchet FG, Kindt R, Legendre P, Minchin PR, O'Hara RB, Simpson GL, Sol ymos P, Stevens HH, Wagner H. 2015. Vegan: community ecology package. Available from: https://cran.r-project.org/web/packages/vegan/index.html
49. Palmer MA, Menninger HL, Bernhard E. 2010. River restoration, habitat heterogeneity and biodiversity: a failure of theory or practice? Freshw Biol. 55:205–222
50. R Development Core Team. 2014. R: a language and environment for statistical computing. Vienna: R Foundation for Statistical Computing. Available from: http://www.R-project.org/
51. Rahel FJ. 2002. Homogenization of freshwater faunas. Annu Rev Ecol Syst. 33:291–315
52. Roth NE, Allan JD, Erickson DL. 1996. Landscape influences on stream biotic integrity assessed at multiple spatial scales. Landsc Ecol. 11:141–156
53. Rudorff BFT, Aguiar DA, Silva WF, Sugawara LM, Adami M, Moreira MA. 2010. Studie s on the rapid expansion of sugarcane for ethanol production in São Paulo State (Brazil) using Landsat data. Remote Sens. 2:1057–1076
54. Sala OE, Chapin FS, Armesto JJ, Berlow E, Bloomfield J, Dirzo R, Huber-Sanwald E, Huenneke LF, Jackson RB, Kinzig A, et al. 2008. Global biodiversity scenarios for the year 2100. Science. 287:1770–1774
55. Santos FB, Ferreira FC, Esteves KE. 2015. Assessing the importance of the riparian zone for stream fish communities in a sugarcane dominated landscape (Piracicaba River Basin, southeast Brazil). Environ Biol Fishes. 98:1895–1912
56. Saunders DL, Meeuwig JJ, Vincent ACJ. 2002. Freshwater protected areas: strategies for conservation. Conserv Biol. 16:30–41
57. Schlosser IJ. 1982. Fish community structure and function along two habitat gradients in a headwater stream. Ecol Monogr. 52:395–41
58. Sheldon AL. 1968. Species diversity and longitudinal succession in stream fishes. Ecology. 49:193–198
59. Sheldon F, Peterson EE, Boone EL, Sippel S, Bunn SE, Harch BD. 2012. Identifying the spatial scale of land use that most strongly influences overall river ecosystem health score. Ecol Appl. 22:2188–2203
60. Shigesada N, Kawasaki K, Takeda Y. 1995. Modeling stratified diffusion in biological invasions. Am Nat. 146:229–251
61. Silva AM, Casatti L, Álvares CA, Leite AM, Martinelli LA, Durrant SF. 2007. Soil loss risk and habitat quality in streams of a mesoscale river basin. Sci Agric. 64:336–343
62. Skalski GT, Gilliam JF. 2000. Modeling diffusive spread in a heterogeneous population: a movement study with stream fish. Ecology. 81:1685–170
63. Strayer R, Beighley E, Thompson LC, Brooks S, Nilsson C, Pinay G, Naiman RJ. 2003. E ffects of land cover on stream ecosystems: roles of empirical models and scaling issues. Ecosystems. 6:407–423
64. Sweeney BW, Newbold JD. 2014. Streamside forest buffer width needed to protect stream water quality, habitat, and organisms: a literature review. J Am Water Resour Assoc. 50:560–584
65. Victor MAM, Cavalli AC, Guillaumon JR, Serra RF. 2005. Cem anos de devastação - Revisitada 30 anos depois [A hundred years of devastation: revisited 30 years later]. Brasília: Ministério do Meio Ambiente.
66. Wang L, Lyons J, Rasmussen P, Seelbach P, Simon T, Wiley M, Kanehl P, Baker E, Nie mela S, Stewart PM. 2003. Watershed, reach, and riparian influences on stream fish assemblages in the Northern lakes and forest ecoregion, U.S.A. Can J Fish Aquat Sci. 60:491–505

67. Warton DI, Hui FKC. 2011. The arcsine is asinine: the analysis of proportions in ecology. Ecology. 92:3–10

# CHAPTER 4

# Acoustic and Visual Courtship Traits in Two Sympatric Marine Gobiidae Species – Pomatoschistus Microps and Pomatoschistus Minutes

**Eva-Lotta Blom & Isabel Mück & Katja Heubel Ola Svensson**

*Department of Biological and Environmental Sciences, University of Gothenburg, Box 463, SE-405 30 Gothenburg, Sweden*

*Animal Evolutionary Ecology, University of Tübingen, Auf der Morgenstelle 28, D-72076 Tübingen, Germany*

*Institute for Zoology, Research Station Grietherbusch, University of Cologne, D-50923 Cologne, Germany*

*The Linnaeus Centre forMarine Evolutionary Biology, University of Gothenburg, Gothenburg, Sweden*

## ABSTRACT

Divergence in courtship traits across species can evolve as adaptations to different environments, and also through avoidance of reproductive interference and character displacement. Differences may also be explained by phylogenetic relationships. We compared different courtship traits, including male courtship sounds, in two sympatric Pomatoschistus species. Both species are characterised by having male and female courtship, and paternal care of eggs in nests under mussel shells and rocks. In addition to presenting novel observations, we reviewed the literature on courtship traits for both species and complemented it with new observations. We found that courting males of the common goby P. microps sing louder and produce sounds of shorter duration than males of the sand goby P. minutus. Furthermore, males of P. microps swim faster towards females during courtship than males of P. minutus. The eyes of P. minutus females turn black during courtship attempts, whereas this is not the case for females of P. microps. Species-specific differences in courtship sounds and behavior may lead to different susceptibility of the two species to environmental change such as noise pollution and turbidity

**Keywords**: Acoustic communication Character displacement .Gobiidae .Matechoice .Sexual selection Sound signal production

# INTRODUCTION

Animals use a wide range of modalities during courtship and the effectiveness of signals co-varies with environmental conditions (Bradbury and Vehrencamp 2011; Candolin and Wong 2012). Therefore, signals (including acoustic signals) often match measures of optimal signal transmission (Bradbury and Vehrencamp 2011; Wilkins et al. 2013). Under turbid conditions, visual courtship signals can be hampered (Järvenpää and Lindström 2004; Seehausen et al. 1997; Heubel and Schlupp 2006) and changes in pH-levels can modify the use of chemical cues (Heuschele and Candolin 2007). Recently, acoustic communication and its interaction with environmental properties and background noise have been of particular interest (Radford et al. 2014; Holt and Johnston 2015). In the so called 'sand goby group' (family Gobiidae, Huyse et al. 2004), at least eight species have been shown to produce low frequency acoustic pulses in a reproductive context (Malavasi et al. 2012; Bolgan et al. 2013; Pedroso et al. 2013; de Jong et al. 2016) two-spotted goby Gobiusculus flavescens (Fabricius, 1779), Adriatic dwarf goby Knipowitschia panizzae (Verga, 1841), Italian spring goby Knipowitschia punctatissima (Canestrini, 1864) canestrini's goby Pomatoschistus c a n e s t r i n i ( N i n n i , 1 8 8 3 ) , ma r b l e d goby Pomatoschistus marmoratus (Risso, 1810), common goby Pomatoschistus microps (Krøyer, 1838) (one individual only), sand goby Pomatoschistus minutes (Pallas, 1770) and painted gobyPomatoschistus pictus (Malm, 1865). The functions of these sounds are still unknown, but they have been proposed to be used in male courtship and female choice (Lugli and Torricelli 1999; Lindström and Lugli 2000; Pedroso et al. 2013), or in species recognition (Pedroso et al. 2013).

P. minutus and P. microps are small marine fish distributed in lagoons, coastal areas and estuaries of the Atlantic, Mediterranean and Baltic region (Miller 1986; Kullander et al. 2012). The two species are sympatric in the study area on the west coast of Sweden, with an extensive overlap between their habitats. However, P. microps is more abundant in very shallow and often muddy areas, whereas P. minutus is usually found on sandy bottoms and in slightly deeper (>0.5 m) water (personal observation, Miller 1986; Nellbring 1986; Nellbring 1993; Kullander et al. 2012). During a single breeding season, these short-lived fishes (1– 2 years) can reproduce repeatedly with different mates (Miller 1975; Miller 1986; Kullander et al. 2012). There is an overlap in the breeding season between species, with spawning peaks occurring in spring and early summer (earlier peak in P. minutus, range March to July, and later peak in P. microps, May to September Kullander et al. 2012). Males of both species generally establish territories and build a nest by covering mussel shells with sand. Both species attract females with visual courtship displays and males provide exclusive paternal care until the eggs hatch (Forsgren 1999; Magnhagen 1999). P. minutus males prepare the surface of the nest ceiling with sperm-containing mucus on which the eggs are laid (Svensson and Kvarnemo 2005). This is most likely also the case for P. microps (Svensson et al. 1998). It has been shown in other gobies that the mucus contains pheromones (Locatello et al. 2002), and has antimicrobial functions (Giacomello et al. 2008). This antimicrobial function can also be important in female choice because egg infections have an effect on egg survival (Lehtonen and Kvarnemo 2015a; Vallon et al. 2016) and influence female choice (Lehtonen and Kvarnemo 2015b). P. minutus and P. microps are highly sympatric and often occur in close proximity, competing over the same nest resources (personal observation). There is a substantial overlap in terms of fish size, depth, and bottom structure and thus interspecific interference is a real concern.

The aim of our paper is to describe and compare the courtship sound and behavior of two highly sympatric goby species.We reviewed the literature on descriptions of previously studied traits and behavior in P. minutes and P. microps and aim to build a foundation and concise, accessible reference for further studies of acoustical and behavioral traits in different species of gobies. Closely related species are expected to avoid reproductive interference (Gröning and Hochkirch 2008) and diverge through character displacement (Pfennig and Pfennig 2009). Therefore, we predict differences in the courtship sounds, and behavioral and morphological courtship traits facilitating discrimination between the two species with extensive habitat overlap.

# METHODS

We conducted the experiment at The Sven Lovén Centre of Marine Infrastructure in Kristineberg on the west coast of Sweden (58°15′ N, 11°27′ E) during the last 2 weeks in June 2013. All aquaria, including experimental aquaria, had continuously renewed surface seawater (salinity 22–34, temperature 12.8–13.4 °C) and a layer of fine sand for the fish to hide in and to use as nest material. Light was provided through large windows and hence followed the natural Swedish summer light regime of 18 h between sunrise and sunset (it is never completely dark).We blocked visual interactions across all tanks. We caught all fish by hand trawling in bays nearby the station and kept them separated by species and sex in different storage tanks (20–50 l) for 2–14 days before the experiment started.We fed the fish once a day with commercial fish food granules (Nutra HP, Skretting).

## Experimental set up

We placed the experimental aquaria (20 l) on top of two marble layers interspaced with cones of silicon for sound insulation. All males (P. microps n = 10, P. minutus n = 10) had access to an artificial nest, made of a piece of polypropylene tube (Ø 56 mm) with a pipe attached like a chimney (Ø 20 mm). To encourage nest building, we introduced a ripe female in a transparent plastic vial in front of the nest in each aquarium. The vial had small holes on the side and a mesh net over the top to allow for water exchange. We inserted a hydrophone into the chimney of the nest (Song Meter SM2+ platform and hydrophone HTI-96-MIN with pre-amplifier, frequency response 2 Hz to 30 KHz, wildlife acoustic, bioacoustics monitoring systems, Maynard, US; 0 dB = 1 V/pa@1 KHz, calibrated hydrophone sensitivity −165 dB re 1 V/μP, gain 48 dB, sampling frequency 16 KHz). In front of the aquarium, we set up a camcorder for video recordings (Canon Legria HF M56, Ōta, Tokyo, Japan). We tested P. minutus (which has an earlier peak in breeding season) during the first week, and P. microps in the following week of the experiment. After 12 h we released the female from the vial and, due to the naturally differing female-biased operational sex ratios reflected in the catches as well as in previous studies (Kvarnemo 1994; Mück 2016), we added two (P. microps) or one (P. minutus) additional females. We recorded visual and acoustic behavior for 60 min. After the trial, we measured the males to the nearest 0.5 mm (LT P. microps: 33.7 ± 2.5 mm; LT P. minutus: 53.8± 3.2 mm).

## Acoustic courtship

An acoustic signal produced by a male contains a number of acoustic pulses. These pulses produce a train, which are then clustered into bursts (Fig. 1). We analyzed the two first trains made by each male, following the protocol of Lindström and Lugli (2000), for (1) number of pulses per train, (2) the total train duration [ms]), (3) pulse repetition rate (pulse per second i.e. number of pulses per train divided by train duration [s]) and (4) pulse period (time [ms] between two pulses) with Matlab R2009b (The Mathworks Inc., Natick, Massachusetts, USA). As the sound files for P. microps were clipped, we measured (5) average pulse amplitude [mV] only for P. minutus.

## Visual courtship

All authors independently examined the films, leading to the synthesis of behaviors presented in the ethogram in Table 2.We quantified and scored behavioral data by watching video clips (without the synchronized sound files) of experimental tanks of both species in random order. Therefore, it was observer-blind in terms of acoustic signaling properties.We collected and analyzed the data on frequency and duration (s) of the behavioral states and events (Table 2) using the event recorder JWatcher 1.0 (D. T. Blumstein et al., University of California, USA & Macquarie University, Australia). We scored the coloration of females and males by rewatching the films with a special emphasis duringmale– female interactions (courtship).We have previously observed that P. microps males move faster than P. minutes males during courtship. To quantify this, we measured courtship movements parallel to the front of the aquaria (the first movement was analyzed). Six males of each species moved in a way, which was possible to analyze. We annotated time (ms) directly form the video recording and measured distances using ImageJ (Rasband, W.S., ImageJ, U. S. National Institutes of Health, Bethesda, Maryland, USA, http://imagej.nih.gov/ij/, 1997–2016).

## Data analysis

In one aquarium we accidently had two P. microps males. However, the subordinate male was dug down in the back corner except for the very first minutes after the release. Thus, we did not use this replicate for the analysis of visual courtship behavior and only included it in the color and sound analyses. The nest-owning male produced the sound inside the nest accompanied by a female.

Sound characteristics were tested with permutation tests (Permutations tester 1.0 Stat Boss Michael J. Lew 2008). We report the median, the range of the sound variables for the five males per species (based on the averages of the two trains produced by each male), as well as the absolute range of all 10 analyzed sounds per species. We used t-tests to test for differences between species in the proportion of the observation time that males spent on courtship, as well as swim speed. We applied square root arcsine transformations to the proportions and a square root transformation to swim speed data prior to testing to fulfill the assumptions for parametric tests. Untransformed means ± SDs are given in the text and Table 2.

# RESULTS

## Acoustic courtship

Five out of ten P. microps males and five out of ten P. minutus males produced sound in a reproductive context. One oscillogram from each species is given in Fig. 1. The males of P. microps only produced sound inside the nest when accompanied by a female while the males of P. minutus also produced sound in the nest opening when a female was close (Table 2). Four P. minutus males produced the trains inside the nest while the females were at the nest entrance, and one male of P. minutus produced the two first trainswhen the female was also inside the nest. The latter two trains fall within the range of all sound parameters of the other eight P. minutus trains, and thus there was no qualitative difference between courtship sounds produced when the female was inside the nest compared to when she was outside the nest.

P. minutus had a longer duration of the train than P. microps (Table 1). P. microps produced fewer pulses per train than P. minutus, although this was not significant. Both species had a similar pulse repetition rate and pulse period (Table 1). All sound files for P. microps males (N = five males, two sounds per male) were clipped, whereas no sound files from P. minutes were clipped (N = five males, two sounds per male) (5/5 vs. 0/5, Fisher exact test, $p = 0.0079$). In both species, all analyzed sounds were produced when the males were lying with their abdomen towards the bottom of the nest and with the same distance from the hydrophone (1 ± 0.5 cm). For P. minutus males (N = 5) we found the amplitude at a range of 1.5–3 mV.

## Visual courtship

In Table 2 we provide references and synthesize previously reported courtship traits and behaviors in P. microps and P. minutus. Novel descriptions of observations that we have not found previously reported are the lack of black eyes in courting P. microps females and that male P. microps swim faster than male P. minutes during courtship.We were able to confirm the previously reported traits and behaviors included in our ethogram and provide our observed range and median

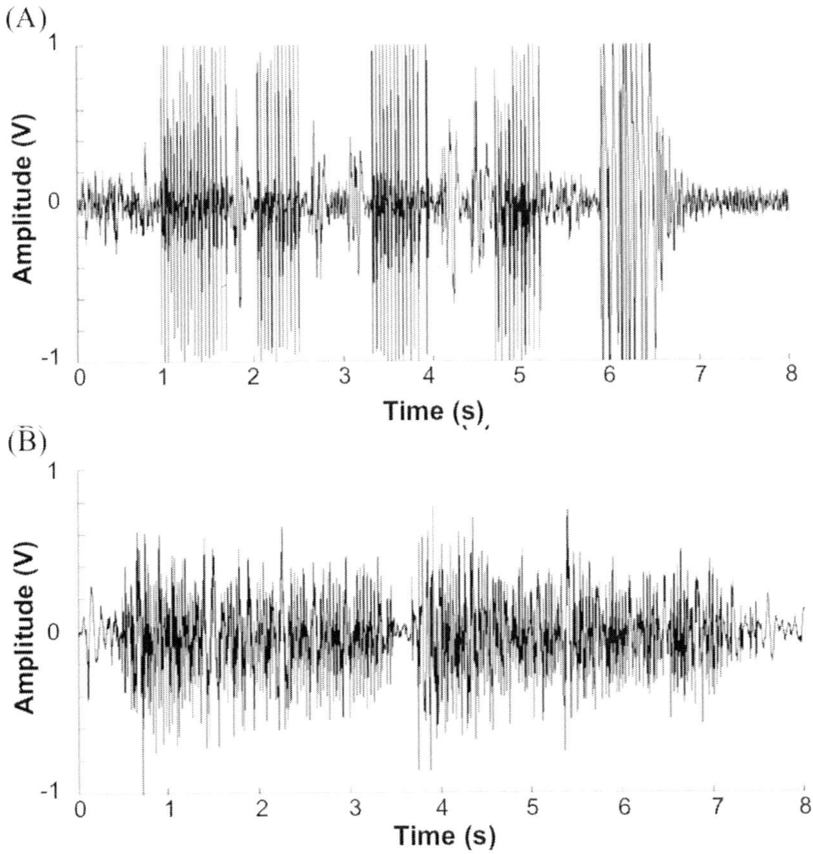


**Figure. 1** a The figure shows a representative sound of five trains clustered into one burst produced by a common goby Pomatoschistus microps male. Amplitudes are clipped at 1 V. b The figure shows a representative sound of two trains clustered into on burst produced by a sand goby Pomatoschistus minutus male. Both oscillograms were made in Matlab R2009b (The Mathworks Inc., Natick, Massachusetts, USA)

**Table 1** Acoustic sound features from males of P. microps and P. minutus produced during courtship

| Acoustic parameters | *P. microps* | | | *P. minutus* | | | Permu-tation test |
|---|---|---|---|---|---|---|---|
| | Median | Range | Range (abs) | Median | Range | Range (abs) | |
| Number of pulses per train | 15.0 | 11.5–17.5 | 9.0–18.0 | 32.5 | 13.0–89.0 | 10.0–121.0 | $p = 0.055$ |
| Train duration (ms) | 535 | 370–390 | 280–840 | 1060 | 625–3450 | 400–4900 | $\mathbf{p = 0.040}$ |
| Pulse repetition rate (pulse/s) | 26.8 | 22.2–37.9 | 21.4–46.7 | 26.1 | 20.5–29.5 | 18.0–31.4 | $p = 0.41$ |
| Pulse period (ms) between two pulses | 39.9 | 38.4–45.1 | 36.7–46.7 | 40.0 | 33.3–50.9 | 30.0–55.0 | $p = 0.95$ |

Descriptive statistics is based on the median of the two first pulse trains in one burst produced by each male (*P. microps* $n = 5$ males, *P. minutus* $n = 5$ males). Range is the range of the means of the two analyses sounds from each male ($n = 5$ for each species) and range (abs) is the range of all measured sounds ($n = 10$ for each species). Significant differences (at alpha = 0.05) are presented in bold

for all behaviors (Table 2). Females of both species courted males. During courtship, the eye, and the area around it, of P. minutus females turns black within seconds (Kvarnemo et al. 1995; Forsgren 1997). This black coloration is absent in P. microps females (Table 2). Females of both species present their bellies during courtship, by hopping up and down in small movements in direct proximity to the male. The two species have a similar male nuptial coloration with a black and blue spot on the anterior dorsal fin, which is strongly colored during a courtship attempt. In particular, P. microps has parallel stripes on the flanks and its facial lines are highly conspicuous during courtship. These facial lines are also present in females and more pronounced in P. microps than in P. minutus. P. microps males lack the blue anal fin typical of P. minutus and have a brown-red throat in contrast to a grey-white throat found in P. minutus (Table 2). Male courtship behavior is similar with both species starting courtship using fast approaches towards females with erected fins. P. microps males do small jumps with their mouth opened and their gill covers raised, including incidences of physical contact with the female, that sometimes escalate to aggression. Typically, the males of both species swim back to the nest in a conspicuous manner, considered a lead display ('lead swim'), as females may choose to follow the courting male into his nest. Other characteristic behaviors observed between courtship attempts were nest building by fanning sand upon the nest with the tail and 'displacement fanning' (fanning in the nest in the absence of eggs) (Table 2). We did not observe any 'eight displays' (the male swims rapidly in front of the female in a figure of eight pattern) which have been described to be typical for P. pictus (Amorim and Neves 2007). The two species did not differ significantly in the proportion of time spent courting (P. microps 0.050 ± 0.06, P. minutus 0.022 ± 0.03; t = 1.26, d.f. = 18, p = 0.22). During courtship, the short distance movements of P. microps were faster than those of P. minutus (t = 2.76, d.f. = 10, p = 0.02, Table 2).

# DISCUSSION

Here we report several courtship traits, which are generally similar but do bear a few intriguingly distinct differences between two closely related, morphologically and ecologically similar goby species (Huyse et al. 2004; Kullander et al. 2012). One important courtship trait that differs between P. microps and P. minutes males are the sounds produced during female attraction. The courtship sound of P. microps males has a shorter duration than the one of P. minutus males. While the settings of the dB gain on the hydrophone worked well for P. minutus, all sounds of P. microps were clipped. Because all sounds were produced inside the nest with the male at the same distance from the hydrophone, we conclude that the courtship sound of P. microps most likely is louder than the sound of P. minutus. Previously, Bolgan et al. (2013) reported sound production in one single individual male of P. microps, and were hence limited in their analyses of the P. microps breeding vocalization. Although our study also suffers from low sample size and methodological issues, we show that there are clear species differences in sound production. By comparing the two species we also highlight that only females of P. minutus display black eyes during courtship. During courtship attempts P. microps males swim faster than P. minutus males. Furthermore, we confirm several previously but separately reported differences (Table 2). We also report for the first time that P. minutus produces courtship sounds inside the nest when the female is outside of the nest.

Species differences in courtship sounds, and other traits involved in mate choice, may reflect the phylogenetic history, but they may also have diverged as a response to differences in the physical environment as well as intra- and inter-specific interactions (Gröning and Hochkirch 2008; Pfennig and Pfennig 2009; Bradbury and Vehrencamp 2011; Wilkins et al. 2013). P. microps and P. minutus belong to two different clades of the 'sand goby group' radiation. The divergence time of the two clades is estimated to be 2–5 Mya (Huyse et al. 2004; Vanhove et al. 2012) and behaviors including sound may have diverged during this time, and hence show a phylogenetic signal (Malavasi et al. 2012). The two other sympatric 'sand goby group' species in shallow bays of Sweden (Kullander et al. 2012), Gobiusculus flavescens (Fabricius, 1779) and P. pictus, as well as the deep water species Pomatoschistus norvegicus (Collett, 1902), appear to belong to the same clade as P. minutus. P. marmoratus, the sister species to P. microps (Huyse et al. 2004; Vanhove et al. 2012), is not found in Sweden (Kullander et al. 2012). G. flavescens, which is semipelagic and not benthic, is at present not in the genus Pomatoschistus, rendering it paraphyletic and in need of revision (Vanhove et al. 2012).

**Table 2** Synthesis of previously described and novel observations of male and female courtship behaviors of P. microps and P. minutes

| Courtship trait | Ethogram/ Description | Result from present study. Properties, range, median (s) for behaviors | | | | Selected references |
|---|---|---|---|---|---|---|
| | | *P. microps* | *P. minutus* | *P. microps* | *P. minutus* | |
| Female | | | | | | |
| Bobbing & Belly display | Up and down bobbing. Display of belly towards the male | Present | Present | 1–11, 3, *n* = 7 | 1–18, 8, *n* = 3 | CG: (Borg et al. 2002; Nyman 1953) SG: (Fonds 1973) |
| Black eyes during courtship | Conspicuous black eyes and lines around the eyes of ready-to-mate females | Absent | Present | | 60–2340, 403, *n* = 5 | SG: (Forsgren 1997; Kvarnemo et al. 1995) |
| Male | | | | | | |
| General male nuptial color | Bands on the sides of the body, anal fin with black edge, a black and blue spot in the first dorsal fin | Present | Present | *n* = 10 | *n* = 10 | CG: (Kullander et al. 2012) SG: (Fonds 1973; Kullander et al. 2012) |
| Blue anal fin | | Absent | Present | *n* = 10 | *n* = 10 | CG: (Kullander et al. 2012) SG: (Kullander et al. 2012) |
| Male throat color | | Brown-red | Grey-white | *n* = 10 | *n* = 10 | CG: (Kullander et al. 2012) SG: (Kullander et al. 2012) |
| Fin display | Erecting all fins while displaying towards female (when still as well as swimming) | Present | Present | <1–84, 8, *n* = 8 | <1–22, 3, *n* = 5 | CG: (Magnhagen 1998) SG: (Forsgren 1997; Forsgren et al. 1996) |
| Jumps | Short distance fast movements | Present | Present | 1–16, 2, *n* = 8 | 1–2, 2, *n* = 3 | CG: (Jones and Reynolds 1999) SG: (Fonds 1973) |
| Swim speed | Swimming speed in vicinity of female | 13.1 ± 9.1 cm/s | 4.3 ± 1.5 cm/s | 5–26, 8, *n* = 6 | 2–6.4, *n* = 6 | Present study |
| Attack | Fast swimming towards the female including bites | Present | Present | <1–11, 2, *n* = 5 | <1–4, 1, *n* = 6 | CG: (Magnhagen 1998) SG: (Fonds 1973) |
| Lead swim | Targeted swimming towards the nest during courtship context | Present | Present | <1–6, 2, *n* = 3 | 2–7, 3, *n* = 5 | CG: (Magnhagen 1998; Nyman 1953) SG: (Fonds 1973; Forsgren 1997) |
| Displacement fanning | Fanning inside the nest also without eggs | Present | Present | 1–84, 8, *n* = 7 | 2–44, 7, *n* = 3 | CG: (Jones and Reynolds 1999) SG: (Forsgren et al. 1996; Pampoulie et al. 2004) |
| Nest building | Nest building during courtship context | Present | Present | <1, 12, 7, *n* = 5 | 1–19, 4, *n* = 5 | SG: (Svensson and Kvarnemo 2005) |
| Pre-spawning sound, female in nest | Low-frequency sound in trains of pulses | Present | Present | *n* = 5 | *n* = 1 | CG: (Bolgan et al. 2013) SG: (Lindström and Lugli 2000; Pedroso et al. 2013) |
| Pre-spawning sound, female outside nest | Low-frequency sound in trains of pulses | Not recorded | Present | | *n* = 4 | Present study |

Range and median (s) are calculated from all recorded behaviors from all individuals that showed the behavior. *n* number of replicates with an individual showing the trait, *CG P. microps* (common goby), *SG P. minutus* (sand goby). Two courtship sounds was analyzed per male i.e. 10 sounds per species

The highly sympatric P. microps and P. minutus have similar habitat preferences with huge habitat overlaps (personal observation, Kullander et al. 2012) and are arguably the most similar species of the Swedish shallow water 'sand goby group' gobies. This suggests that courtship traits may show parallel evolution. However, the differences that do occur in habitat choice may explain courtship sounds and courtship repertoires. The three main differences that may affect courtship and hence result in divergence are nest substrate, bottom substrate and spawning depth. P. microps often build nests using empty shells of Cerastoderma cockles whereas P. minutus use larger shells e.g. from Mya clams (personal observation). Because the nest construction affects the sound (Lugli 2013), sound production could be adapted to the nest construction. The two species also prefer different bottom substrates, with P. microps being more prevalent in soft bottom areas whereas P. minutus is more common at sandy beaches (Nellbring 1986; Nellbring 1993; Magnhagen 1999). The preferred habitats can be only meters apart or be in completely different bays (personal observation). In addition, bottomsubstrate affects turbidity, and visibility has been shown to affect courtship and mate choice in fishes including P. minutus (Seehausen et al. 1997; Järvenpää and Lindström 2004; Engstrom-Öst and Candolin 2007; Michelangeli et al. 2015). Because P. microps prefer to spawn in even shallower water than P. minutus (less than one meter compared to a few meters, Nellbring 1986; Nellbring 1993), the range of light conditions differs as well as the distance from the shore. Hence, the visual and acoustic environment may correlate with spawning habitat. Furthermore, there are differences between P. minutus populations as well as between P. minutus and P. microps in the rhodopsin (RH1) genes, which are most likely involved in the spectral tuning mechanism (Larmuseau et al. 2009; Larmuseau et al. 2010), potentially causing diverging sensory biases (Ryan and Keddy-Hector 1992). These differences coincide with the general photic conditions (Larmuseau et al. 2009; Larmuseau et al. 2010). An alternative, but not mutually exclusive, mechanism for the divergence in courtship traits is character displacement (Pfennig and Pfennig 2009). Character displacement may decrease fitness loss due to reproductive interference (Gröning and Hochkirch 2008), and has been suggested as a potential mechanism for the divergence in courtship sound between P. minutus and P. pictus (Pedroso et al. 2013). Because mate choice takes time and prolonged conspicuous courtship increases exposure to predators (Magnhagen 1990; Magnhagen 1991), both males and females should be selected to avoid spending time on inspecting and courting the wrong species.

Changes in the environment due to human activities, such as aquatic noise and turbidity, have the potential to affect signal transmission, and are of increasing concern (Candolin and Wong 2012; Popper and Hawkins 2012; Chivers et al. 2013). Species with different acoustic courtship strategies may react differently to noise pollution. For example, P. minutus has a longer song repertoire with a lower volume than P. microps while also producing sound when the female is outside the nest (i.e. further away from the source of the sound). It has recently been shown that acoustic noise affect goby be havior (de Jong et al. 2016). Furthermore, increased levels of aquatic noise may increase reproductive interference. In gobies, it has been shown that behavior is affected by turbidity (Michelangeli et al. 2015) which can cause a mating system breakdown (Järvenpää and Lindström 2004). We show that P. minutus and P. microps differ in several acoustic and visual courtship traits, and hence females would still have enough available cues even in a perturbed environment.

To conclude, by comparing two species of Pomatoschistus, we highlight differences in courtship signals and behaviors such as train duration of courtship sounds and swimming speed during courtship. These differences may be consequential for how specific species react to a changing environment, including species identification and reproductive interference.

**Acknowledgments** M. Celander, K. de Jong and J. Wilson helped in the field and the Sven Lovén Centre of Marine Infrastructure in Kristineberg for use of their facilities, technical support and hospitality.We thank CalumNinnes for valuable comments on the manuscript. The manuscript was greatly improved by the helpful comments from two anonymous reviewers. Financial support was provided by the Graduate School in Marine Environmental Research at the Gothenburg Centre forMarine Research, Helge Ax:son Johnssons stiftelse, Herbert & Karin Jacobssons stiftelse, (to ELB), ASSEMBLE grant agreement no 227799 (to IM and KH), Volkswagen foundation (to KH), the Linnaeus Centre for Marine Evolutionary Biology at the University of Gothenburg which is supported by a Linnaeus-grant from the Swedish Research Councils VR and Formas (to OS).

## COMPLIANCE WITH ETHICAL STANDARDS

**Ethics** This study complies with the Swedish law and was

performed under the ethical permits 135–2010 and 143–2012.

All fish were released into their natural environment.

## REFERENCES

1. Amorim MCP, Neves ASM (2007) Acoustic signalling during courtship in the painted goby, Pomatoschistus pictus. J Mar Biol Assoc UK 87(4):1017–1023. doi:10.1017/s0025315407056822
2. Bolgan M, Pedroso SS, Picciulin M, Fonseca PJ, Amorim MCP (2013) Differential investment in acoustic communication during social interactions in two closely-related sand goby species. Behaviour 150(2):133–152. doi:10.1163/1568539x-00003041
3. Borg ÅA, Forsgren E, Magnhagen C (2002) Plastic sex-roles in the common goby - the effect of nest availability. Oikos 98(1):105–115
4. Bradbury JW, Vehrencamp SL (2011) Principles of animal communication, 2nd edn. Sinauer Associates, Inc., Sunderland
5. Candolin U,Wong BBM (eds) (2012) Behavioural responses to a changing world: mechanisms and consequences. Oxford University Press, Oxford
6. Chivers DP, Al-Batati F, Brown GE, Ferrari MCO (2013) The effect of turbidity on recognition and generalization of predators and non-predators in aquatic ecosystems. Ecol Evol 3(2):268–277. doi:10.1002/ece3.454
7. de Jong K, Amorim MCP, Fonseca PJ, Klein A, Heubel KU (2016) Noise affects acoustic courtship behavior similarly in two species of gobies. Proc Mtgs Acoust. 27: 010018. doi: 10.1121/2.0000272
8. Engstrom-Öst J, Candolin U (2007) Human-induced water turbidity alters selection on sexual displays in sticklebacks. Behav Ecol 18(2):393–398. doi:10. 1093/beheco/arl097
9. Fonds M (1973) Sand gobies in the Dutch Wadden Sea (Pomatoschistus, Gobiidae, Pisces). Neth J Sea Res 6(4): 417–478. doi:10.1016/0077 +7579(73)90001-x
10. Forsgren E (1997) Mate sampling in a population of sand gobies. Anim Behav 53:267–276

11. Forsgren E (1999) Sexual selection and sex roles in the sand goby. In: Almada VC, Oliveira RF, Gonçalves EJ (eds) Behaviour and conservation of littoral fishes. ISPA, Lisboa, pp 249–274
12. Forsgren E, Karlsson A, Kvarnemo C (1996) Female sand gobies gain direct benefits by choosing males with eggs in their nests. Behav Ecol Sociobiol 39(2):91–96
13. Giacomello E, Marri L, Marchini D, Mazzoldi C, Rasotto MB (2008) Sperm-duct gland secretion of the grass goby Zosterisessor ophiocephalus exhibits antimicrobial activity. J Fish Biol 73(7):1823–1828. doi:10.1111/j.1095- 8649.2008.02069.x
14. Gröning J, Hochkirch A (2008) Reproductive interference between animal species. Q Rev Biol 83(3):257–282
15. Heubel KU, Schlupp I (2006) Turbidity affects association behavior in male Poecilia latipinna. J Fish Biol 68(2):555–568. doi:10.1111/j.0022-1112.2006.00941.x
16. Heuschele J, Candolin U (2007) An increase in pH boosts olfactory communication in sticklebacks. Biol Lett 3(4):411–413. doi:10.1098/rsbl.2007.0141
17. Holt DE, Johnston CE (2015) Traffic noise masks acoustic signals of freshwater stream fish. Biol Conserv 187:27–33. doi:10.1016/j.biocon.2015.04.004
18. Huyse T, Van Houdt J, Volckaert FAM (2004) Paleoclimatic history and vicariant speciation in the Bsand goby^ group (Gobiidae, Teleostei). Mol Phylogenet Evol 32(1):324–336
19. JärvenpääM, Lindström K (2004)Water turbidity by algal blooms causes mating system breakdown in a shallow-water fish, the sand goby Pomatoschistus minutus. Proc R Soc B Biol Sci 271(1555):2361–2365
20. Jones JC, Reynolds JD (1999) Oxygen and the trade-off between egg ventilation and brood protection in the common goby. Behaviour 136:819–832
21. Kullander SO, Nyman L, Jilg K, Delling B (2012) Nationalnyckeln till Sveriges flora och fauna. Strålfeninga fiskar. Actinopterygii. Artdatabanken, SLU, Uppsala
22. Kvarnemo C (1994) Temperature differentially affects male and female reproductive rates in the sand goby: consequences for operational sex-ratio. Proc R Soc B Biol Sci 256(1346):151– 156
23. Kvarnemo C, Forsgren E, Magnhagen C (1995) Effects of sex ratio on intra- and inter-sexual behaviour in sand gobies. Anim Behav 50:1455–1461
24. Larmuseau MHD, Raeymaekers JAM, Ruddick KG, Van Houdt JKJ, Volckaert FAM (2009) To see in different seas: spatial variation in the rhodopsin gene of the sand goby (Pomatoschistus minutus). Mol Ecol 18(20):4227–4239. doi:10.1111/j.1365-294X.2009.04331.x
25. Larmuseau MHD, Huyse T, Vancampenhout K, Van Houdt JKJ, Volckaert FAM (2010) High molecular diversity in the rhodopsin gene in closely related goby fishes: a role for visual pigments in adaptive speciation? Mol Phylogenet Evol 55(2): 689–698. doi:10.1016/j.ympev.2009.10.007
26. Lehtonen TK, Kvarnemo C (2015a) Infections may select for filial cannibalism by impacting egg survival in interactions with water salinity and egg density. Oecologia 178(3):673–683. doi:10.1007/s00442-015-3246-1
27. Lehtonen TK, Kvarnemo C (2015b) Odour cues from suitors' nests determine mating success in a fish. Biol Lett 11(5): 20150021. doi:10.1098/rsbl.2015.0021
28. Lindström K, Lugli M (2000) A quantitative analysis of the courtship acoustic behaviour and sound patterning in male sand goby, Pomatoschistusminutus. Environ Biol Fish 58(4): 411–424
29. Locatello L, Mazzoldi C, Rasotto MB (2002) Ejaculate of sneaker males is pheromonally inconspicuous in the black goby, Gobius niger (Teleostei, Gobiidae). J Exp Zool 293(6):601–605
30. Lugli M (2013) Sand pile above the nest amplifies the sound emitted by the male sand goby. Environ Biol Fish 96(8): 1003–1012. doi:10.1007/s10641-012-0097-z
31. Lugli M, Torricelli P (1999) Prespawning sound production in Mediterranean sand-gobies. J Fish Biol 54(3):691–694. doi:10.1006/jfbi.1998.0898

32. Magnhagen C (1990) Reproduction under predation risk in the sand goby, Pomatoschistus minutus, and the black goby, Gobius niger: the effect of age and longevity. Behav Ecol Sociobiol 26(5):331–335
33. Magnhagen C (1991) Predation risk as a cost of reproduction. Trends Ecol Evol 6(6):183–185
34. Magnhagen C (1998) Alternative reproductive tactics and courtship in the common goby. J Fish Biol 53(1):130–137
35. Magnhagen C (1999) Alternative mating tactics, nest defence and courtship in gobies - just a matter of size? In: Almada VC, Oliveira RF, Gonçalves EJ (eds) Behaviour and conservation of littoral fishes. ISPA, Lisboa, pp 275–300
36. Malavasi S, Gkenas C, Leonardos I, Torricelli P, McLennan DA (2012) The phylogeny of a reduced 'sand goby' group based on behavioural and life history characters. Zool J Linnean Soc 165(4):916–924. doi:10.1111/j.1096-3642.2012.00832.x
37. Michelangeli M, Tuomainen U, Candolin U, Wong BBM (2015) Habitat alteration influences male signalling effort in the Australian desert goby. Behav Ecol 26(4):1164–1169. doi:10.1093/beheco/arv060
38. Miller PJ (1975) Age-structure and life-span in the common goby. J Zool 177:425–448
39. Miller PJ (1986) Gobiidae. In: Whitehead PJP (ed) Fishes of the north-eastern Atlantic and the Mediterranean Vol 3. Unesco, Paris
40. Mück I (2016) The role of an environmental gradient in driving population divergence in common gobies (Pomatoschistus microps). Universität Tübingen
41. Nellbring S (1986) A quantitative and experimental stufy of goby (Pomatoschistus spp.) nests on a shallow soft bottom, northern Baltic proper. Ophelia 25(1):15–23
42. Nellbring S (1993) Spawning of two Pomatoschistus species (Gobiidae) in relation to nest availability and depth: a field experiment. Neth J Sea Res 31(2):173–179. doi:10.1016 /0077-7579(93)90007-f
43. Nyman K-J (1953) Observations on the behaviour of Gobius microps. Acta Soc pro Fauna et Flora Fenn 65:1–11 Pampoulie C, Lindström K, St Mary CM (2004) Have your cake and eat it too: male sand gobies show more parental care in the presence of female partners. Behav Ecol 15(2):199–204
44. Pedroso SS, Barber I, Svensson O, Fonseca PJ, Amorim MCP (2013) Courtship sounds advertise species identity and male quality in sympatric Pomatoschistus spp. gobies. PLoS One 8(6):e64620. doi:10.1371/journal.pone.0064620
45. Pfennig KS, Pfennig DW (2009) Character displacement: ecological and reproductive responses to a common evolutionary problem. Q Rev Biol 84(3):253–276
46. Popper AN, Hawkins A (eds) (2012) The effects of noise on aquatic life. Springer, New York
47. Radford AN, Kerridge E, Simpson SD (2014) Acoustic communication in a noisy world: can fish compete with anthropogenic noise? Behav Ecol 25(5):1022–1030. doi:10.1093 /beheco/aru029
48. Ryan MJ, Keddy-Hector A (1992) Directional patterns of female mate choice and the role of sensory biases. Am Nat 139:S4– S35
49. Seehausen O, vanAlphen JJM, Witte F (1997) Cichlid fish diversity threatened by eutrophication that curbs sexual selection. Science 277(5333):1808–1811
50. Svensson O, Kvarnemo C (2005) The importance of sperm competition risk and nest appearance for male behavior and female choice in the sand goby, Pomatoschistus minutus. Behav Ecol 16:1042–1048. doi:10.1093/beheco/ari085
51. Svensson O, Magnhagen C, Forsgren E, Kvarnemo C (1998) Parental behaviour in relation to the occurrence of sneaking in the common goby. Anim Behav 56:175–179
52. Vallon M, Anthes N, Heubel KU (2016)Water mold infection but not paternity induces selective filial cannibalism in a goby. Ecol Evol. doi:10.1002/ece3.2403
53. Vanhove MPM et al (2012) Phylogenetics and biogeography of the Balkan sand gobies' (Teleostei: Gobiidae): vulnerable species in need of taxonomic revision. Biol J Linn Soc 105(1):73–91. doi:10.1111/j.1095-8312.2011.01781.x

54. Wilkins MR, Seddon N, Safran RJ (2013) Evolutionary divergence in acoustic signals: causes and consequences. Trends Ecol Evol 28(3):156–166. doi:10.1016/j.tree.2012.10.002

# CHAPTER 5

# Environmental Factors Affecting Large-Bodied Coral Reef Fish Assemblages in the Mariana Archipelago

**Benjamin L. Richards[1,3]* , Ivor D. Williams[2], Oliver J. Vetter[3], Gareth J. Williams[4]**

[1] *Department of Zoology, University of Hawai'i at Ma¯noa, Honolulu, Hawaii, United States of America,*

*2 Joint Institute for Marine and Atmospheric Research, University of Hawai'i at Ma¯noa, Honolulu, Hawaii, United States of America,*

*3 NOAA Fisheries Service, Pacific Islands Fisheries Science Center, Honolulu, Hawaii, United States of America,*

*4 Scripps Institution of Oceanography, University of California - San Diego, San Diego, California, United States of America*

## ABSTRACT

Large-bodied reef fishes represent an economically and ecologically important segment of the coral reef fish assemblage. Many of these individuals supply the bulk of the reproductive output for their population and have a disproportionate effect on their environment (e.g. as apex predators or bioeroding herbivores). Large-bodied reef fishes also tend to be at greatest risk of overfishing, and their loss can result in a myriad of either cascading (direct) or indirect trophic and other effects. While many studies have investigated habitat characteristics affecting populations of small-bodied reef fishes, few have explored the relationship between large-bodied species and their environment. Here, we describe the distribution of the large-bodied reef fishes in the Mariana Archipelago with an emphasis on the environmental factors associated with their distribution. Of the factors considered in this study, a negative association with human population density showed the highest relative influence on the distribution of large-bodied reef fishes; however, depth, water temperature, and distance to deep water also were important. These findings provide new information on the ecology of large-bodied reef fishes can inform discussions concerning essential fish habitat and ecosystem-based management for these species and highlight important knowledge gaps worthy of additional research.

**Funding:** The National Marine Fisheries Service Office of Habitat Conservation and National Oceanic and Atmospheric Administration's Coral Reef Conservation Program provided funding for this work. The funders had no role in study design, data collection and analysis, decision to publish, or preparation of the manuscript.

## INTRODUCTION

In the terrestrial world, studies of habitat use and resource selection form much of our understanding of wildlife habitat requirements [1], and such studies can potentially provide the basis for effective, spatially based ecosystem management in the marine world as well. Agardy et al. [2] state that marine spatial planning should, at a minimum, include the identification of priority areas as a key element in the design process, and investigations of the relationships between reef fishes and their environment are key in identifying such priority areas. Indeed, many reef fishes have evolved adaptations suiting them to particular reef zones [3], and many studies have described ways in which habitat characteristics affect assemblages of small-bodied reef fishes living in close association with the reef framework and how knowledge of these associations can inform management decisions [4]–[12]. Several authors have also investigated the effects of environmental variables on the distribution of small and large-bodied pelagic [13] and estuarine [14]–[16] fishes. However, less work has focused on the relationships between environmental variables and large-bodied fishes associated with coral reefs [17], especially those in the Mariana Archipelago. Many of these large-bodied fishes are capable of traveling great distances, allowing them to regularly move among various different habitat types. Meyer et al. [18] found that giant trevally (*Caranx ignobilis*) in the Northwestern Hawaiian Islands, made periodic atoll-wide excursions of up to 29 km. These same authors found that green jobfish (*Aprion virescens*) were seasonally attached to core areas of up to 12 km in length. These fish were able to range 19 m across atolls with daily round-trip excursions of up to 24 km [19]. McKibben and Nelson [20] found that grey reef sharks (*Carcharhinus amblyrhynchos*) at Enewetak inhabited home ranges of up to 53 $km^2$ and 16 km in length. Finally, Holland et al. [21] found that Bluefin trevally (Caranx melampygus) on patch reefs in Kaneohe Bay, Hawai'i, appear to inhabit fairly stable home ranges for periods of at least a few weeks and possibly up to a year. During the day, tagged fish would patrol back and forth along the face of the patch reef, traveling distances of several hundred meters, and often changing direction at the same point each along the reef. With the exception of most of the pelagic studies and work done on the Great Barrier Reef [22]–[24], most research has focused at relatively small scales—a single estuary, bay, or island—and not at the scale of an entire archipelago. While it is important to investigate small-scale relationships that may differ from island to island and even between different sites within an island, it is no less important to investigate higher-level relationships operating at the archipelagic scale and which may not be apparent from analyses carried out at the site or even island scale [10], [25].

Although tourism is now a key industry, a healthy nearshore coral reef ecosystem able to provide key ecological, economic, and social benefits is no less important [26]. Fish represent the primary natural resource in the Mariana Archipelago, and large-bodied individuals including some species of surgeonfishes (Acanthuridae), jacks (Carangidae), emperors (Lethrinidae), snappers (Lutjanidae), groupers (Serranidae), wrasses (Labridae), and parrotfishes (Scaridae) tend to be preferentially harvested both for local consumption and export [27]–[29]. Large-bodied reef fishes also are of particular importance for a variety of ontogenetic and phylogenetic reasons. Within a species, the largest individuals tend to supply the bulk of the reproductive output for their population [30]. For example, Sudekum et al. [31] found that fecundity in *Caranx melampygus* ranged from approximately 50,000 mature ova for a fish of 760 g (32.8 cm SL) to over 4 million for an individual of 6490 g (64.0 cm SL). Sequential hermaphroditism is common in many teleost fishes including many parrotfishes (e.g. *Scarus rubroviolaceus*) and wrasses (e.g. *Cheilinus undulatus*), with the largest individuals often being disproportionately male or female [32]. Hence, the removal of the largest individuals can have a disproportionate impact on a single gender with associated effects on the reproductive potential of the population. Large-bodied species also tend to have protracted spawning periods [33], and can have a disproportionate impact on their environment, often as apex predators [32], [34]or primary agents of bioerosion [35], [36]. Furthermore, the loss of large-bodied species can have a cascading (direct) or indirect effect on lower trophic levels and ecosystem balance [37]–[39]. Unfortunately, large-bodied individuals tend to be preferentially targeted by fishers, and many large-bodied species are among the most vulnerable to overfishing because they share a suite of life-history characteristics including large size, slow growth rates, and delayed sexual maturity [40]–[42]. Many species of sharks are additionally vulnerable due to their low reproductive rates.

Currently, several fishery targets including snappers, groupers, jacks, surgeonfish, sharks, and large emperors show higher relative abundances in the remote northern islands of the Mariana Archipelago compared to the populated islands of the south, and it has been suggested that abundance and biomass of some taxa has declined in recent decades [28]. Guam, the southernmost major island in the archipelago, has experienced a decline in nearshore reef-associated fish populations accompanied by a sharp decline in catch per unit effort (CPUE) [43]–[45]. Large-bodied species including bumphead parrotfish, humphead wrasse, stingrays, parrotfish, jacks, emperors, and groupers are considered rare in Guam, and it has been suggested that this may be the result of heavy fishing [26]. Technological improvements, which have facilitated the expansion of activities like nighttime SCUBA spearfishing, have resulted in a reappearance of larger species in fishery catch statistics [46]. While it is possible that this resurgence in catches indicates greater targeting of "healthy" populations, it would seem more likely that such increases are related to the use of new technology, representing the next step in the serial depletion cycle seen in many fisheries [47], [48].

At local, regional, and international scales, marine resource management is moving toward a suite of ecosystem-based management approaches including spatial closure and the protection of Essential Fish Habitat [49]. The Western Pacific Regional Fishery Management Council is beginning to integrate ecosystem approaches to management in Guam and the CNMI [26]. However, their current Fishery Ecosystem Plan for the Mariana Archipelago report also states that little is known about the life history, habitat utilization, food habits, or spawning behavior of most of the resident coral reef species. Agardy et al. [2] further argue that true integrated marine protected area planning has yet to be achieved and that the "blind faith" many have placed in often poorly planned and inadequately thought-out marine protected areas carries with it great risk.

Here we test the relative influence of various anthropogenic, physical, oceanographic, and biological environmental factors on the distribution of large-bodied reef fishes in the Mariana Archipelago. Our results show that human populations likely affect the distribution of large-bodied reef fishes but that assemblages are highly variable and other factors including depth, temperature, distance to deep water and many others likely play a role. These results provide additional insight on the relationship between these reef fishes and their environment, can inform discussions of essential fish habitat for these large-bodied reef fishes, and can provide information for managers in their efforts to implement ecosystem-based approaches to fishery management.

# METHODS

## Study Area

The Mariana Archipelago (politically the CNMI and the US Territory of Guam) is an elongate string of islands stretching 950 km northward in an arc from the island of Guam (13.44°N, 144.76°E) to the Farallon de Pajaros (20.54°N, 144.89°E) (Fig. 1). Islands range in age from 1 to 1.5 million years in the north to >30 million years in the south [50]. The northern islands are volcanically active and topographically complex. The archipelago contains ~230 $km^2$ of shallow-water (<18 m) coral reef habitat, and is the second largest reef area under U.S. jurisdiction in the Pacific [28], [51]. Each island is surrounded by diverse shallow-water coral reefs ranging in size from 0.8 $km^2$ around Farallon de Pajaros to 108 $km^2$ surrounding Guam [51]. Humans have occupied the Mariana Archipelago since about 3500 B.P. [52], and the present day human population generally declines along a south-to-north gradient from ~150,000 in Guam and ~62,000 in Saipan to near zero in the northern islands (U.S. Census Bureau; Census 2000).

## Survey Protocols

Between 2003 and 2009, the Coral Reef Ecosystem Division (CRED) of the NOAA Pacific Islands Fisheries Science Center (PIFSC) conducted 508 large-scale towed-diver surveys for large-bodied (>50 cm Total Length) reef fishes, covering 1038 ha (10.4 $km^2$) of the shallow-water (<30 m), coral reef environment around 20 islands and offshore banks in the Mariana Archipelago. These surveys were conducted as part of an ecosystem-wide long-term monitoring program and the lower size threshold of 50 cm TL was chosen to enable observers to concentrate their efforts on the large-bodied, patchily distributed, rare and more mobile species and individuals that were not as effectively sampled by other methods [53]. The 50 cm TL threshold also effectively captured the majority of species often referred to as "apex predators" (e.g., carcharhinids, carangids), as well as the larger herbivores (e.g., scarids) and some species of particular concern (i.e., *C. undulatus*). With the exception of sharks, this size threshold also generates data on the largest of the sexually mature individuals (e.g. Lm for *Caranx melampygus*=35 cm), which, as mentioned earlier, are expected to have the greatest effect on their environment or population.

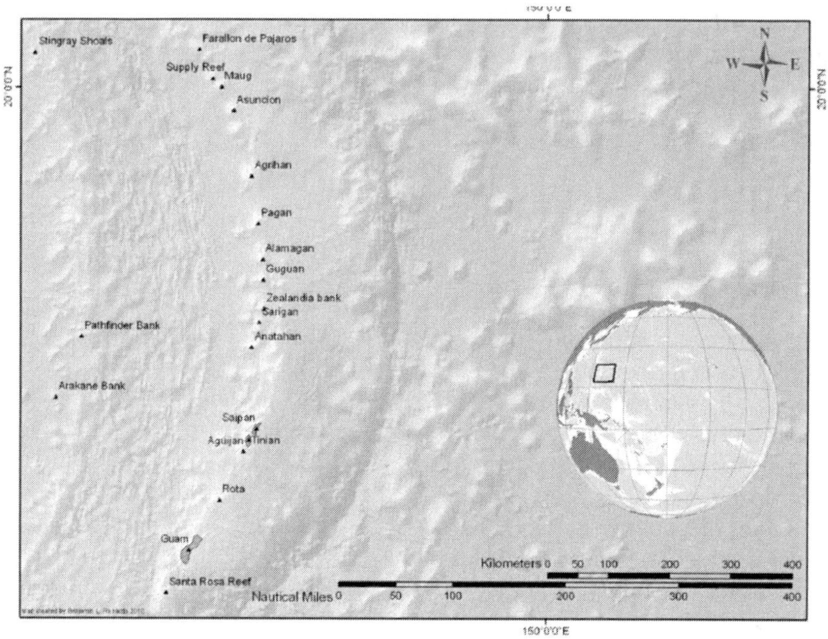


**Figure 1.** Study area, The Mariana Archipelago showing the location of all islands and banks.

Within each year, the entire archipelago was surveyed within a span of 1–2 months. The details of the methodology and its utility for surveying assemblages of large-bodied reef fishes are provided by Richards et al. [53]. In brief, surveys were conducted by means of SCUBA diver-controlled towboards equipped with still and video camera equipment and SeaBird™ SBE39 high-resolution temperature/depth recorders. Divers were towed side-by-side at the end of 60 m lines behind a small survey launch along a depth contour (~15 m) at a speed of ~1.5 kt for a duration of 50 minutes per survey. Each survey was partitioned into ten 5-minute survey segments. A global positioning system (GPS) unit mounted on the survey launch recorded survey track points at 5-second intervals, and a custom layback algorithm allowed the diver track to be plotted in a geographic information system (GIS) and merged with spatial data sets (e.g. benthic habitat maps, wave energy estimates). During each survey, one diver recorded, to the lowest possible taxon, all large-bodied reef fishes (>50 cm TL) observed within a 10-m wide lane, focusing observations ahead in a 10-m long moving window. The total length of each fish was estimated to the nearest 5 cm. Cryptic species and fish observed behind the diver were not recorded. Fishes seen leaving and then reentering the survey area were recorded only once. The second diver simultaneously recorded benthic habitat information (e.g. % cover of coral, algae, habitat complexity) as detailed by Kenyon et al. [54]. Habitat complexity is recorded on a six-point scale and is a subjective measure of the topographic diversity or amount of "roughness" of the substrate. Examples of low complexity include sand flat or rubble plains. Areas of steep spur and groove, canyon, pinnacles and walls would be classified as high or very high complexity.

## Data Assimilation

Fish abundance and size data were converted to biomass density (hereafter "biomass") using the allometric length-weight conversion: $W=aTL^b$, where parameters a and b are species-specific constants, TL is total length in mm, and W is weight in grams. Length-weight fitting parameters that most closely matched they survey location and fish size class were obtained from FishBase [55] and Kulbicki et al. [56]. When length-weight fitting parameters used other than total length, length-length conversions were made using formulas from the same sources. A data subset was created that contained species-level biomass values. A relational database and ArcGIS 9.3.1 were used to generate a variety of *in situ* and remotely sensed environmental variables having demonstrated potential to influence the distribution of reef fishes (Table 1). While measures of primary productivity are important to include in ecological models, we have excluded such measures from this analysis for several reasons. Firstly, existing data are unlikely to provide an accurate measure of nearshore primary productivity patterns in this region and at the scale of our study as the standard algorithms are based on the assumption of "optically deep" (i.e. no bottom reflectance) waters. This assumption is violated in the clear, nearshore, shallow-water areas from which the data for this study was derived. Secondly, terrigenous nutrient loading from natural and anthropogenic sources is likely to greatly exceed and mask any open ocean patterns. Hence, remotely sensed primary productivity values derived form open ocean sources would not reflect actual nearshore patterns in primary productivity. Because many of the data layers did not completely overlap with one another or with our towed-diver surveys, it was necessary to further subset the data to create a data matrix that contained only those records with values for all variables.

## Benthic Habitat

Reef area was calculated based on hard-bottom habitat present within the 0–30 m depth range. Benthic habitat was classified using a combination of *in situ* and remotely sensed data. *In situ*data included % cover of coral, macroalgae, crustose coralline algae (CCA), sand, and rubble and the habitat complexity index we described in the survey protocols. Remotely sensed benthic habitat map layers depicting benthic habitat structure and cover classes, based on IKONOS™ satellite imagery [57] were converted to 5-m resolution raster layers in ArcGIS 9.3.1. Major reef structure (e.g. Rock/Boulder, Pavement, Aggregate Reef) and cover (e.g. Coral, Macroalgae, Uncolonized) categories were defined by NOAA's National Centers for Coastal Ocean Science (NCCOS), using a standard hierarchical classification scheme in which the term benthic cover refers only to biological cover type (i.e. coral, algae) and structure refers to only substrate type, denoting geomorphologic structure (i.e. pavement, sand, reef). The dominant benthic cover and structure were calculated for each survey. The dominant habitat was defined as that assigned to the majority of grid cells within a given survey area. Cover and structure richness values were also calculated based on the "variety" or number of different cover or structure types contained within each survey area.

**Table 1.** Environmental predictor variables included in the boosted regression tree analysis.

| Variable Name | Source | Units | Range |
|---|---|---|---|
| Moon phase | calculation from date | days from new | 0–29 |
| Time of Day | Index from sequential dive # | # | 1–6 |
| Temperature | *in situ* (SBE39) | °C | 23.43–30.35 |
| Wave Energy | calculation from NOAA WW3[2] | kW/m | 13.76–517.99 |
| Depth | *in situ* (SBE39) | m | 1.6–26.5 |
| Distance to 50 m contour | NOAA PIBHMC[3] bathymetry | m | 1–500 |
| Quadrant | calculation | categorical | NE, SE, SW, NW |
| Human Population/Reef Area | Calculation (2000 US Census) | #/km$^2$ | 0–2445 |
| Complexity | benthic towed-diver visual estimate | index | 0–6 |
| Reef Structure | NOAA BHM (Benthic Habitat Maps)[4] | categorical | Reef, Sand, etc |
| Benthic Cover | NOAA BHM[4] | categorical | Coral, Algae, etc |
| % Cover – Coral | benthic towed-diver visual estimate | % | 0–68.8 |
| % Cover – Macroalgae | benthic towed-diver visual estimate | % | 0–87.5 |
| % Cover – Crustose coralline algae | benthic towed-diver visual estimate | % | 0–68.75 |
| % Cover – Sand | benthic towed-diver visual estimate | % | 8.0–87.5 |
| % Cover – Rubble | benthic towed-diver visual estimate | % | 0–56.3 |
| Variety of Cover Types | Calculation from BHM[4] data | # | 1–4 |
| Variety of Structure Types | Calculation from BHM[4] data | # | 1–4 |
| **Variables removed due to multi-collinearity** | | **Correlated with:** | |
| Mean Sea Surface Temperature | NODC Pathfinder[1] | Human Population/Reef Area | |
| Latitude | *in situ* | Human Population/Reef Area | |
| Longitude | *in situ* | Human Population/Reef Area | |
| Human Population | 2000 US Census | Human Population/Reef Area | |
| Reef Area | calculation : 0–30 m hardbottom | Human Population/Reef Area | |

[1]Pathfinder satellite data - http://www.nodc.noaa.gov/ Satellite Data/ pathfinder4km/.

[2]NOAA WaveWatch 3 - http://polar.ncep.noaa.gov/waves/main_int.html.

[3]Pacific Islands Benthic Habitat Mapping Center - http://www.soest.hawaii.edu/pibhmc.

[4]NOAA Benthic Habitat Maps- http://ccma.nos.noaa.gov /ecosystems /coralreef/us_pac_mapping.html.

https://doi.org/10.1371/journal.pone.0031374.t001

An attempt was made to incorporate quantitative information on benthic slope and complexity from existing shallow-water multibeam bathymetric maps [58]. However, the level of overlap (~20%) between the multibeam data (for which the shallowest extent is ~30 m) and our biological surveys (where the deepest extent is 30 m) was insufficient for analysis. In an effort to nonetheless incorporate information of this type, measures of benthic habitat complexity (as recorded by the benthic towed diver) and distance between each survey and the 50-m isobath were used.

## Oceanography and other Variables

Sea water temperature and diver depth were recorded during each survey using SeaBird™ SBE39 temperature and pressure loggers. Larger-scale sea surface temperature at the island-scale was derived from Pathfinder 4.1 Advanced Very High Resolution Radiometer (AVHRR) Global Climatology [59]. SST values were averaged over a 9 km grid spacing, using nighttime monthly means, averaged over five years of data (2006–2010). NOAA Wave Watch III (WW3) data were used as a proxy for overall wave energy impinging on the shallow reef. WW3 output is a time series (six hour interval) of offshore (deepwater) directional wave energy spectra. Twelve years of WW3 data (1997–2009) were used to calculate average energy flux for each frequency and direction, using deepwater linear wave theory (Equation 1). Direction was binned in 10-degree increments from 0–360 degrees. Period was binned in 2-second increments from 4 seconds to 20 seconds. Frequency is defined as 1/Period yielding a frequency range of 0.05 Hz to 0.25 Hz.

Equation 1. Wave energy flux in watts per meter of wave front.

$$E_{flux} = (\rho g^2 / 64\pi) H_{sig.o}^2 T$$

$E_{flux}$ = Wave energy flux in watts per m (W/m)

$\rho$ = Density of seawater (1024 kg/m$^3$)

$g$ = Gravitation alacceleration (9.81 m/s$^2$)

$H_{sig.o}$ = Offshore significant wave height (m)

$T$ = Wave period (s)

The frequency range with the maximum calculated energy flux was used as an approximation of maximum, and therefore most significant, offshore wave stress. This yielded a maximum offshore wave energy flux (watts per meter of wave front), or maximum power, for each directional bin for each island. The maximal wave energy bin in each quadrant was then used as input for the model. Island quadrant (i.e., NE, SE, SW, NW) was also included as a descriptive variable and was calculated based on the previously calculated directional bins. The time of day for each survey was recorded and moon phase was calculated based on the survey date.

## Anthropogenic Impacts

In the absence of specific data on human impacts such as fishing intensity, human population density per unit reef area was used as a proxy for anthropogenic pressure. Local population estimates came from the 2000 United States Census. Reef area was calculated using a range of GIS layers on bathymetry and bottom type to represent all hardbottom area <30 m deep. Islands were also classified into human population categories of High (>100,000 individuals), Mid (50–1,000 individuals), Low (1–50 individuals), and None. Islands in the Saipan-Tinian-Aguijan complex were each classified as Mid-level based on likely impact level given their close proximity to one another (Tinian is 5 km from Saipan, Aguijan is 8 km from Tinian and 30 km from Saipan), reports that fishers frequently travel among these islands, and information that shore-based fishing has occurred on Aguijan by operations based on the other islands (Trianni, M., pers. comm.). It should be noted that, while we do not have data on the percentage of the population that regularly engages in fishing activities, a large number of immigrant workers reside on Saipan, and these workers do not typically engage in fishing. There are fewer of these workers on Tinian and Rota and therefore, while the total population is lower, the percentage of fishers on these islands is likely greater than on Saipan.

## Data Analysis

The final merged data set (with fish and environmental variables) contained 3711 individual survey segments from 445 surveys conducted around 12 of the 20 islands/banks: Agrihan, Aguijan, Alamagan, Asuncion, Guam, Guguan, Maug, Pagan, Rota, Saipan, Sarigan, and Tinian (Fig. 1). The offshore banks (Stingray Shoals, Pathfinder, and Arakane) and Anatahan were excluded from the analysis, as they were surveyed only during the first year of sampling. Farallon de Pajaros (FDP) was excluded because of the low number of replicates that included all variables. A two-stage analysis was employed using *PRIMER* v6.1.13 [60], followed by Boosted Regression Trees (*BRT*) [61] in the R statistical language v2.6.1 [62]. Data were analyzed at two scales: a larger scale in which data were pooled by island and a smaller scale using data from each towed-diver survey.

*PRIMER* was used for the initial, broad-scale multivariate analysis involving all large-bodied reef fish species at both the island and survey-segment scale to determine the following: 1) similarities in large-bodied reef fish assemblages among islands across the archipelago, 2) which large-bodied reef fish species contributed primarily to any observed patterns and 3) which of the available environmental variables best predict the patterns in the large-bodied reef fish assemblages. The results of the *PRIMER* analyses were further used to help define the scope of the smaller-scale *BRT* analysis (e.g. which large-bodied reef fish species would be analyzed). Overall, *PRIMER* was deemed better able to handle the full multivariate data set, whereas *BRT* was better able to handle the combination of categorical and continuous variables and produced a more detailed description of the relationship between environmental predictor variables and fish response variables.

## Multivariate analyses

Species-level biomass densities were transformed using species-specific dispersion weighting to down-weight highly variable or 'clumped' species and to reduce similarity between sampled subregions [63]–[66]. This transformation also down-weighted particularly high biomass species and the need for further transformation was not indicated.

Data were pooled by island to assess large-scale patterns across the archipelago and the relationship among islands. A similarity profile test (*SIMPROF*; [67] was used to determine if significant, interpretable structure in the data existed and, where structure was evident, resemblance matrices were constructed using Bray-Curtis similarity measures [68]. The results of the *SIMPROF* analysis were visualized using hierarchical cluster analysis and nonmetric multidimensional scaling (nMDS) [69]. The *SIMPER* [69] routine was used to determine the relative influence of individual species on the dissimilarity among groups.

## Environmental Correlations

At the smaller scale, Spearman rank correlation was used to identify covarying explanatory variables and, where pairwise rank correlation coefficients were greater than |0.75|, we selected a single variable to represent each variable group. This variable was selected based on data range, interpretability, and ecological theory. Variables estimating % cover were log transformed to reduce skew and all variables were normalized to account for different ranges and measurement scales. The *RELATE* routine [60] was used to test for a relationship between segment-level species biomass and environmental variables and the *BEST* routine [70] was used to select a subset which best explained the overall structure in the large-bodied assemblage.

## Relative Impacts

To measure the relative impact of candidate environmental predictor variables on fish response variables and to assess the specific shape of each relationship, 3711 individual survey segments were modeled simultaneously against 18 continuous and categorical predictor variables (Table 1) using Boosted Regression Trees (*BRT*s) [61]. *BRTs* were constructed using the *gbm* (v1.5-7) [71] and *gbm.step* [61] packages in the *R* statistical language v2.6.1 [62](Table 2). Regression trees have many desirable properties, including 1) the ability to handle various types of response and predictor variables including both continuous and categorical, 2) the invariance of trees to monotonic transformations of predictors, 3) the ability to model complex interactions in a simple form, and 4) the ability to easily manage missing predictor values with minimal information loss [72]. The two main weaknesses of trees – that they are poor predictors and that large trees can be difficult to interpret – are largely overcome through the use of boosting and consequently, boosted trees are increasingly used in ecological studies [72]–[75].

*BRT*s do not assign real probabilities (i.e. *p*-values) and instead use a cross-validation process, which requires some of the data to be held back for model development and validation. Nevertheless, the full data set is still used to fit the model. We used cross validation deviance (CVD) and standard error (SE) as the measure of model performance, where lower values indicate a better model (e.g. a CVD of zero indicates that the model is able to predict new data without any predictive error, and larger CVDs indicate increasing amounts of error). Model optimization was achieved by varying the model parameters: tree complexity, learning rate, and bag-fraction. Tree complexity determines the number of nodes in a tree, while the learning rate is used to shrink the contribution of each tree as it is added to the model. The bag-fraction determines the proportion of the data to be selected at each step in model development and therefore affects stochasticity. We coded a loop routine which sought to minimize model CVD, among models with a minimum of 1000 trees, by adjusting all possible combinations of tree complexity (1, 2, 3, 4, 5), learning rate (0.05, 0.01, 0.001, 0.0001), and bag fraction (0.1, 0.5, 0.75). The combination with the lowest CVD was used to create the final *BRT* model. Relative importance values for each environmental variable were calculated based on the number of times each variable was selected for splitting [72]. Higher relative importance values for a given environmental variable indicate a stronger influence on either total large-bodied reef fish biomass or species distribution. Partial dependency plots were used to interpret the relationship between the environmental predictors and the fish response variables.

**Table 2.** Boosted regression tree (BRT) analysis: Optimal parameter settings, predictive performance, and relative influence of environmental variables on total large-bodied reef fish biomass and presence/absence of key species.

| Booted Regression Tree Output (optimized) | | | | | | | | | |
|---|---|---|---|---|---|---|---|---|---|
| **Model Parameters** | **Total Biomass** | **CAAB** | **LUBO** | **MASP** | **TROB** | **NAHE** | **SCRU** | **CAME** | **CHUD** |
| **Tree complexity** | 3 | 5 | 5 | 1 | 4 | 5 | 4 | 5 | 1 |
| **Learning rate** | 0.010 | 0.001 | 0.010 | 0.010 | 0.001 | 0.01 | 0.010 | 0.001 | 0.010 |
| **Bag fraction** | 0.50 | 0.75 | 0.75 | 0.50 | 0.75 | 0.75 | 0.75 | 0.75 | 0.75 |
| **# of trees** | 1400 | 5750 | 1150 | 2500 | 3050 | 3100 | 1350 | 3750 | 1350 |
| **Mean Total Deviance** | 0.004 | 0.289 | 0.835 | 0.424 | 0.390 | 0.273 | 0.459 | 0.335 | 0.141 |
| **CV Deviance**[1] | 0.003 | 0.207 | 0.595 | 0.309 | 0.341 | 0.228 | 0.394 | 0.300 | 0.131 |
| **SE**[1] | 0.0001 | 0.006 | 0.016 | 0.008 | 0.008 | 0.008 | 0.007 | 0.005 | 0.005 |
| **Percent deviance explained (%)** | 25 | 28 | 29 | 27 | 13 | 16 | 14 | 10 | 7 |
| **Environmental Variables** | **Relative influence (%)** | | | | | | | | |
| Human Population | 26.4 | 19.9 | 22.6 | 24.2 | 30.8 | 11 | 7.2 | 12.7 | 8.6 |
| Temperature | 15.7 | 17.4 | 13.5 | 17.9 | 12.1 | 8.1 | 21.1 | 18.4 | 6.8 |
| Depth | 11.5 | 9.8 | 5.8 | 6.7 | 17 | 14.2 | 19.2 | 13.4 | 49.4 |
| Distance to Deep Water | 9.4 | 12.8 | 7.8 | 29.5 | 6.7 | 23.6 | 11.1 | 13.2 | 2.8 |
| Wave Energy | 6.5 | 4.1 | 4.6 | 1.5 | 11.9 | 6.6 | 11.7 | 6.9 | 1.4 |
| Benthic Habitat Structure | 4.7 | 4.6 | 17.9 | 6.9 | 3.3 | 2.4 | 3.4 | 6.1 | 10.7 |
| Topographic Complexity | 4.7 | 1.7 | 2.3 | 7 | 0.7 | 1.5 | 5.8 | 2.3 | 1.9 |
| % Macroalgae | 4.1 | 4.6 | 2.6 | 0.4 | 1.6 | 3 | 3.1 | 3.1 | 2.4 |
| Moon Phase | 3.6 | 5.5 | 4.1 | 0.5 | 5.5 | 6.7 | 5 | 3.2 | 2.1 |
| Island Quadrant | 3 | 4.6 | 4.6 | 0.7 | 0.9 | 0.9 | 2.1 | 6.1 | 0.6 |
| Time of Day | 2.4 | 3.9 | 4.1 | 0.9 | 1.4 | 3.6 | 2.8 | 3 | 2 |
| % Sand | 1.8 | 1.2 | 2.2 | 0.1 | 1.2 | 2.3 | 1.5 | 2.7 | 0.6 |
| % Coral | 1.7 | 1.9 | 3.4 | 2.2 | 4 | 4.2 | 2 | 3 | 0.7 |
| Benthic Habitat Cover | 1.6 | 0.5 | 0.8 | 0 | 0.4 | 1.6 | 0.9 | 0.4 | 4.5 |
| % Crustose Coralline Algae (CCA) | 1.5 | 3.2 | 2.9 | 1.2 | 0.7 | 5.7 | 1.7 | 1.9 | 0.5 |
| % Rubble | 0.8 | 3.8 | 0.9 | 0.4 | 1.8 | 3.5 | 0.8 | 2.7 | 0 |
| Variety in Benthic Habitat Cover | 0.4 | 0.2 | 0.2 | 0 | 0.2 | 1.1 | 0.4 | 0.9 | 0.5 |
| Variety in Benthic Habitat Structure | 0.1 | 0.3 | 0 | 0 | 0 | 0.1 | 0 | 0.1 | 4.6 |

Note: Total Biomass values were double-log+1 transformed to achieve pseudo-normality, species were analyzed based on presence/absence. Species selection was based on PRIMER BEST analysis, maximum data density, and management importance. CAAB = *Carcharhinus amblyrhynchos*, LUBO = *Lutjanus bohar*, MASP = *Macolor* spp., TROB = *Triaenodon obesus*, NAHE = *Naso hexacanthus*, SCRU = *Scarus rubroviolaceus*, CAME = *Caranx melampygus*, CHUD = *Cheilinus undulatus*.
[1]Cross-validation (CV) deviance and standard error (SE) is shown as the measure of model performance (the lower the value the better the model performance).
doi:10.1371/journal.pone.0031374.t002

Analyses were carried out for total biomass of large-bodied reef fishes and for the occurrence (presence-absence) of eight individual species (Table 3). We were unable to use biomass data at the species level, as assumptions of normality were not valid even after extreme transformation. Species were primarily chosen based on their prevalence in the data, but also because of their relative contribution to overall patterns, and/or their ecological, economic, or management importance (e.g. as apex predators [39], [76], bioeroding herbivores [35], [77], fishery targets, or as IUCN "red list" species) (Table 3). *Macolor niger* and *M. macularis* were grouped as species of *Macolor* because of their ecological similarity and the difficulty distinguishing between the two species in the field. Total biomass values were double (log+1) transformed to achieve normality (or pseudo-normality) [66] and were analyzed based on a Gaussian distribution as required by the technique. Species-level counts were converted to presence-absence by survey segments and were analyzed using a binomial distribution, because the large proportion of zero values precluded normality even after extreme transformation. For each taxonomic group, we also quantified interaction effects among the various predictors (the collinearity and synergistic effect upon predicting the response in question) using the *gbm.interactions* routine [61]. In this routine, the relative strength of interaction fitted by BRT is quantified by the residual variance from a linear model, and the value indicates the relative degree of departure from a purely additive effect, with zero indicating no interaction effects. One can also think of the interaction value as the relative contribution of the interaction between the two predictors towards the overall predictive performance of the individual model (the cv deviance value). For each taxonomic group, we have chosen to report the top two interactions based on interaction value.

**Table 3.** List of taxonomic groups and reasons for inclusion in the analysis.

| Taxonomic Group | Data prevalence (# of records) | Contribution to overall pattern | Significance | | |
|---|---|---|---|---|---|
| | | | Ecological | Economic | Management |
| Total Biomass | X (1537) | X | X | X | X |
| *Carcharhinus amblyrhynchos* (CAAB) | X (122) | | X | X | X |
| *Lutjanus bohar* (LUBO) | X (545) | X | | | |
| *Macolor* spp. (MASP) | X (203) | X | | | |
| *Triaenodon obesus* (TROB) | X (181) | X | X | | X |
| *Naso hexacanthus* (NAHE) | X (113) | | X | | |
| *Scarus rubroviolaceus* (SCRU) | X (226) | X | X | X | X |
| *Caranx melampygus* (CAME) | X (148) | X | X | X | X |
| *Cheilinus undulates* (CHUD) | (49) | | X | | X |

Species were chosen based on their prevalence in the data, their relative contribution to overall patterns in the all-species multivariate PRIMER analysis, and/or their ecological, economic, or management importance.
doi:10.1371/journal.pone.0031374.t003

# RESULTS

## General Patterns

A total of 6280 individual large-bodied reef fish from 67 species were encountered during the survey period. While we were able to reject a null hypothesis of no spatial autocorrelation using Spearman rank correlation between resemblance matrices based on multivariate species-level biomass and the geographic coordinates for each survey segment, the effect size was weak (ρ=0.113, $p$=0.001, 999 permutations). The predominant pattern is that of a latitudinal gradient in mean total biomass of large-bodied reef fish (all species and years pooled), ranging from 0.10 g/100 $m^2$ (SE 0.02) at Guam to 2.13 g/100 $m^2$ (SE 0.39) at Farallon de Pajaros (Fig. 2). The northern (FDP – Sarigan) section of the archipelago has a mean total biomass level nearly twice that of the southern (Saipan – Guam) section (1.81 g/100 $m^2$ [SE 0.10] vs. 0.97 g/100 $m^2$[SE 0.05]). This separation between the northern and southern portion of the archipelago was maintained in the restricted data set that incorporates the environmental data.

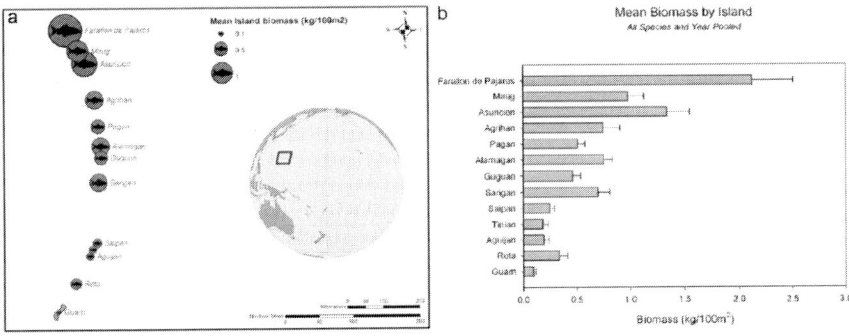


**Figure 2.** Biomass of large-bodied reef fish (all species and years pooled) in the Mariana Archipelago.

This pattern was evident in the hierarchical *CLUSTER* and n*MDS* analysis, based on species-level biomass density, and the *SIMPROF* test, which showed a distinct separation between the northern islands, which showed absent (None) to Low human population levels, and those in the south, which showed Mid to High populations (Global R=0.044, $p$=0.001) (Fig. 3). This dissimilarity was driven primarily by five major taxa (*Carcharhinus amblyrhynchos*, *Lutjanus bohar*, species of *Macolor*, *Triaenodon obesus*, and *Naso hexacanthus*), all of which showed higher biomass values at the northern, less populated islands (Table 4). *Carcharhinus amblyrhynchos* and species of *Macolor* were observed only in the northern islands, with the highest biomass densities observed at Asuncion Island (0.57 g/100 $m^2$ [SE 0.16] and 0.21 g/100 $m^2$ [SE 0.08], respectively). Although *Lutjanus bohar* and *Naso hexacanthus* were observed at both the northern and southern islands, biomass densities were higher at the northern islands, with the highest levels of both observed also at Asuncion (0.28 g/100 $m^2$ [SE 0.04] and 0.11 g/100 $m^2$ [SE 0.03], respectively). *Triaenodon obesus* were observed at both the northern and southern islands; however, biomass densities were higher in the northern islands with the highest levels at Maug and Alamagan Islands (0.14 g/100 $m^2$ [SE 0.02] and 0.12 g/100 $m^2$ [SE 0.03], respectively).

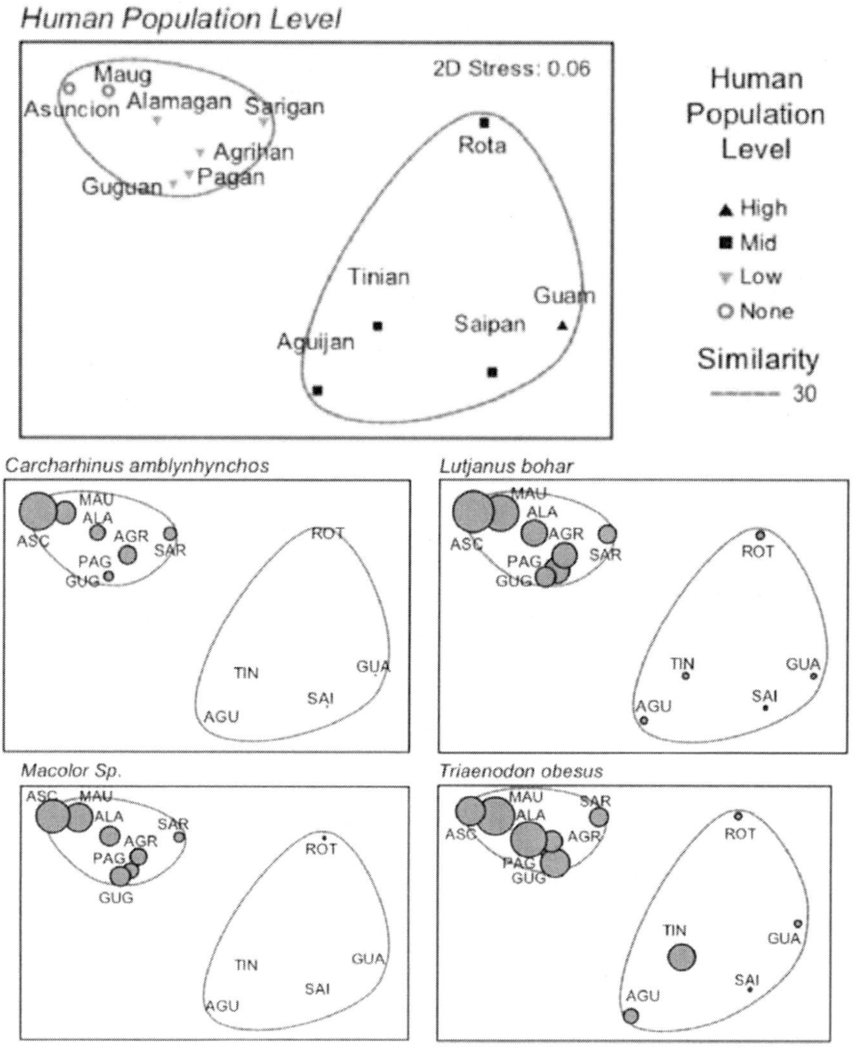

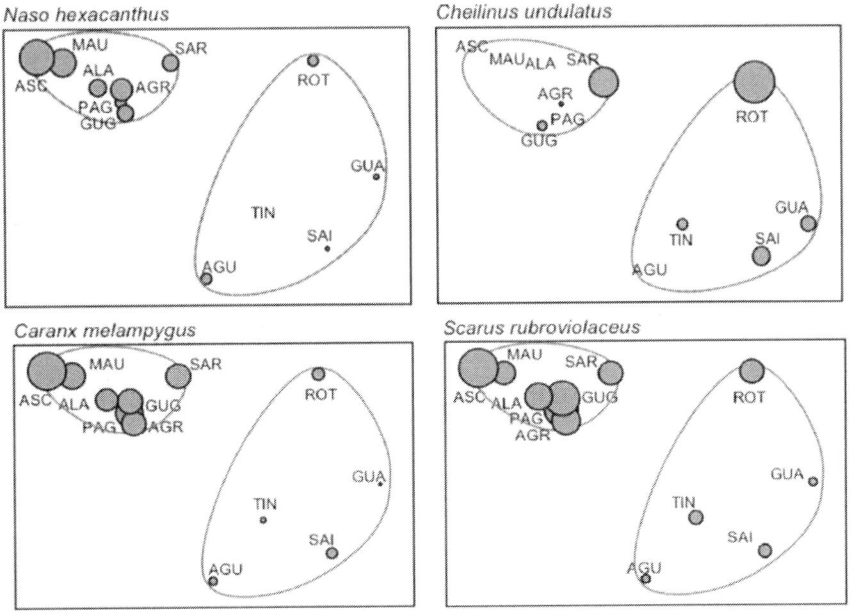


**Figure 3. Two-dimensional nMDS ordination plots of islands in the Mariana Archipelago based on multivariate species-level dispersion-weighted biomass of large-bodied (>50 cm TL) reef fishes and Bray-Curtis similarities.** Cluster contours represent significant SIMPROF groups (~30% similarity within groups). In the first panel, symbols reflect level of human population. Aguijan is categorized as "Mid" human population due to the influence of nearby Tinian and Saipan as well as reported visitation by fishers from Guam. Subsequent panels are the same nMDS as panel 1, with symbols representing relative biomass of each species.

**Table 4.** *SIMPER* analysis results showing species contribution to the dissimilarity between significant *SIMPROF* groupings based on species-level biomass (dispersion-weighted) and human population.

| | Average Biomass (disp) | | | | | |
|---|---|---|---|---|---|---|
| Species | (+) Humans | (−) Humans | Average Dissimilarity | SD | % Contribution | Cummulitave % contribution |
| *Carcharhinus amblyrhynchos* | 0.04 | 0.21 | 9.75 | 1.84 | 18.74 | 18.74 |
| *Lutjanus bohar* | 0.09 | 0.24 | 9.27 | 4.72 | 17.81 | 36.55 |
| *Macolor* spp. | 0.05 | 0.18 | 7.61 | 4.66 | 14.62 | 51.17 |
| *Triaenodon obesus* | 0.06 | 0.11 | 3.67 | 1.32 | 7.05 | 58.22 |
| *Naso hexacanthus* | 0.03 | 0.09 | 3.56 | 3.33 | 6.84 | 65.06 |

High-Mid human population group is designated as (+) Humans while Low-None group is designated as (−) Humans.
doi:10.1371/journal.pone.0031374.t004

## Fish-Habitat Relationships

The null hypothesis of no relationship between large-bodied reef fish species distribution and the predictor variables was rejected (ρ=0.109, $p$=0.001). Of the 15 continuous predictors, human population density (human population per reef area) provided the *BEST* match (corr.=0.197). Excluding human population density from the analysis resulted in a *BEST* match that, while having a lower correlation (0.106), indicated four environmental variables as significant predictors (Moon phase, % Coral Cover, % Sand Cover, and Variety in Habitat Structure). We were able to quantify the relative influence of each of the 18 continuous and categorical environmental variables using Boosted Regression Trees (Fig. 4 & Table 2) and were able to interpret the specific nature of each relationship in more detail using the resulting partial dependency plots (Fig. 5).

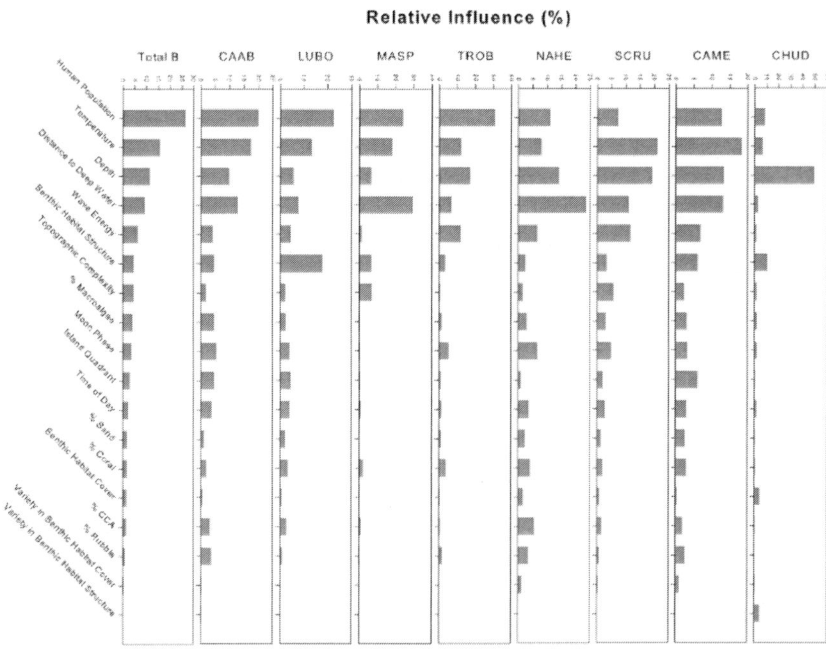


**Figure 4.** Relative importance plots for boosted regression tree analysis of total large-bodied reef fish biomass (all species pooled) and presence/absence of key species.

Species selection was based on maximum data density, and management and ecological importance. CAAB=*Carcharhinus amblyrhynchos*, LUBO= *Lutjanus bohar*, MASP=*Macolor* spp., TROB=*Triaenodon obesus*, NAHE=*Naso hexacanthus*, SCRU=*Scarus rubroviolaceus*, CAME=*Caranx melampygus*, CHUD=*Cheilinus undulatus*.

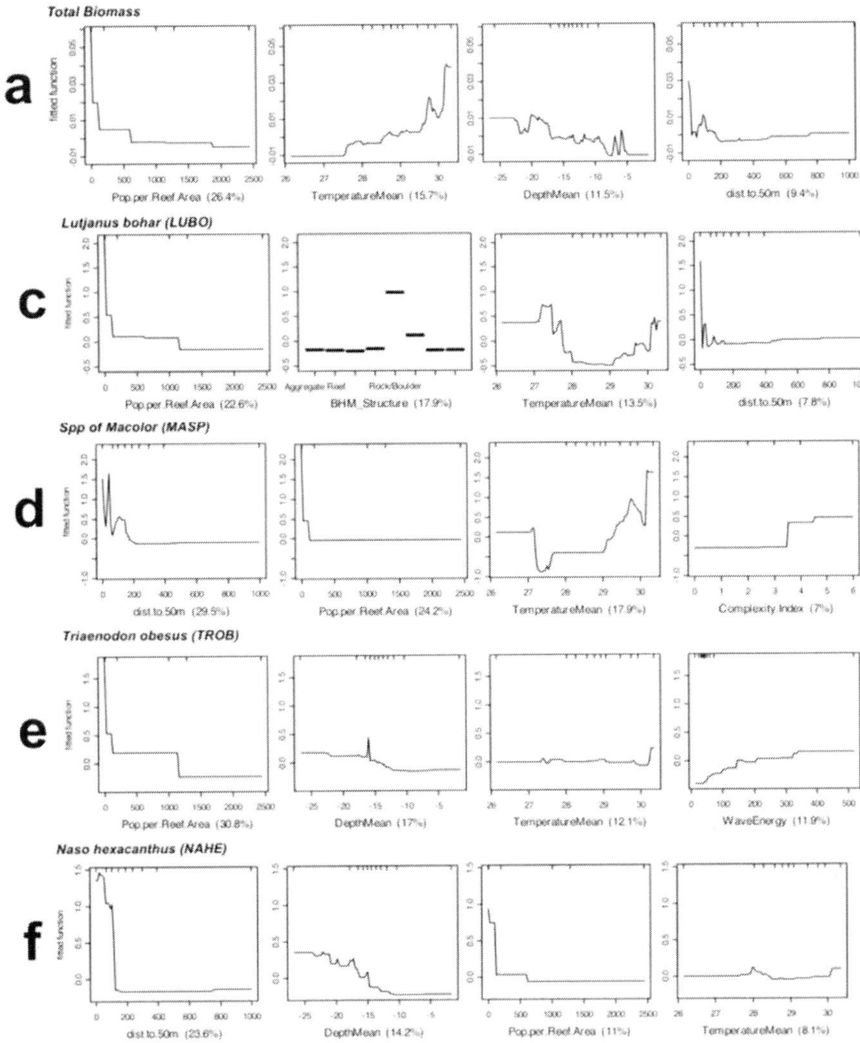

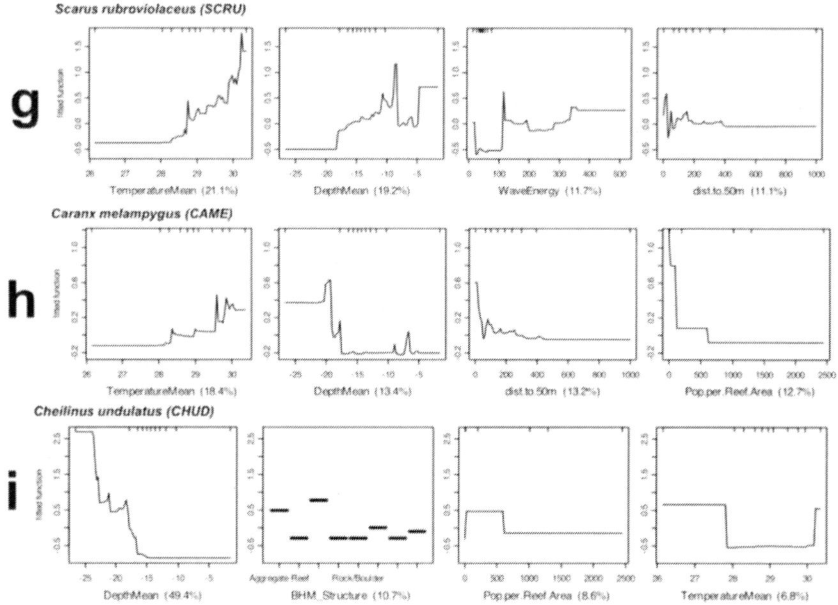


**Figure 5.** Partial dependency plots for boosted regression tree analysis of total large-bodied reef fish biomass (all species pooled) and presence/ absence of key species.

Species selection was based on maximum data density, and management and ecological importance. CAAB=*Carcharhinus amblyrhynchos*, LUBO=*Lutjanus bohar*, MASP=*Macolor* spp., TROB=*Triaenodon obesus*, NAHE=*Naso hexacanthus*, SCRU=*Scarus rubroviolaceus*, CAME=*Caranx melampygus*, CHUD=*Cheilinus undulatus*.

## Total large-bodied reef fish

Three relationships contributed most strongly to predicting total large-bodied reef fish biomass (Figs. 4 & 5a). Biomass was highest in areas with low human population density (relative influence [RI]=26.4%) and decreased as human population increased. Biomass also gradually increased as *in situ* water temperature increased (RI=15.7%), rose steeply once water temperature neared 30°C, and peaked at just over 30°C. Total biomass density of large-bodied reef fish also increased with increasing depth (RI=11.5%) and with proximity to deep water (RI=9.4%). Overall model CVD was 0.003 with a second order interaction between depth and temperature (Table 5). The effect of depth was exacerbated in the warmest areas (near 30°C) and the effect of temperature was similarly exacerbated in the deepest areas (deeper than ~20 m). In general biomass is highest at deeper depths (>20 m) and at higher temps (above 30°) (Fig. 6). However, biomass is also higher below 20 m across the entire temperature range.

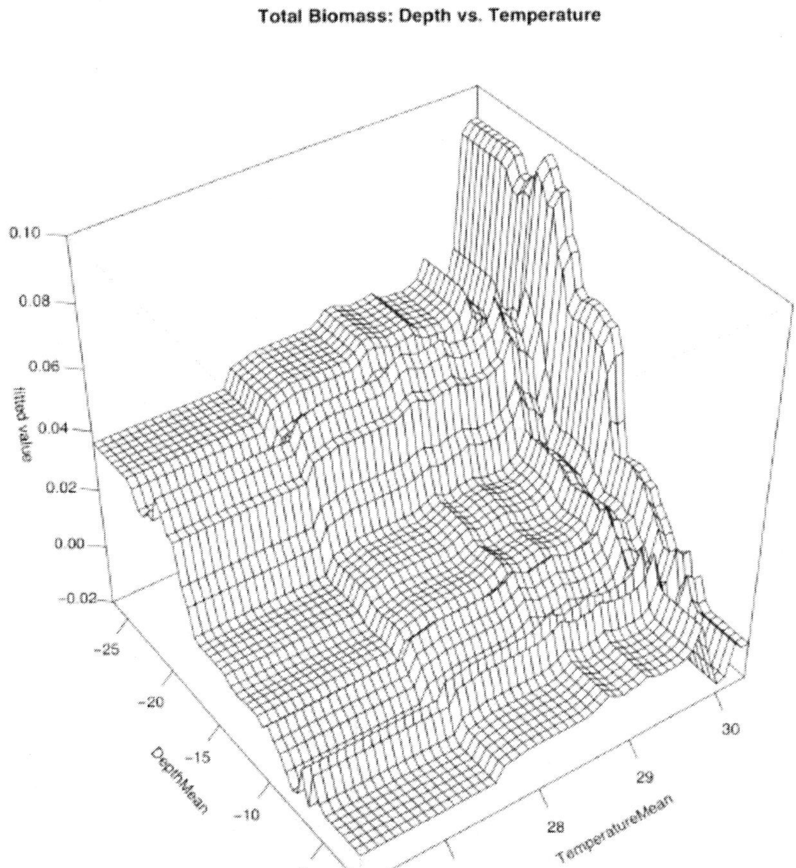


**Figure 6. Pairwise interaction between depth and temperature with respect to total biomass of large-bodied reef fish.**The effect of temperature on total large-bodied reef fish biomass is magnified in areas of warm water and the effect of temperature is magnified at deeper depths.

**Table 5.** Pairwise interactions between predictor variables used to relate total biomass and occurrence of each taxa to environment.

| Taxa | Predictor | Predictor | Interaction value | Interaction Summary |
|---|---|---|---|---|
| Total Biomass | Depth | Temperature | 0.03 | Deeper Depth+Higher Temp = Higher Biomass |
| CAAB | (BHM_Structure) | Temperature | 32.31 | |
| | Depth | Human Population | 28.25 | Depth >15 m+0 people = More CAAB |
| CAME | % CCA | Depth | 11.78 | Depth >20 m+CCA >20% = More CAME |
| | (BHM_Structure) | Depth | 7.64 | |
| CHUD | NA | NA | NA | *To few samples to model interactions* |
| LUBO | Wave Energy | Temperature | 25.12 | Temp <27 or >29+Higher Wave Energy = More LUBO |
| | Depth | (Island Quardrant) | 9.17 | |
| MASP | NA | NA | NA | *To few samples to model interactions* |
| NAHE | Distance to Deep Water | (BHM_Structure) | 20.82 | |
| | Distance to Deep Water | (BHM_Cover) | 6.62 | |
| SCRU | Wave Energy | Depth | 27.61 | Shallow depth+low to moderate Wave Energy = More SCRU |
| | % Sand | Wave Energy | 11.92 | % Sand <20+Wave Energy between 100 & 200 = More SCRU |
| TROB | Depth | Human Population | 20.99 | Depth >15 m+0 people = More TROB |
| | Wave Energy | Human Population | 4.17 | |

*Interactions displayed are the top two for each taxonomic group (based on value) that involved the 8 predictors offering the highest contribution to the model displayed in Figure 5. Only one interaction is displayed for Total Biomass as none of the lower interactions had values greater than 0.01. Interaction value indicates the relative degree of departure from a purely additive effect; with a value of zero indicating that no interaction is present. A summary description is given for the association of the peak in each of 9 taxonomic groups and the pairwise interactions for those predictor variables showing a clear relationship (for example positive, negative, or modal) with the taxa in Figure 5. Categorical variables are noted with "()".*
*NA indicates sample sizes insufficient to model interactions.*
doi:10.1371/journal.pone.0031374.t005

## Carcharhinus amblyrhynchos

Three relationships contributed most strongly to predicting the occurrence (presence/absence) of *Carcharhinus amblyrhynchos* (Figs. 4 & 5b). These were a steep negative relationship with human population density (RI=19.9%), a somewhat variable relationship with *in situ* water temperature (RI=17.4%), and a negative relationship with distance to deep water (RI=12.8%). Sightings were most frequent where water temperatures were just below 28°C and decreased gradually with a smaller peak at 29°C. Overall model CVD was 0.207 with second order interactions between habitat structure and temperature as well as between depth and human population (Table 5). While the partial dependency plots show that both human population density and depth are correlated with distribution of *C. amblyrhynchos* (with higher occurrences in areas of low human population density and in deep water), the correlation with depth is partially dependent on local human population density. The effect of depth is weaker (or almost non-existent) where human population density is high and stronger (or only seems to matter) where there are fewer (to zero) people (Fig. 7). Similarly, the effect of local human population density is much reduced in shallower depths.

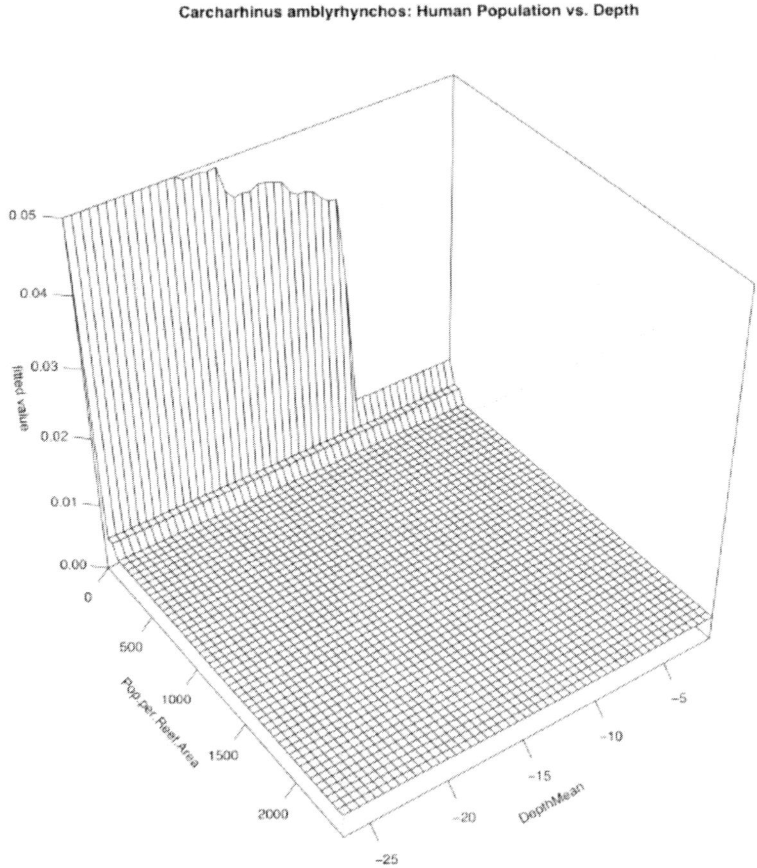


**Figure 7. Pairwise interaction between human population density and depth with respect to occurrence of *Carcharhinus amblyrhynchos*.** Depth is only influential in areas of low human population density and the effect of human population density is reduced in shallow waters.

## Lutjanus bohar

Two relationships contributed most strongly to predicting the occurrence of *Lutjanus bohar*(Figs. 4 & 5c). A negative relationship with human population density was the strongest predictor (RI=22.6%). This was followed by a relationship with benthic habitat structure (RI=17.9%) for which *Lutjanus bohar* sightings were highest in areas classified as "Rock/Boulder". Sightings decreased as water temperatures warmed from 27°C to 28°C and increased again once water temperatures reached 29°C (RI=13.5%). Sightings also dropped steeply as distance to deep water increased (RI=7.8%). Overall model CVD was 0.595 with second order interactions between wave energy and temperature as well as between depth and island quadrant (Table 5). *L. bohar* occurred most frequently in low temperature (<27.5°C) high wave energy waters (>250 kW/m). The effect of wave energy was stronger at these lower water temperatures and temperature appeared to have more of an effect at higher wave energies (Fig. 8). The interaction between depth and island quadrant suggests that the effect of depth was not universal across island quadrants, but this type of interaction (between categorical and continuous variables) is difficult to interpret.

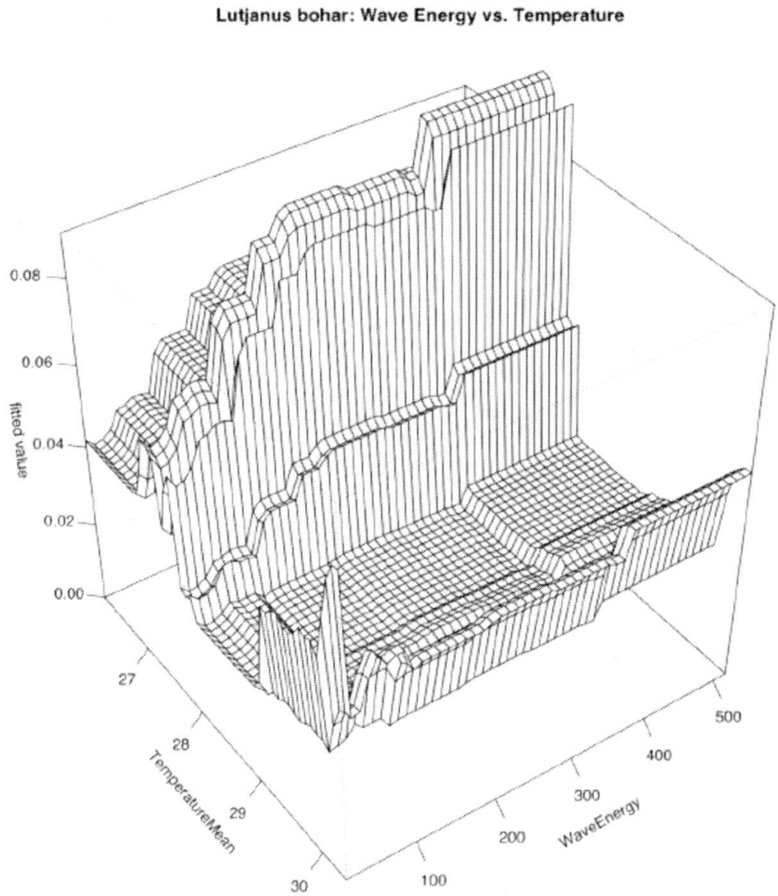


**Figure 8. Pairwise interaction between wave energy and temperature with respect to occurrence of *Lutjanus bohar*.** The effect of temperature is magnified by wave energy while wave energy is primarily influential at temperatures below 28°C.

## Macolor spp

Three relationships contributed most strongly to predicting the occurrence of species of *Macolor*(Figs. 4 & 5d). An overall negative relationship with distance to deep water was the strongest predictor (RI=29.5%). This was followed by a steep negative relationship with human population density (RI=24.2%). Sightings of species of *Macolor* also decreased as water temperatures warmed from 27°C to 29°C and increased again once water temperatures exceeded 29°C (RI=17.9%). Sightings of species of *Macolor* also increased with increases in observed benthic habitat complexity (RI=7%). Overall model CVD was 0.309 and sample sizes were insufficient to model secondary interactions.

## Triaenodon obesus

A negative relationship with human population density was the strongest predictor of sightings of *Triaenodon obesus* (RI=30.8%) with a steep decline in presence between 0 and approximately 200 humans/km$^2$ of reef and another drop once human population density reached 1000 individuals/km$^2$ (Figs. 4 & 5e). A slightly positive relationship existed between sightings of *Triaenodon obesus* and depth (RI=17%), with a spike in sightings at depths between 15 and 20 m. Slightly positive relationships also existed between sightings of this species and temperature (RI=12.1%) and wave energy (RI=11.9%). Overall model CVD was 0.341 with second order interactions between human population density and depth and wave energy (Table 5). In a pattern reminiscent of that seen in *C. amblyrhynchos*, the relationship between *T. obesus* and depth is partially dependent on local human population density. The effect of depth is weaker (or almost non-existent) where human population density is high and stronger (or only seems to matter) where there are fewer (to zero) people (Fig. 9). Similarly, the effect of local human population density is much reduced in shallower depths. Similarly, wave energy was only influential when human population density was low and the effect of human population density appeared to be reduced in areas with low wave energy (Fig. 10).

## Naso hexacanthus

A steep negative relationship with distance to deep water most strongly predicted sightings of *Naso hexacanthus* (RI=23.6%), with fitted values dropping to near zero as distances to the 50 m isobath exceeded 200 m (Figs. 4 & 5f). A positive relationship also existed with depth (RI=14.2%), and there was a steep negative relationship with human population density (RI=11%). Overall model CVD was 0.228 with second order interactions between distance to deep water and benthic structure and cover (Table 5). The interaction between distance to deep water and benthic structure and cover (both categorical variables) suggests that the effect of distance to deep water was stronger in certain structure or cover classes as opposed to others, and that the relationship with cover and structure is partially dependent on proximity to deep water. Again however, this type of interaction (between categorical and continuous variables) is difficult to interpret.

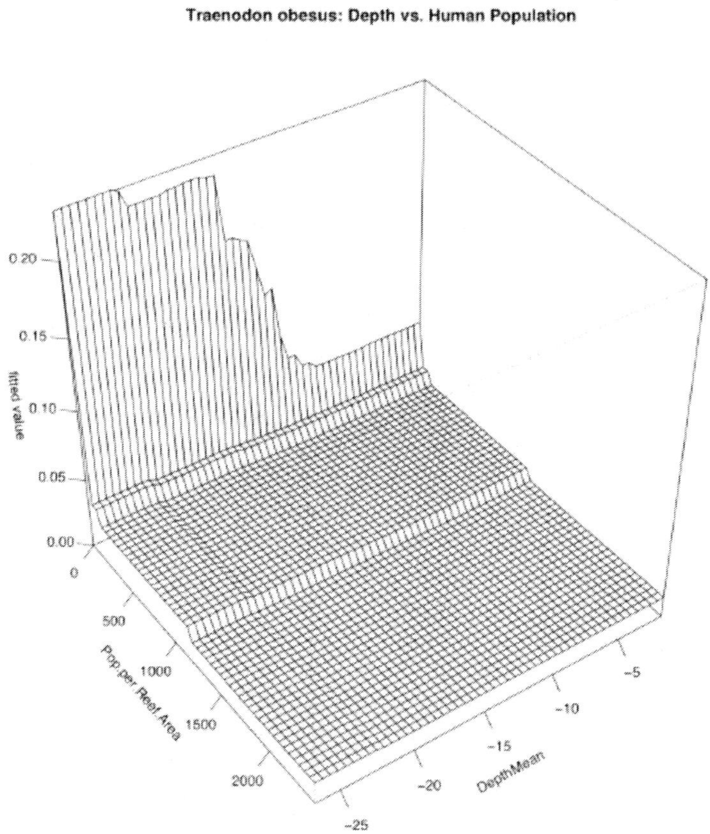


**Figure 9. Pairwise interaction between human population density and depth with respect to occurrence of *Triaenodon obesus*.**The effect of depth is influential only at the lowest human population densities and the effect of human population density is reduced in shallow areas.

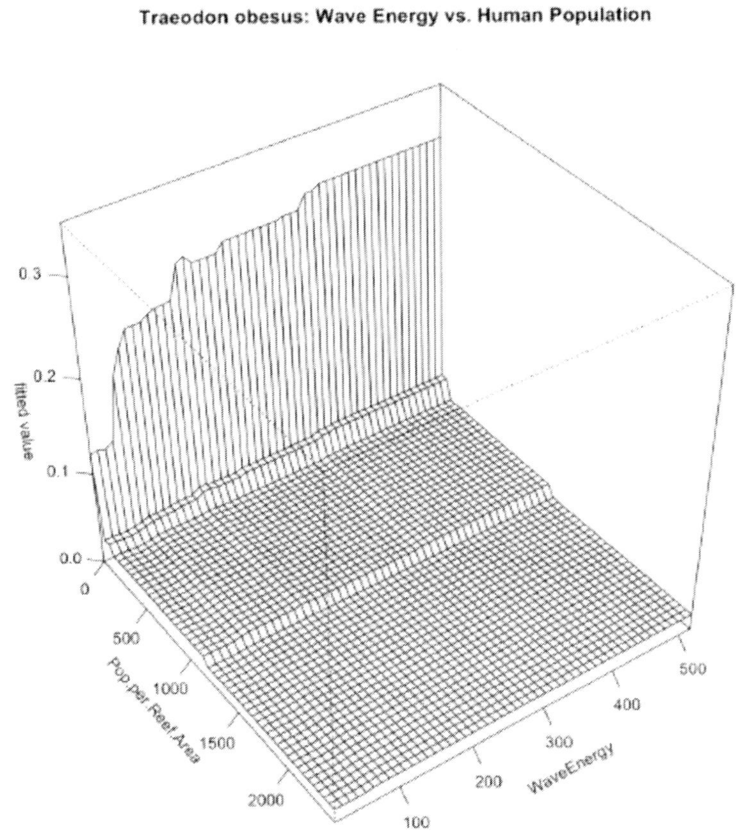


**Figure 10. Pairwise interaction between human population density and wave energy with respect to occurrence of *Triaenodon obesus*.**The effect of wave energy is influential only at the lowest human population densities and the effect of human population density is magnified in higher wave energy areas.

## Scarus rubroviolaceus

Relationships with temperature and depth contributed most strongly to predicting the occurrence of *Scarus rubroviolaceus* (RI=21.1% and 19.2%, respectively) (Figs. 4 & 5g The relationship with temperature was generally positive but was highly variable. The relationship with depth was also highly variable but sighting frequency was highest in depths near 10 m or less than 5 m. Highly variable relationships also existed with wave energy (RI=11.7%) and distance to deep water (RI=11.1%). There was not a strong relationship between the occurrence of this species and local human population density. Overall model CVD was 0.394 with second order interactions between wave energy and depth and sand cover (Table 5). The interaction between depth and wave energy is complicated. The relationship between *S. rubroviolaceus* and depth appears to be most pronounced in low wave energy environments (Fig. 11). As one might expect, the relationship between *S. rubroviolaceus* and wave energy is apparent only at depths shallower than 15 m and is strongest at depths shallower than 5 m. The effect of wave energy is also much reduced in areas of high sand cover. The effect of sand is highest at wave energies of between 100 and 200 kW/m (Fig. 12).

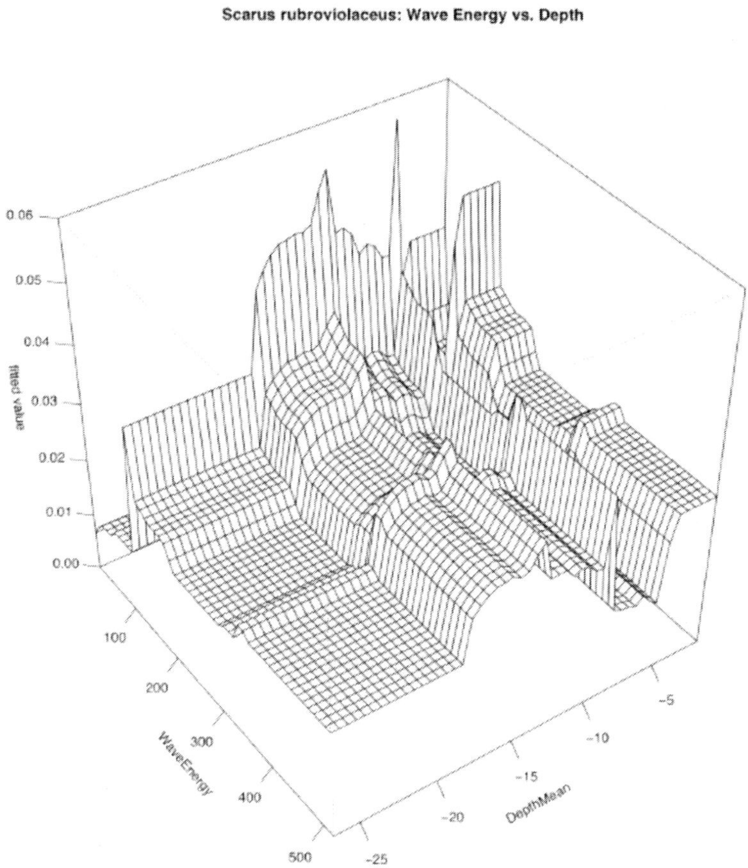


**Figure 11. Pairwise interaction between depth and wave energy with respect to occurrence of *Scarus rubroviolaceus*.** The effect of wave energy is influential only in depths shallower than 15 m. The effect of depth is somewhat magnified in lower wave energy areas.

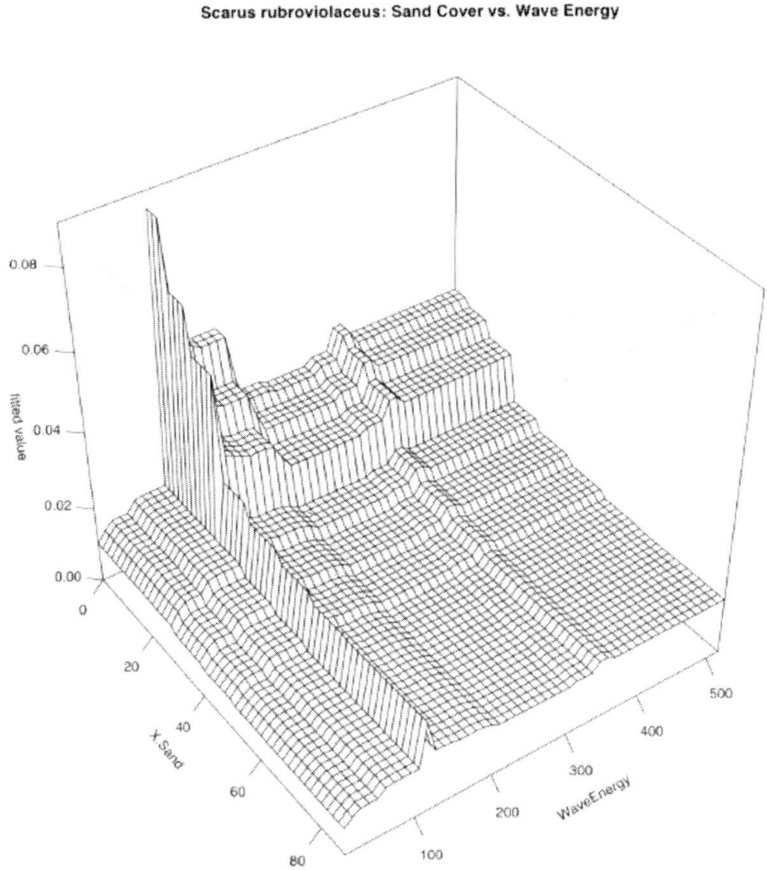


**Figure 12. Pairwise interaction between wave energy and sand cover with respect to occurrence of *Scarus rubroviolaceus*.**The effect of wave energy was highest in areas with low sand cover and the effect of sand cover was greatest in areas were wave energy was between 100 and 200 kW/m.

## Caranx melampygus

Temperature, depth, distance to deep water, and human population density all contributed to predicting the occurrence of *Caranx melampygus* (RI=18.4%, 13.4%, 13.2%, 12.7%, respectively; Figs. 4 & 5h). The relationship with temperature and depth were positive but highly variable, while the relationship with distance to deep water was negative. Sighting frequency declined steeply as human population density increased from zero. Overall model CVD was 0.300 with second order interactions between depth and percent cover of crustose coralline algae (CCA) and benthic structure (Table 5). The effect of percent cover of CCA was almost non-existent and depth shallower than 17 m and the effect of depth seems to have been reduced in areas with low CCA cover (Fig. 13). The interaction between depth and benthic structure (another categorical variable) suggests that the effect of depth was stronger in certain structure classes as opposed to others, and that the relationship with structure is partially dependent on depth.

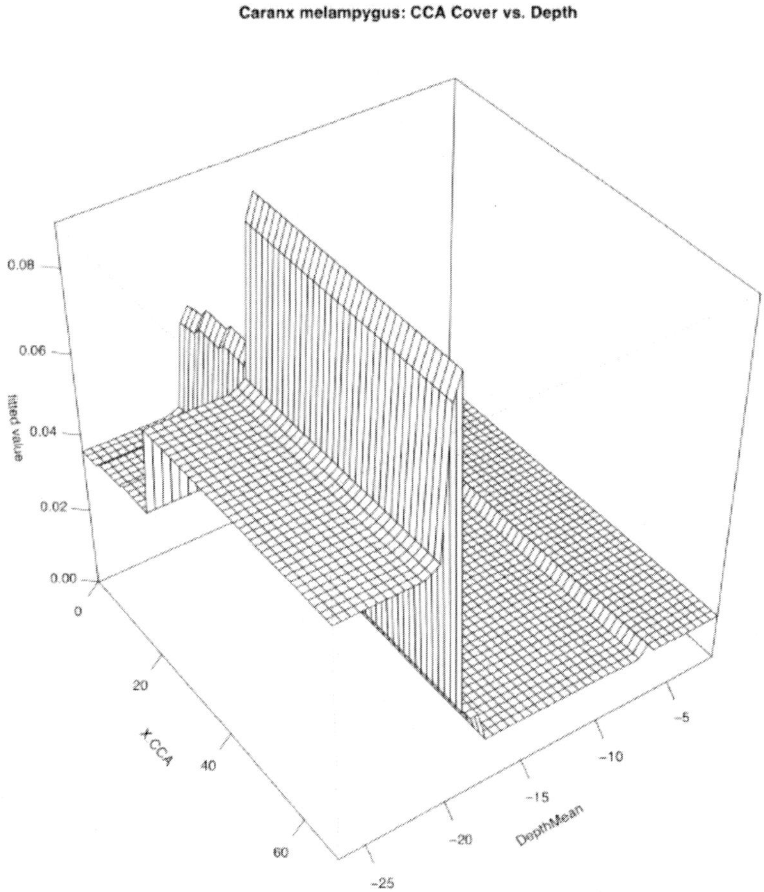


**Figure 13. Pairwise interaction between depth and percent cover of crustose coralline algae (CCA) with respect to occurrence of *Caranx melampygus*.**The effect of CCA was greatest at depths greater than 15 m while the effect of depth was reduced in areas with lower CCA cover.

## Cheilinus undulatus

Depth most strongly contributed to predicting the occurrence of *Cheilinus undulatus* with a greater number of sightings at depths greater than 15 m (RI=49.4%) (Figs. 4 & 5i). Sightings were also higher in areas classified as pavement and aggregate reef (RI=10.7%). Human population density had a relative influence of 8.6%, with higher occurrences of *Cheilinus undulatus* in areas where human population density was between 0 and 750 individuals per km$^2$ of reef. Overall model CVD was 0.131 and sample sizes were insufficient to model secondary interactions.

# DISCUSSION

The islands of the Mariana Archipelago span a wide range of geographic, environmental, and anthropogenic gradients. The archipelago spans nearly ten degrees of latitude and ranges from relatively large, heavily populated, carbonate islands in the south to small, remote, unpopulated, volcanic islands in the north. While one might expect significant spatial structure in this type of data, a test for spatial autocorrelation indicates that our data are highly variable across space and that any effect of spatial autocorrelation is weak and likely overshadowed by the environmental variables used out model. Nevertheless, a number of the environmental and anthropogenic variables selected for analysis covaried, thus limiting our ability to draw definitive conclusions about the effect of any one variable to the exclusion of others. For example, the north-south gradient in human population density followed similar gradients in latitude, mean sea surface temperature, open ocean primary productivity and island size. Consequently, while we feel that human populations density is the most likely causative factor, and a substantial body of literature exists that would support such a conclusion [32], [38], [78]–[84], we are nevertheless, not able to say with certainty that human population density is the causative factor to the exclusion of all others and our results should be used to indicate factors that appear to be important in shaping the spatial distribution of large-bodied reef fishes in this geographic area.

Measurements of *in situ* temperature factored highly in our analysis of relative influence and many studies have focused on the importance of temperature in the ontogeny of fishes as well as in structuring marine and aquatic communities. Francis [85] discusses the importance of sea surface temperature in determining the year class strength of New Zealand snapper while Worm et al. [86] show that patterns of diversity in top predators can be correlated with thermal fronts and patterns in the distribution of dissolved oxygen. In their 1980 study of habitat preference and fisheries oceanography, Magnuson et al. [87] showed that sea surface temperature was the best predictor of catch per unit effort (CPUE). These authors do, however, suggest that, as with any environmental correlation, the causal hypothesis regarding temperature needs laboratory verification as the finding that fish distributions following temperature gradients does not necessarily mean that temperature itself is the forcing function. Rather, it is possible that higher trophic-level fishes may be responding to the availability of prey species that are the ones responding to temperature. Furthermore, it is clear that fish assemblages are not responding directly to human population density, but are rather responding to a variety of direct and indirect factors (e.g. fishing, sedimentation, habitat degradation, etc.) related to human population density. It should also be noted that the coral reef system is enormously complex, with myriad interacting factors affecting species distributions through a variety of direct and indirect ways. Furthermore, the assemblages of large-bodied reef fishes we describe are rare, highly mobile, and patchily distributed. Hence, even our most complete model leaves the majority of deviance unexplained. That being said, the key findings of our study, such as the high relative importance of human population density and the effect of depth and distance to deep water remain important considerations for management of the marine resources of Guam and the Commonwealth of the Northern Mariana Islands and are consistent with research on other taxa from other locations [32], [34], [80], [81], [88]. While the low relative influence of small-scale environmental factors (i.e. benthic cover and habitat heterogeneity) that have been previously identified as important in the distribution of small-bodied reef fishes [10], [89]–[91] is intuitive given their life history, to our knowledge this has seldom been demonstrated in the literature regarding this area. It stands to reason that small-scale factors are of prime importance to those species that interact most directly with the reef either for food or shelter [92]. Chabanet et al. [90] found no relationship between habitat cover and the abundance of planktivores or carnivores at Reunion Island, and suggest that the relationship between the abundance of fish and coverage of living coral may be stronger in shallow water where fish remain in closer physical proximity to the substrate. However, as Levins [93] suggests, the perceived “grain size” or “resolution” of habitat depends on the body size of the individual animal. Hence, for wide-ranging large-bodied species small-scale differences in benthic cover are likely less important.

## Anthropogenic Impacts

Community structure, relative biomass, and species occurrence differed substantially between the heavily populated southern islands and the more remote and unpopulated islands in the north. Overall biomass of large-bodied reef fish in the northern islands was nearly twice that found in the southern islands. Several taxa such as *Carcharhinus amblyrhynchos* and species of *Macolor* were abundant in the northern islands and nearly absent in the southern islands (Fig. 3). Other species such as *Lutjanus bohar, Triaenodon obesus, Scarus rubroviolaceus,* and *Caranx melampygus* were found both north and south, but biomass densities were much lower at the southern islands. At present, we do not know of a robust direct measure of fishing pressure in the Mariana Archipelago, so the evident relationship between many of the taxa in our study and local human population density cannot be directly attributed to the effects of fishing. While anthropogenic habitat destruction cannot be ruled out as a possible cause, we would expect anthropogenic habitat destruction to manifest itself through changes in biological benthic cover (e.g. shifts from coral to algae) and possibly changes in small-scale substrate complexity (e.g. sedimentation, coral death and erosion). As these habitat factors were of low relative influence in our study, it seems unlikely that anthropogenic habitat destruction is the basis for the high relative influence of human population density on the distribution of large-bodied reef fishes in this study. It should be noted that, with the exception of the two shark species, each of the chosen species do appear, in varying levels, in commercial and recreation catch data from Guam and the Commonwealth of the Northern Mariana Islands. While *L. bohar*is considered ciguatoxic in the Mariana Archipelago, and is therefore not commonly considered a "fishery target", the species does appear—albeit in modest amounts—in both boat-based and shore-based creel survey data from Guam, as well as in commercial landings data, where it is classified under the the common name "Tagafi" and possibly as a component of the overall "snapper" grouping (unpublished data collected and processed by Guam Division of Aquatic Resources and provided to the authors via the WPacFIN program [94]).

Interestingly, overall biomass density of *Cheilinus undulatus*, an IUCN Red List species, was higher in the southern islands than in the north, with the highest densities found around the islands of Rota and Sarigan, which have moderate to low human population densities. Initially we thought this might be due to a higher number of smaller individuals in the southern islands compared with a lower number of larger individuals in the north. However, this does not appear to be the case, as few individuals of any size were sighted in the northern portion of the archipelago. Of the environmental factors evaluated in the present study, depth had the highest relative influence with higher occurrence of *C. undulatus* at deeper depths. However, the waters around the southern islands tend to be shallower than those in the north, so this does not explain our findings. It is possible that *C. undulatus* distribution reflects the availability of certain habitats or conditions conducive to this species at Rota and Sarigan that we failed to measure, or the lack of the same in the northern islands. *Cheilinus undulatus* is an IUCN Red List species and is, therefore, of high management importance. However, it should be noted that *C. undulatus* was observed on only 49 of our surveys compared to a minimum of 100 survey observations for each of the other taxonomic groups. Hence, our results may be an artifact of low sample size, may not be representative of true distribution patterns, and should be interpreted with caution. Further, targeted research on this low-density species is needed to fully resolve the factors affecting its distribution.

All *in situ* surveys likely influence their study subjects, causing fishes to either aggregate around or flee from the observers or instrument platform [95]. We feel that the towed-diver methods used in this study are not as prone to bias as other diver-based techniques [53]. However, it should be noted that the relative difference between remote and human-inhabited islands could be exaggerated if fishes aggregate around divers in remote areas and flee from divers in areas frequented by people. That being said, our findings are consistent with the preliminary findings of Schroeder et al. [28], statements such as those by the Western Pacific Regional Fishery Management Council [26] suggesting that many large bodied species "became rare on shallow reefs [around Guam] due to heavy fishing ...", and the substantial literature linking anthropogenic impacts to decreases in fish populations [27], [32], [34], [38], [47], [48], [80]–[82], [96]–[106]. Fish and shellfish dominate middens dated to the period of early human habitation in the Mariana Islands and those from early settlement sites contain the remains of a wide range of fishes from coastal reefs, lagoons, and deep water (Green 1997). During this time, fishing appears to have been concentrated on coral reef and lagoon species [107] and stable isotope data indicate that marine foods constituted nearly 40% of the prehistoric diet [108]. Shortcomings in fishery statistics make it virtually impossible to assess the total harvest of contemporary coral reef fisheries in the Mariana Archipelago [26]. While most fisheries are limited to nearshore areas off Saipan, Rota, and Tinian, the accuracy of reporting is suspect in many areas and virtually no information is available for the inshore subsistence and recreational fisheries. At least six of the northern islands have been commercially fished to some extent [26], and poaching by foreign vessels has been documented [46]. While poaching is difficult to quantify, it is a legitimate concern as poachers tend to target high-value, rare, or otherwise heavily fished resources [26] – qualities that describe many large-bodied reef fish populations.

The evidence that early human populations had at least some impact on the local reef environment and that contemporary anthropogenic impacts may be greatly underestimated should be to be taken into account when discussing relationships between reef fishes and their habitat, as it is likely that contemporary relationships that include human influences may not necessarily be representative of healthy let alone "pristine" systems. Nevertheless, contemporary fish-habitat relationships can reveal the importance of certain physical, oceanographic, and biological factors in structuring the distribution of large-bodied reef fishes. As we discuss below, these relationships between fishes and their environment can inform discussions of important or "essential" fish habitats and their roles in ecosystem-based management.

## Fish - Habitat Relationships

The Boosted Regression Tree analysis, based on species presence/absence, was able to assess the relative influence of each environmental factor adjusting for the impacts of each of the other variables. Human population density was still the most influential factor, but others appear to play a role (Figs. 4 & 5). In certain cases, such as with the two shark species *Carcharhinus amblyrhynchos* and *Triaenodon obesus*, a look at prominent pairwise interactions among variables shows that variables including depth and wave energy are only influential in areas with the lowest human population densities. While the absolute influence of each variable was relatively low in this highly variable data set, oceanographic factors such as depth, temperature, and proximity to deep water (quantified as distance to the 50-m depth contour) consistently the most influential of the factors we analyzed. In almost all cases, species occurrence increased in deeper waters and in areas of higher temperature. The effects of temperature and depth were synergistic with respect to total large-bodied reef fish biomass, with the effect of each being magnified by the other. Depth was particularly important for *Cheilinus undulatus*, which was only observed on surveys in greater than 15 m. *Caranx melampygus* and *Scarus rubroviolaceus* displayed an inverse relationship with respect to depth, which highlights their differing niches. Carnivorous C. melampygus were primarily found in deeper waters (>15 m) while herbivorous *S. rubroviolaceus* were primarily found in the shallower areas (<15 m) more conducive to algae growth. Proximity to deep water was particularly important for *Carcharhinus amblyrhynchos*, species of *Macolor*, *Caranx melampygus* and *Naso hexacanthus*, which are all found near escarpments or walls. These findings are consistent with those of Friedlander et al. [88] who found depth and proximity to the pelagic environment to predictably alter the structure of reef fish assemblages at Kingman Reef in the northern Line Islands. Specifically, these authors found that both abundance and biomass increased with proximity to the pelagic environment and with increasing depth on the forereef. Our findings are also consistent with those of Wetherbee et al. [109] and Papastamatiou et al. [110] who found that CPUE or the relative number of grey reef sharks (*Carcharhinus amblyrhynchos*) caught in Hawai'i generally increased with depth. Friedlander et al. [88] also found that apex predators, and the planktivorous snappers, on which they prey, were more abundant on the deeper forereef and reef walls, presumably because currents consistently deliver higher concentrations of plankton to these areas [111], [112]. These findings are consistent with the relationship we describe regarding the planktivorous species of *Macolor* and *Naso hexacanthus*. It is possible that the relationship between the apex predators (*Carcharhinus amblyrhynchos, Caranx melampygus*) and proximity to deep water is a secondary relationship related to the distribution of their prey species. This conclusion is consistent with Wetherbee et al. [109] and Papastamatiou et al. [110], who found that teleost fishes (e.g., holocentrids, monacanthids, and acanthurids) dominated the gut contents of *Carcharhinus amblyrhynchos* in the Hawaiian Islands.

*In situ* temperature also factored highly in the distribution of large-bodied reef fish, with overall biomass and the occurrence of many taxa positively correlated with temperature. In certain taxa including *Carcharhinus amblyrhynchos* and *Lutjanus bohar*, temperature showed high relative influence, but the relationship did not have a clear linear trend. In the case of *L. bohar*, there was an interaction between temperature and wave energy in which the effect of temperature was magnified in areas of high wave energy. The thermal environment of coral reefs is highly stable and thermal gradients can play a large role in life history and distribution both on and off coral reefs [113]–[116]; hence it is not surprising that fish react to temperature and that the relationship can be quite complex. Meyer et al. [19] found that short-term movements of *Aprion virescens* at Pearl and Hermes Reef in the NWHI were oriented to tidal cycles, with fish moving along the barrier reef on the flooding tide and returning on the ebbing tide, and tidal rhythmicity has been previously documented in acoustic monitoring studies of other fishes. We did not account for tidal cycle in our analysis, and it is possible that the importance of temperature is magnified by a correlation with tidal cycling. There may also be a relationship between temperature and primary productivity that would follow the north-south human population gradient. However, unlike remotely sensed mean sea surface temperature, *in situ* temperature measurements did not clearly correlate with other environmental gradients measured in this study (max $\rho$=0.41 with % sand cover) and the difference in mean *in situ* temperature measurements between remote and populated sections of the archipelago (the strongest overall pattern) was only 0.4°C.

Topographic complexity, benthic habitat structure, and benthic cover had relatively little influence in our analysis of large-bodied reef fish. In contrast, these factors have a strong influence on the distribution and relative abundance of smaller reef fishes [5], [6], [8], [11], [89], [91], [117], [118]. It is likely that our estimates of topographic complexity, which are based on a six-point scale visually estimated by divers, are more subjective or variable than quantitative measures such as multibeam SONAR or LIDAR. These latter types of data might provide a better means for assessing this relationship, but, unfortunately, they are presently unavailable for these areas. Organisms interact with their environment at a range of scales, and the relative heterogeneity of the environment depends on the size of the individual [93]. It is possible that the resolution at which we measured these variables is not ideal for the large-bodied portion of the reef fish assemblage.

Benthic habitat structure appeared to be important for *Lutjanus bohar* and *Cheilinus undulates*, with higher occurrences of these species in areas characterized as "Rock/Boulder" and "Pavement" or "Aggregate Reef," respectively. While a majority of species showed a positive relationship with benthic complexity, the relative influence of complexity was not high compared to the other measured variables. This may be because many large-bodied reef fishes are not as closely associated with small-scale benthic structures (but see Lindberg et al. [17], tend to have wider ranges, and interact with their environment in a more "coarse-grained" manner (sensu Levins [93]). It is also possible that we failed to measure habitat characteristics in sufficient detail or at the scales important to this portion of the assemblage. As noted above, the ability to incorporate high-resolution bathymetric information for large areas, such as that provided by multibeam SONAR or LIDAR technologies, would likely allow for a more detailed investigation of these relationships.

Fisheries management in coral reef ecosystems is moving towards ecosystem-based approaches with a priority placed on the identification of "essential fish habitat" – 'those waters and substrate necessary to fish for spawning, feeding, breeding, and growth to maturity' [7], [26]. However, logistical constraints have limited much of the previous research to small-bodied coral reef fishes and often to areas adjacent to human populations [8], [119]–[121]. As many fisheries tend to preferentially target large-bodied species and large individuals [48], [122], [123], determination of the relative importance of environmental factors for this segment of the assemblage is essential to improving fisheries and ecosystem-based management. Our results show that proximity to areas of high human population had the highest influence relative to the other environmental factors we measured and likely influences large-bodied reef fish assemblages in the Mariana Islands. Whether this is a result of fishing or other anthropogenic impacts for which we do not yet have accurate data is unknown, but extraction is commonly assumed to be the most important impact associated with human population centers [27], [80], [101], [124].

While the negative relationship with human population was the most important predictor variable in this study, the human population gradient was correlated with latitudinal and temperature gradients. Large-scale oceanographic factors such as depth, temperature, and proximity to deep water were also important. Shallow-water wave stress may be significant locally, but this is difficult to quantify accurately without bathymetry data. Large-scale habitat structure was also shown to be important for certain species. Conveniently, data on many of these factors may be easily obtained at large scales through remote sensing, allowing managers and researchers to develop predictive models delineating areas conducive to large-bodied reef fishes using methods similar to those outlined by Pittman et al. [12], De'ath [72] and Elith et al. [61]. Such information can be used in ecosystem-based spatial planning and management. For example, even for cases in which visual or other population survey data are lacking, areas likely capable of supporting species of particular concern can be identified as a areas for further research or consideration as marine protected areas. The reader will note that, as with other studies focusing on low density and patchily distributed assemblages, our models explain only a small portion of the overall deviance. There are likely many factors that contribute to the spatial and temporal distribution of individual species. Our aim has been to evaluate the relative influence of a specific suite of environmental variables that have been identified as important in previous studies. By gaining a better understanding of the relationships between large-bodied reef fishes and their environment, especially with those environmental variables that showed high relative influence, management agencies can better conserve large-bodied reef fish populations. While many of the environmental variables identified as important in this study cannot be directly manipulated, such as proximity to human population or distance to deep water, our results can highlight the importance of certain areas, such as remote and uninhabited islands with steep walls, that may serve as important refuges for these fishes and are therefore worthy of protection and monitoring. With appropriate caution, model predictions resulting from this kind of research can be used in an adaptive management framework to adjust the boundaries of existing marine protected areas that are not meeting their management objectives and to target further detailed research.

## ACKNOWLEDGMENTS

This work is part of an interdisciplinary effort by the NOAA Pacific Islands Fisheries Science Center, Coral Reef Ecosystem Division (PIFSC CRED) to assess, understand, and monitor coral-reef ecosystems in the U.S. Pacific. We thank the many CRED towed divers who, over the past 10 years, have been instrumental in collecting these data. Special thanks to F. Parrish, E. DeMartini and R. Boland of the PIFSC who initially developed the CRED towed-diver method and to the CRED benthic team who have worked tirelessly to collate the benthic towed-diver data set. We would also like to thank the officers and crews of the NOAA Ships *Townsend Cromwell, Hi'ialakai* and *Oscar Elton Sette* for logistic support and field assistance. Permission to work in the Mariana Archipelago was granted by the Commonwealth of the Northern Mariana Islands (CNMI) Division of Fish and Wildlife, Coastal Resources Management Office, Division of Environmental Quality and the Guam Division of Aquatic and Wildlife Resources. The use of trade, firm, or corporation names in this publication is for the convenience of the reader and does not constitute an official endorsement or approval of any product or service to the exclusion of others that may be suitable.

# REFERENCES

1. Long RA, Muir JD, Rachlow JL, Kie J (2009) A comparison of two modeling approaches for evaluating wildlife-habitat relationships. Journal Of Wildlife Management 73: 294–302.RA LongJD MuirJL RachlowJ. Kie2009A comparison of two modeling approaches for evaluating wildlife-habitat relationships.Journal Of Wildlife Management73294302
2. Agardy T, Notarbartolo di Sciara G, Christie P (2011) Mind the gap: Addressing the shortcomings of marine protected areas through large scale marine spatial planning. Marine Policy 35: 226–232.T. AgardyG. Notarbartolo di SciaraP. Christie2011Mind the gap: Addressing the shortcomings of marine protected areas through large scale marine spatial planning.Marine Policy35226232
3. Friedlander AM (2001) Essential fish habitat and the effective design of marine reserves: application for marine ornamental fishes. Aquarium Sciences and Conservation 3: 135–150.AM Friedlander2001Essential fish habitat and the effective design of marine reserves: application for marine ornamental fishes.Aquarium Sciences and Conservation3135150
4. Jones RS (1968) Ecological relationships in Hawaiian and Johnston Island Acanthuridae (Surgeonfishes). Micronesica 4: 309–361.RS Jones1968Ecological relationships in Hawaiian and Johnston Island Acanthuridae (Surgeonfishes). Micronesica4309361
5. Hixon MA, Beets JP (1993) Predation, prey refuges, and the structure of coral-reef fish assemblages. Ecological Monographs 63: 77–101.MA HixonJP Beets1993 Predation, prey refuges, and the structure of coral-reef fish assemblages.Ecological Monographs6377101
6. Friedlander AM, Parrish JD (1998) Habitat characteristics affecting fish assemblages on a Hawaiian coral reef. Journal of Experimental Marine Biology and Ecology 224: 1–30.AM FriedlanderJD Parrish1998Habitat characteristics affecting fish assemblages on a Hawaiian coral reef.Journal of Experimental Marine Biology and Ecology224130
7. Lange M (2003) What is essential in fish habitat? Environmental Biology of Fishes 66: 99–101.M. Lange2003What is essential in fish habitat?Environmental Biology of Fishes6699101
8. Friedlander AM, Brown EK, Jokiel PL, Smith WR, Rodgers KS (2003) Effects of habitat, wave exposure, and marine protected area status on coral reef fish assemblages in the Hawaiian archipelago. Coral Reefs 22: 291–305.AM FriedlanderEK BrownPL JokielWR SmithKS Rodgers2003Effects of habitat, wave

exposure, and marine protected area status on coral reef fish assemblages in the Hawaiian archipelago.Coral Reefs22291305

9. Pittman SJ, McAlpine CA, Pittman KM (2004) Linking fish and prawns to their environment: a hierarchical landscape approach. Marine Ecology Progress Series 283: 233–254.SJ PittmanCA McAlpineKM Pittman2004Linking fish and prawns to their environment: a hierarchical landscape approach.Marine Ecology Progress Series283233254
10. Chittaro PM (2004) Fish-habitat associations across multiple spatial scales. Coral Reefs 23: 235–244.PM Chittaro2004Fish-habitat associations across multiple spatial scales.Coral Reefs23235244
11. Friedlander AM, Brown EK, Monaco MA, Clark A (2005) Fish habitat utilization patterns and evaluation of the efficacy of marine protected areas in Hawaii: integration of NOAA digital benthic habitat mapping and coral reef ecological studies. NOAA Technical Memorandum NOS NCCOS 23: 210.AM FriedlanderEK BrownMA MonacoA. Clark2005Fish habitat utilization patterns and evaluation of the efficacy of marine protected areas in Hawaii: integration of NOAA digital benthic habitat mapping and coral reef ecological studies.NOAA Technical Memorandum NOS NCCOS23210
12. Pittman SJ, Christensen JD, Caldow C, Menza C, Monaco MA (2007) Predictive mapping of fish species richness across shallow-water seascapes in the Caribbean. Ecological Modeling 204: 9–21.SJ PittmanJD ChristensenC. CaldowC. MenzaMA Monaco2007Predictive mapping of fish species richness across shallow-water seascapes in the Caribbean.Ecological Modeling204921
13. Bigelow KA, Boggs CH, HE XI (1999) Environmental effects on swordfish and blue shark catch rates in the US North Pacific longline fishery. Fisheries Oceanography 8: 178–198.KA BigelowCH BoggsXI HE1999Environmental effects on swordfish and blue shark catch rates in the US North Pacific longline fishery.Fisheries Oceanography8178198
14. Hopkins TE, Cech JJ (2003) The Influence of Environmental Variables on the Distribution and Abundance of Three Elasmobranchs in Tomales Bay, California. Environmental Biology of Fishes 66: 279–291.TE HopkinsJJ Cech2003The Influence of Environmental Variables on the Distribution and Abundance of Three Elasmobranchs in Tomales Bay, California.Environmental Biology of Fishes66279291
15. Simpfendorfer C, Freitas G, Wiley T, Heupel M (2005) Distribution and habitat partitioning of immature bull sharks (*Carcharhinus leucas*) in a Southwest Florida estuary. Estuaries and Coasts 28: 78–85.C. SimpfendorferG. FreitasT. WileyM. Heupel2005Distribution and habitat partitioning of immature bull sharks (*Carcharhinus leucas*) in a Southwest Florida estuary.Estuaries and Coasts287885
16. Heithaus MR, Delius BK, Wirsing AJ, Dunphy-Daly MM (2009) Physical factors influencing the distribution of a top predator in a subtropical oligotrophic estuary. Limnology and Oceanography 54: 472–482.MR HeithausBK DeliusAJ WirsingMM Dunphy-Daly2009Physical factors influencing the distribution of a top predator in a subtropical oligotrophic estuary.Limnology and Oceanography54472482
17. Lindberg WJ, Frazer TK, Portier KM, Vose F, Loftin J, et al. (2006) Density-dependent habitat selection and performance by a large mobile reef fish. Ecological Applications 16: 731–746.WJ LindbergTK FrazerKM PortierF. VoseJ. Loftin2006 Density-dependent habitat selection and performance by a large mobile reef fish.Ecological Applications16731746
18. Meyer CG, Holland KN, Papastamatiou YP (2007) Seasonal and diel movements of giant trevally *Caranx ignobilis* at remote Hawaiian atolls: implications for the design of Marine Protected Areas. Marine Ecology Progress Series 333: 13–25.CG MeyerKN HollandYP Papastamatiou2007Seasonal and diel movements of giant trevally *Caranx ignobilis* at remote Hawaiian atolls: implications for the design of Marine Protected Areas.Marine Ecology Progress Series3331325

19. Meyer CG, Papastamatiou YP, Holland KN (2007) Seasonal, diel, and tidal movements of green jobfish (*Aprion virescens*, Lutjanidae) at remote Hawaiian atolls: implications for marine protected area design. Marine Biology 151: 2133–2143.CG MeyerYP PapastamatiouKN Holland2007Seasonal, diel, and tidal movements of green jobfish (*Aprion virescens*, Lutjanidae) at remote Hawaiian atolls: implications for marine protected area design.Marine Biology15121332143
20. McKibben JN, Nelson DR (1986) Patterns of movement and grouping of gray reef sharks, *Carcharhinus amblyrhynchos*, at Enewetak, Marshall Islands. Bulletin of Marine Science 38: 89–110.JN McKibbenDR Nelson1986Patterns of movement and grouping of gray reef sharks, *Carcharhinus amblyrhynchos*, at Enewetak, Marshall Islands.Bulletin of Marine Science3889110
21. Holland KN, Lowe CG, Wetherbee BM (1996) Movements and dispersal patterns of blue trevally (*Caranx melampygus*) in a fisheries conservation zone. Fisheries Science 25: 279–292.KN HollandCG LoweBM Wetherbee1996Movements and dispersal patterns of blue trevally (*Caranx melampygus*) in a fisheries conservation zone.Fisheries Science25279292
22. Russ GR (1984) Distribution and abundance of herbivorous grazing fishes in the central Great Barrier Reef. I. Levels of variability across the entire continental shelf. Marine Ecology Progress Series 20: 23–34.GR Russ1984Distribution and abundance of herbivorous grazing fishes in the central Great Barrier Reef. I. Levels of variability across the entire continental shelf.Marine Ecology Progress Series202334
23. Newman SJ, Williams DM (1996) Variation in reef associated assemblages of the Lutjanidae and Lethrinidae at different distances offshore in the central Great Barrier Reef. Environmental Biology of Fishes 46: 123–138.SJ NewmanDM Williams1996Variation in reef associated assemblages of the Lutjanidae and Lethrinidae at different distances offshore in the central Great Barrier Reef.Environmental Biology of Fishes46123138
24. Beger M, Possingham HP (2008) Enrironmental factors that influence the distribution of coral reef fishes: modeling occurrence data for broad-scale conservation and management. Marine Ecology Progress Series 361: 1–13.M. BegerHP Possingham2008Enrironmental factors that influence the distribution of coral reef fishes: modeling occurrence data for broad-scale conservation and management.Marine Ecology Progress Series361113
25. Caselle JE, Warner RR (1996) Variability in Recruitment of Coral Reef Fishes: The Importance of Habitat at Two Spatial Scales. Ecology 77: 2488–2504.JE CaselleRR Warner1996Variability in Recruitment of Coral Reef Fishes: The Importance of Habitat at Two Spatial Scales.Ecology7724882504
26. Western Pacific Regional Fishery Management Council (2009) 231 p.Western Pacific Regional Fishery Management Council2009Fishery Ecosystem Plan for the Mariana Archipelago. 231 Fishery Ecosystem Plan for the Mariana Archipelago.
27. Russ GR, Alcala AC (1989) Effects of intense fishing pressure on an assemblage of coral-reef fishes. Marine Ecology Progress Series 56: 13–27.GR RussAC Alcala1989Effects of intense fishing pressure on an assemblage of coral-reef fishes.Marine Ecology Progress Series561327
28. Schroeder RE, Trianni MS, Moots KA, Laughlin JL, Zgliczynski BJ, et al. (2003) Status of target species on coral reefs of the Marianas Archipelago. Proceedings of the 10th International Coral Reef Symposium 1016–1027.RE SchroederMS TrianniKA MootsJL LaughlinBJ Zgliczynski2003Status of target species on coral reefs of the Marianas Archipelago.Proceedings of the 10th International Coral Reef Symposium10161027
29. Birkeland C, Dayton PK (2005) The importance in fishery management of leaving the big ones. Trends in Ecology and Evolution 20: 356–358.C. BirkelandPK Dayton2005The importance in fishery management of leaving the big ones.Trends in Ecology and Evolution20356358
30. Coleman FC, Koenig CC, Huntsman GR, Musick JA, Eklund AM, et al. (2000) Long-lived reef fishes: The grouper-snapper complex. Fisheries 25: 14–21.FC ColemanCC

KoenigGR HuntsmanJA MusickAM Eklund2000Long-lived reef fishes: The grouper-snapper complex.Fisheries251421

31. Sudekum AE, Parrish JD, Radtke RL, Ralson S (1991) Life History and Ecology of Large Jacks in Undisturbed, Shallow, Oceanic Communities. Fishery Bulletin 89: 493–513.AE SudekumJD ParrishRL RadtkeS. Ralson1991Life History and Ecology of Large Jacks in Undisturbed, Shallow, Oceanic Communities.Fishery Bulletin89493513
32. DeMartini EE, Friedlander AM, Sandin SA, Sala E (2008) Differences in fish-assemblage structure between fished and unfished atolls in the northern Line Islands, central Pacific. Marine Ecology Progress Series 365: 199–215.EE DeMartiniAM FriedlanderSA SandinE. Sala2008Differences in fish-assemblage structure between fished and unfished atolls in the northern Line Islands, central Pacific.Marine Ecology Progress Series365199215
33. Francis RC, Hixon MA, Clarke ME, Murawski SA, Ralson S (2007) Perspective: Fisheries management. Fisheries 32: 217–233.RC FrancisMA HixonME ClarkeSA MurawskiS. Ralson2007Perspective: Fisheries management.Fisheries32217233
34. Friedlander AM, DeMartini EE (2002) Contrasts in density, size, and biomass of reef fishes between the northwestern and the main Hawaiian islands: the effects of fishing down apex predators. Marine Ecology Progress Series 230: 253–264.AM FriedlanderEE DeMartini2002Contrasts in density, size, and biomass of reef fishes between the northwestern and the main Hawaiian islands: the effects of fishing down apex predators.Marine Ecology Progress Series230253264
35. Bellwood DR, Choat JH (1990) A functional analysis of grazing in parrotfishes (family Scaridae): the ecological implications. Environmental Biology of Fishes 28: 189–214.DR BellwoodJH Choat1990A functional analysis of grazing in parrotfishes (family Scaridae): the ecological implications.Environmental Biology of Fishes28189214
36. Bruggemann H, van Kessel AM, van Rooij JM, Breeman AM (1996) Bioerosion and sediment ingestion by the Caribbean parrotfish *Scarus vetula* and *Sparisoma viride*: implications of fish size, feeding mode and habitat use. Marine Ecology Progress Series 134: 59–71.H. BruggemannAM van KesselJM van RooijAM Breeman 1996Bioerosion and sediment ingestion by the Caribbean parrotfish *Scarus vetula* and *Sparisoma viride*: implications of fish size, feeding mode and habitat use.Marine Ecology Progress Series1345971
37. Hixon MA (1997) Effects of reef fishes on corals and algae. In: Birkeland C, editor. Life and Death of Coral Reefs. New York: Chapman & Hall. pp. 230–248.MA Hixon1997Effects of reef fishes on corals and algae.C. BirkelandLife and Death of Coral ReefsNew YorkChapman & Hall230248
38. Jackson JBC, Kirby MX, Berger WH, Bjorndal KA, Botsford LW, et al. (2001) Historical overfishing and the recent collapse of coastal ecosystems. Science 293: 629–638.JBC JacksonMX KirbyWH BergerKA BjorndalLW Botsford2001Historical overfishing and the recent collapse of coastal ecosystems.Science293629638
39. Sandin SA, Walsh SM, Jackson JBC (2010) Prey release, trophic cascades, and phase shifts in tropical nearshore ecosystems. In: Terborgh J, Estes JA, editors. Trophic cascades: predators, prey, and the changing dynamics of nature. Washington DC: Island Press. pp. 71–90.SA SandinSM WalshJBC Jackson2010Prey release, trophic cascades, and phase shifts in tropical nearshore ecosystems.J. TerborghJA EstesTrophic cascades: predators, prey, and the changing dynamics of natureWashington DCIsland Press7190
40. Jennings S, Reynolds JD, Mills SC (1998) Life history correlates of responses to fisheries exploitation. Proceedings of the Royal Society B: Biological Sciences 265: 333–339.S. JenningsJD ReynoldsSC Mills1998Life history correlates of responses to fisheries exploitation.Proceedings of the Royal Society B: Biological Sciences265333339
41. Russ GR, Alcala AC (1998) Natural fishing experiments in marine reserves 1983–1993: roles of life history and fishing intensity in family responses. Coral Reefs 17:

399–416.GR RussAC Alcala1998Natural fishing experiments in marine reserves 1983–1993: roles of life history and fishing intensity in family responses.Coral Reefs17399416
42. Jennings S, Greenstreet PR, Reynolds JD (1999) Structural change in an exploited fish community: a consequence of differential fishing effects on species with contrasting life histories. Journal of Animal Ecology 68: 617–627.S. JenningsPR GreenstreetJD Reynolds1999Structural change in an exploited fish community: a consequence of differential fishing effects on species with contrasting life histories.Journal of Animal Ecology68617627
43. Flores T (2006) Offshore Fisheries Program Annual Program Report. T. Flores2006Offshore Fisheries Program Annual Program Report.Government of Guam Department of Agriculture Annual Report. Government of Guam Department of Agriculture Annual Report.
44. Flores T (2006) Inshore Fisheries Participation, Effort, and Harvest Surveys Program Annual Program Report FY 2006. T. Flores2006Inshore Fisheries Participation, Effort, and Harvest Surveys Program Annual Program Report FY 2006. Government of Guam Department of Agriculture Annual Report. Government of Guam Department of Agriculture Annual Report.
45. Burdick D, Brown V, Asher J, Caballes C, Gawel M, et al. (2006) 76 p.D. BurdickV. BrownJ. AsherC. CaballesM. Gawel2006Status of the Coral Reef Ecosystems of Guam. 76 Status of the Coral Reef Ecosystems of Guam.
46. Green A (1997) An assessment of the status of the coral reef resources, and their patterns of use, in the U.S. Pacific Islands. Final report prepared for the Western Pacific Regional Fishery Management Council 278: A. Green1997An assessment of the status of the coral reef resources, and their patterns of use, in the U.S. Pacific Islands.Final report prepared for the Western Pacific Regional Fishery Management Council278
47. Ault JS, Bohnsack JA, Meester GA (1998) A retrospective (1979–1996) multispecies assessment of coral reef fish stocks in the Florida Keys. Fishery Bulletin 96: 395–414.JS AultJA BohnsackGA Meester1998A retrospective (1979–1996) multispecies assessment of coral reef fish stocks in the Florida Keys.Fishery Bulletin96395414
48. Pauly D, Christensen V, Dalsgaard J, Froese R Jr, FT (1998) Fishing down marine food webs. Science 279: 860–863.D. PaulyV. ChristensenJ. DalsgaardR. Froese JrFT1998Fishing down marine food webs.Science279860863
49. United States Congress (1996) Magnuson-Stevens Fishery Conservation and Management Act. Public Law 94-265. United States Congress1996Magnuson-Stevens Fishery Conservation and Management Act.Public Law 94-265
50. Asakura A, Ohba T, Miyano S, Furuki T, Kurozumi T, et al. (1994) Outline of the biological expedition to the Northern Mariana Islands, Micronesia. Natural History Research Special Issue 1: 1–11.A. AsakuraT. OhbaS. MiyanoT. FurukiT. Kurozumi1994Outline of the biological expedition to the Northern Mariana Islands, Micronesia.Natural History Research Special Issue1111
51. Rohmann SO, Hayes JJ, Newhall RC, Monaco ME, Grigg RW (2005) The area of potential shallow-water tropical and subtropical coral ecosystems in the United States. Coral Reefs 24: 370–383.SO RohmannJJ HayesRC NewhallME MonacoRW Grigg2005The area of potential shallow-water tropical and subtropical coral ecosystems in the United States.Coral Reefs24370383
52. Rainbird P (1994) Prehistory in the northwest tropical Pacific: The Caroline, Mariana, and Marshall Islands. Journal of World Prehistory 8: 293–349.P. Rainbird1994Prehistory in the northwest tropical Pacific: The Caroline, Mariana, and Marshall Islands.Journal of World Prehistory8293349
53. Richards BL, Williams ID, Nadon MO, Zgliczynski BJ (2011) A towed-diver survey method for mesoscale fishery-independent assessment of large-bodied reef fishes. Bulletin of Marine Science 87: 55–74.BL RichardsID WilliamsMO NadonBJ Zgliczynski2011A towed-diver survey method for mesoscale fishery-independent assessment of large-bodied reef fishes.Bulletin of Marine Science875574

54. Kenyon JC, Brainard RE, Hoeke RK, Parrish FA, Wilkinson CB (2006) Towed-diver surveys, a method for mesoscale spatial assessment of benthic reef habitat: a case study at Midway Atoll in the Hawaiian Archipelago. Coastal Management 34: 339–349.JC KenyonRE BrainardRK HoekeFA ParrishCB Wilkinson2006Towed-diver surveys, a method for mesoscale spatial assessment of benthic reef habitat: a case study at Midway Atoll in the Hawaiian Archipelago.Coastal Management34339349
55. Froese R, Pauly D (2009) FishBase. R. FroeseD. Pauly2009 Fish Base. Available: www.fishbase.org via the Internet. Available: www.fishbase.org via the Internet.
56. Kulbicki M, Guillemot N, Amand M (2005) A general approach to length-weight relationships for New Caledonian lagoon fishes. Cybium 29: 235–252.M. KulbickiN. GuillemotM. Amand2005A general approach to length-weight relationships for New Caledonian lagoon fishes.Cybium29235252
57. NOAA National Centers for Coastal Ocean Science [NCCOS] (2005) Shallow-Water Benthic Habitats of American Samoa, Guam, and the Commonwealth of the Northern Mariana Islands (CD-ROM). NOAA Technical Memorandum NOS NCCOS 8: NOAA National Centers for Coastal Ocean Science [NCCOS]2005Shallow-Water Benthic Habitats of American Samoa, Guam, and the Commonwealth of the Northern Mariana Islands (CD-ROM).NOAA Technical Memorandum NOS NCCOS8
58. Pacific Islands Benthic Habitat Mapping Center (2010) Commonwealth of Northern Mariana Islands (CNMI) and Guam. Pacific Islands Benthic Habitat Mapping Center2010Commonwealth of Northern Mariana Islands (CNMI) and Guam.http://wwwsoesthawaiiedu/pibhmc/pibhmc_cnmihtm via the Internet.http://wwwsoesthawaiiedu/pibhmc/pibhmc_cnmihtm via the Internet.
59. Kilpatrick KA, Podestá GP, Evans RH (2001) Overview of the NOAA/NASA Pathfinder algorithm for Sea Surface Temperature and associated Matchup Database. Journal of Geophysical Research 106: 9179–9198.KA KilpatrickGP PodestáRH Evans2001Overview of the NOAA/NASA Pathfinder algorithm for Sea Surface Temperature and associated Matchup Database.Journal of Geophysical Research10691799198
60. Clarke KR, Gorley RN (2006) PRIMER v6: User Manual/Tutorial. KR ClarkeRN Gorley2006PRIMER v6: User Manual/Tutorial.vv6113. PRIMER-E. vv6113. PRIMER-E.
61. Elith J, Leathwick JR, H T (2008) A working guide to boosted regression trees. Journal of Animal Ecology 77: 802–813.J. ElithJR LeathwickT. H2008A working guide to boosted regression trees.Journal of Animal Ecology77802813
62. R Development Core Team (2009) R: A Language and Environment for Statistical Computing. R Development Core Team2009R: A Language and Environment for Statistical Computing.v261. R Foundation for Statistical Computing. v261. R Foundation for Statistical Computing.
63. Plotkin JB, Muller-Landau HC (2002) Sampling the species composition of a landscape. Ecology 83: 3344–3356.JB PlotkinHC Muller-Landau2002Sampling the species composition of a landscape.Ecology8333443356
64. Valesini FJ, Potter IC, Clarke KR (2004) To what extent are the fish compositions at nearshore sites along a heterogeneous coast related to habitat type? Estuarine, Coastal and Shelf Science 60: 737–754.FJ ValesiniIC PotterKR Clarke2004To what extent are the fish compositions at nearshore sites along a heterogeneous coast related to habitat type?Estuarine, Coastal and Shelf Science60737754
65. Clarke KR, Chapman MG, Somerfield PJ, Needham HR (2006) Dispersion-based weighting of species counts in assemblage analyses. Marine Ecology Progress Series 320: 11–27.KR ClarkeMG ChapmanPJ SomerfieldHR Needham2006Dispersion-based weighting of species counts in assemblage analyses.Marine Ecology Progress Series3201127
66. Williams GJ, Cameron MJ, Turner JR, Ford RB (2008) Quantitative characterization of reef fish diversity among nearshore habitats in a northeastern New Zealand marine reserve. New Zealand Journal of Marine and Freshwater Research 42: 33–

46.GJ WilliamsMJ CameronJR TurnerRB Ford2008Quantitative characterization of reef fish diversity among nearshore habitats in a northeastern New Zealand marine reserve.New Zealand Journal of Marine and Freshwater Research423346

67. Clarke KR, Somerfield PJ, Gorley RN (2008) Testing of null hypotheses in exploratory community analyses: similarity profiles and biota-environment linkage. Journal of Experimental Marine Biology and Ecology 366: 56–69.KR ClarkePJ SomerfieldRN Gorley2008Testing of null hypotheses in exploratory community analyses: similarity profiles and biota-environment linkage.Journal of Experimental Marine Biology and Ecology3665669

68. Clarke KR, Somerfield PJ, Chapman MG (2006) On resemblance measures for ecological studies, including taxonomic dissimilarities and a zero-adjusted Bray–Curtis coefficient for denuded assemblages. Journal of Experimental Marine Biology and Ecology 330: 55–80.KR ClarkePJ SomerfieldMG Chapman2006On resemblance measures for ecological studies, including taxonomic dissimilarities and a zero-adjusted Bray–Curtis coefficient for denuded assemblages.Journal of Experimental Marine Biology and Ecology3305580

69. Clarke KR (1993) Non-parametric multivariate analyses of changes in community structure. Austral Ecology 18: 117–143.KR Clarke1993Non-parametric multivariate analyses of changes in community structure.Austral Ecology18117143

70. Clarke KR, Ainsworth M (1993) A method of linking multivariate community structure to environmental variables. Marine Ecology Progress Series 92: 205–219.KR ClarkeM. Ainsworth1993A method of linking multivariate community structure to environmental variables.Marine Ecology Progress Series92205219

71. Ridgeway G Generalized boosted regression models. Documentation on the R package "gbm" v15-7.Ridgeway G Generalized boosted regression models.Documentation on the R package "gbm"v157

72. De'ath G (2007) Boosted trees for ecological modeling and prediction. Ecology 88: 243–251.G. De'ath2007Boosted trees for ecological modeling and prediction.Ecology88243251

73. Cappo M, De'ath G, Boyle S, Aumend J, Olbrich R, et al. (2005) Development of a robust classifier of freshwater residence in barramundi (*Lates calcarifer*) life histories using elemental ratios in scales and boosted regression trees. Marine and Freshwater Research 56: 713–723.M. CappoG. De'athS. BoyleJ. AumendR. Olbrich2005Development of a robust classifier of freshwater residence in barramundi (*Lates calcarifer*) life histories using elemental ratios in scales and boosted regression trees.Marine and Freshwater Research56713723

74. Leathwick JR, Elith J, Francis MP, Hastie T, Taylor P (2006) Variation in demersal fish species richness in the oceans surrounding New Zealand: an analysis using boosted regression trees. Marine Ecology Progress Series 321: 267–281.JR LeathwickJ. ElithMP FrancisT. HastieP. Taylor2006Variation in demersal fish species richness in the oceans surrounding New Zealand: an analysis using boosted regression trees.Marine Ecology Progress Series321267281

75. Williams GJ, Aeby GS, Cowie ROM, Davy SK (2010) Predictive Modeling of Coral Disease Distribution within a Reef System. PLoS ONE 5: e9264.GJ WilliamsGS AebyROM CowieSK Davy2010Predictive Modeling of Coral Disease Distribution within a Reef System.PLoS ONE5e9264

76. Myers RA, Baum JK, Shepherd TD, Powers SP, Peterson CH (2007) Cascading Effects of the Loss of Apex Predatory Sharks from a Coastal Ocean. Science 315: 1846–1850.RA MyersJK BaumTD ShepherdSP PowersCH Peterson2007Cascading Effects of the Loss of Apex Predatory Sharks from a Coastal Ocean.Science31518461850

77. Bellwood DR (1995) Direct estimate of bioerosion by two parrotfish species, *Chlorurus gibbus* and *C. sordidus*, on the Great Barrier Reef, Australia. Marine Biology 121: 419–429.DR Bellwood1995Direct estimate of bioerosion by two parrotfish species, *Chlorurus gibbus* and *C. sordidus*, on the Great Barrier Reef, Australia.Marine Biology121419429

78. Maragos JE (1993) Impact of coastal construction on coral reefs in the U.S.-affiliated Pacific Islands. Coastal Management 21: 235–269.JE Maragos1993Impact of coastal construction on coral reefs in the U.S.-affiliated Pacific Islands.Coastal Management21235269

79. Roberts CM (2007) The Unnatural History of the Sea. Island Press. Washington, D.C..CM Roberts2007The Unnatural History of the SeaIsland PressWashington, D.C.

80. Williams ID, Walsh WJ, Schroeder RE, Friedlander AM, Richards BL, et al. (2008) Assessing the importance of fishing impacts on Hawaiian coral reef fish assemblages along regional-scale human population gradients. Environmental Conservation 35: 261–272.ID WilliamsWJ WalshRE SchroederAM FriedlanderBL Richards2008Assessing the importance of fishing impacts on Hawaiian coral reef fish assemblages along regional-scale human population gradients.Environmental Conservation35261272

81. Sandin SA, Smith JE, DeMartini EE, Dinsdale EA, Donner SD, et al. (2008) Baselines and degradation of coral reefs in the Northern Line Islands. PLoS ONE 3: e1548.SA SandinJE SmithEE DeMartiniEA DinsdaleSD Donner2008Baselines and degradation of coral reefs in the Northern Line Islands.PLoS ONE3e1548

82. Stallings CD (2009) Fishery-Independent Data Reveal Negative Effect of Human Population Density on Caribbean Predatory Fish Communities. PLoS ONE 4: e5333.CD Stallings2009Fishery-Independent Data Reveal Negative Effect of Human Population Density on Caribbean Predatory Fish Communities.PLoS ONE4e5333

83. Williams ID, Richards BL, Sandin SA, Baum JK, Schroeder RE, et al. (2011) Differences in Reef Fish Assemblages between Populated and Remote Reefs Spanning Multiple Archipelagos Across the Central and Western Pacific. Journal of Marine Biology 14: ID WilliamsBL RichardsSA SandinJK BaumRE Schroeder2011Differences in Reef Fish Assemblages between Populated and Remote Reefs Spanning Multiple Archipelagos Across the Central and Western Pacific.Journal of Marine Biology14

84. Mora C, Aburto-Oropeza O, Ayala B, Arturo , Ayotte PM, Banks S, et al. (2011) Global Human Footprint on the Linkage between Biodiversity and Ecosystem Functioning in Reef Fishes. PLoS Biology 9: e1000606.C. MoraO. Aburto-OropezaB. AyalaArturoPM AyotteS. Banks2011Global Human Footprint on the Linkage between Biodiversity and Ecosystem Functioning in Reef Fishes.PLoS Biology9e1000606

85. Francis MP (1993) Does water temperature determine year class strength in New Zealand snapper (*Pagrus auratus*, Sparidae)? Fisheries Oceanography 2: 65–72.MP Francis1993Does water temperature determine year class strength in New Zealand snapper (*Pagrus auratus*, Sparidae)?Fisheries Oceanography26572

86. Worm B, Sandow M, Oschlies A, Lotze HK, Myers RA (2005) Global Patterns of Predator Diversity in the Open Oceans. Science 309: 1365–1369.B. WormM. SandowA. OschliesHK LotzeRA Myers2005Global Patterns of Predator Diversity in the Open Oceans.Science30913651369

87. Magnuson JJ, Brandt SB, Stewart DJ (1980) Habitat Preference and Fisheries Oceanography. In: Bardach JE, Magnuson JJ, May RC, Reinhart JM, editors. Fish behavior and its use in the capture and culture of fishes. Manilia: International center for living and aquatic resources management. pp. 371–382.JJ MagnusonSB BrandtDJ Stewart1980Habitat Preference and Fisheries Oceanography.JE BardachJJ MagnusonRC MayJM ReinhartFish behavior and its use in the capture and culture of fishesManiliaInternational center for living and aquatic resources management371382

88. Friedlander AM, Sandin SA, DeMartini EE, Sala E (2010) Spatial patterns of the structure of reef fish assemblages at a pristine atoll in the central Pacific. Marine Ecology Progress Series 410: 219–231.AM FriedlanderSA SandinEE DeMartiniE. Sala2010Spatial patterns of the structure of reef fish assemblages at a pristine atoll in the central Pacific.Marine Ecology Progress Series410219231

89. Luckhurst BE, Luckhurst K (1978) Analysis of the influence of substrate variables on coral reef fish communities. Marine Biology 49: 317–323.BE LuckhurstK. Luckhurst1978Analysis of the influence of substrate variables on coral reef fish communities.Marine Biology49317323
90. Chabanet P, Ralambondrainy H, Amanieu M, Faure G, Galzin R (1997) Relationships between coral reef substrata and fish. Coral Reefs 16: 93–102.P. ChabanetH. RalambondrainyM. AmanieuG. FaureR. Galzin1997Relationships between coral reef substrata and fish.Coral Reefs1693102
91. Friedlander AM, Brown EK, Monaco ME (2007) Defining reef fish habitat utilization patterns in Hawaii: comparisons between marine protected areas and areas open to fishing. Marine Ecology Progress Series 351: 221–233.AM FriedlanderEK BrownME Monaco2007Defining reef fish habitat utilization patterns in Hawaii: comparisons between marine protected areas and areas open to fishing.Marine Ecology Progress Series351221233
92. Sale PF (1972) Effect of Cover on Agonistic Behavior of a Reef Fish, A Possible Spacing Mechanism. Ecology 53: 753–758.PF Sale1972Effect of Cover on Agonistic Behavior of a Reef Fish, A Possible Spacing Mechanism.Ecology53753758
93. Levins R (1968) Evolution in Changing Environments: Some Theoretical Explorations. Princeton, New Jersey: Princeton University Press. 120 p.R. Levins1968Evolution in Changing Environments: Some Theoretical ExplorationsPrinceton, New JerseyPrinceton University Press120
94. WPacFIN (2011) Western Pacific Fisheries Information Network. WPacFIN2011Western Pacific Fisheries Information Network. Available: http:/ /www. pifsc.noaa.gov/wpacfin/ via the Internet. Available: http:/ /www. pifsc.noaa.gov/wpacfin/ via the Internet.
95. Biro PA, Dingemanse NJ (2009) Sampling bias resulting from animal personality. Trends in Ecology and Evolution 24: 66–67.PA BiroNJ Dingemanse2009Sampling bias resulting from animal personality.Trends in Ecology and Evolution246667
96. Koslow JA, Hanley F, Wicklund R (1988) Effects of fishing on reef fish communities at Pedro Bank and Port Royal Cays, Jamaica. Marine Ecology Progress Series 43: 201–212.JA KoslowF. HanleyR. Wicklund1988Effects of fishing on reef fish communities at Pedro Bank and Port Royal Cays, Jamaica.Marine Ecology Progress Series43201212
97. Hughes TP (1994) Catastrophes, phase shifts, and large-scale degradation of a Caribbean coral reef. Science 265: 1547–1551.TP Hughes1994Catastrophes, phase shifts, and large-scale degradation of a Caribbean coral reef.Science26515471551
98. Sheppard C (1995) The Shifting Baseline Syndrome. Marine Pollution Bulletin 30: 766–767.C. Sheppard1995The Shifting Baseline Syndrome.Marine Pollution Bulletin30766767
99. Dayton PK, Thrush SF, Agardy MT, Hofman RJ (1995) Environmental effects of marine fishing. Aquatic Conservation: Marine and Freshwater Ecosystems 5: 205–232.PK DaytonSF ThrushMT AgardyRJ Hofman1995Environmental effects of marine fishing.Aquatic Conservation: Marine and Freshwater Ecosystems5205232
100. Myers RA, Hutchings JA, Barrowman NJ (1997) Why do fish stocks collapse? The example of cod in Atlantic Canada. Ecological Applications 7: 91–106.RA MyersJA HutchingsNJ Barrowman1997Why do fish stocks collapse? The example of cod in Atlantic Canada.Ecological Applications791106
101. Jackson JBC (1997) Reefs since Columbus. Coral Reefs 16: S23–S32.JBC Jackson1997Reefs since Columbus.Coral Reefs16S23S32
102. Jackson JBC (2001) What was natural in the coastal oceans? Proceedings of the National Academy of Sciences of the United States of America 98: 5411–5418.JBC Jackson2001What was natural in the coastal oceans?Proceedings of the National Academy of Sciences of the United States of America9854115418
103. Bohnsack JA (2003) Shifting Baselines, marine reserves, and Leopold's biotic ethic. Gulf and Caribbean Research 14: 1–7.JA Bohnsack2003Shifting Baselines, marine reserves, and Leopold's biotic ethic.Gulf and Caribbean Research1417

104. Bellwood DR, Hughes TP, Nyström M (2004) Confronting the coral reef crisis. Nature 429: 827–833.DR BellwoodTP HughesM. Nyström2004Confronting the coral reef crisis.Nature429827833

105. Jennings S, Blanchard JL (2004) Fish abundance with no fishing: predictions based on macroecological theory. Journal of Animal Ecology 73: 632–642.S. JenningsJL Blanchard2004Fish abundance with no fishing: predictions based on macroecological theory.Journal of Animal Ecology73632642

106. Halpern BS, Walbridge S, Selkoe KA, Kappel CV, Micheli F, et al. (2008) A Global Map of Human Impact on Marine Ecosystems. Science 319: 948–952.BS HalpernS. WalbridgeKA SelkoeCV KappelF. Micheli2008A Global Map of Human Impact on Marine Ecosystems.Science319948952

107. Amesbury J, Hunter-Anderson R, Wells E (1989) Native fishing rights and limited entry in the CNMI. J. AmesburyR. Hunter-AndersonE. Wells1989Native fishing rights and limited entry in the CNMI.Western Pacific Regional Fishery Management Council. Western Pacific Regional Fishery Management Council.

108. Pate FD, Craib JL, Heathcote GM (2001) Stable isotopic analysis of prehistoric human diet in the Mariana Islands, western Pacific. Australian Archaeology 52: 1–4.FD PateJL CraibGM Heathcote2001Stable isotopic analysis of prehistoric human diet in the Mariana Islands, western Pacific.Australian Archaeology5214

109. Wetherbee BM, Crow GL, Lowe CG (1997) Distribution, reproduction and diet of the gray reef shark *Carcharhinus amblyrhynchos* in Hawaii. Marine Ecology Progress Series 151: 181–189.BM WetherbeeGL CrowCG Lowe1997Distribution, reproduction and diet of the gray reef shark *Carcharhinus amblyrhynchos* in Hawaii.Marine Ecology Progress Series151181189

110. Papastamatiou YP, Wetherbee BM, Lowe CG, Crow GL (2006) Distribution and diet of four species of carcharhinid shark in the Hawaiian Islands: evidence for resource partitioning and competitive exclusion. Marine Ecology Progress Series 320: 239–251.YP PapastamatiouBM WetherbeeCG LoweGL Crow2006Distribution and diet of four species of carcharhinid shark in the Hawaiian Islands: evidence for resource partitioning and competitive exclusion.Marine Ecology Progress Series320239251

111. Hobson ES, Chess JR (1978) Trophic relationships among fishes and plankton in the lagoon at Enewetok atoll, Marshall Islands. Fisheries Bulletin 76: 133–153.ES HobsonJR Chess1978Trophic relationships among fishes and plankton in the lagoon at Enewetok atoll, Marshall Islands.Fisheries Bulletin76133153

112. Thresher RE, Colin PL (1986) Trophic structure, diversity and abundance of fishes of the deep reef (30–300 m) at Enewetak, Marshall Islands. Bulletin of Marine Science 38: 410–426.RE ThresherPL Colin1986Trophic structure, diversity and abundance of fishes of the deep reef (30–300 m) at Enewetak, Marshall Islands.Bulletin of Marine Science38410426

113. Meekan MG, Carleton JH, McKinnon AD, Flynn K, Furnas M (2003) What determines the growth of tropical reef fish larvae in the plankton: food or temperature? Marine Ecology Progress Series 256: 193–204.MG MeekanJH CarletonAD McKinnonK. FlynnM. Furnas2003What determines the growth of tropical reef fish larvae in the plankton: food or temperature?Marine Ecology Progress Series256193204

114. Humston R, Ault JS, Larkin MF, Luo J (2005) Movements and site fidelity of the bonefish Albula vulpes in the northern Florida Keys determined by acoustic telemetry. Marine Ecology Progress Series 291: 237–248.R. HumstonJS AultMF LarkinJ. Luo2005Movements and site fidelity of the bonefish Albula vulpes in the northern Florida Keys determined by acoustic telemetry.Marine Ecology Progress Series291237248

115. Sponaugle S, Grorud-Colvert K, Pinkard D (2006) Temperature-mediated variation in early life history traits and recruitment success of the coral reef fish *Thalassoma bifasciatum* in the Florida Keys. Marine Ecology Progress Series 308: 1–15.S. SponaugleK. Grorud-ColvertD. Pinkard2006Temperature-mediated variation in early life history traits and recruitment success of the coral reef fish *Thalassoma bifasciatum* in the Florida Keys.Marine Ecology Progress Series308115

116. Song L, Zhou J, Zhou Y, Nishida T, Jiang W, et al. (2009) Environmental preferences of bigeye tuna, Thunnus obesus, in the Indian Ocean: an application to a longline fishery. Environmental Biology of Fishes 85: 153–171.L. SongJ. ZhouY. ZhouT. NishidaW. Jiang2009Environmental preferences of bigeye tuna, Thunnus obesus, in the Indian Ocean: an application to a longline fishery.Environmental Biology of Fishes85153171
117. Carpenter KE, Miclat RI, Albaladejo VD, Corpus VT (1981) The influence of substrate structure on the local abundance and diversity of Philippine reef fishes. Proceedings of the 4th International Coral Reef Symposium 2: 497–502.KE CarpenterRI MiclatVD AlbaladejoVT Corpus1981The influence of substrate structure on the local abundance and diversity of Philippine reef fishes.Proceedings of the 4th International Coral Reef Symposium2497502
118. DeMartini EE, Anderson TW (2007) Habitat associations and aggregations of recruit fishes on Hawaiian coral reefs. Bulletin of Marine Science 81: 139–152.EE DeMartiniTW Anderson2007Habitat associations and aggregations of recruit fishes on Hawaiian coral reefs.Bulletin of Marine Science81139152
119. Kemp JM (1998) Zoogeography of the coral reef fishes of the Socotra Archipelago. Journal of Biogeography 25: 919–933.JM Kemp1998Zoogeography of the coral reef fishes of the Socotra Archipelago.Journal of Biogeography25919933
120. Nakamura Y, Sano M (2004) Overlaps in habitat use of fishes between a seagrass bed and adjacent coral and sand areas at Amitori Bay, Iriomote Island, Japan: importance of the seagrass bed as juvenile habitat. Fisheries Science 70: 788–803.Y. NakamuraM. Sano2004Overlaps in habitat use of fishes between a seagrass bed and adjacent coral and sand areas at Amitori Bay, Iriomote Island, Japan: importance of the seagrass bed as juvenile habitat.Fisheries Science70788803
121. Pittman SJ, Caldow C, Hile SD, Monaco ME (2007) Using seascape types to explain the spatial patterns of fish in the mangroves of SW Puerto Rico. Marine Ecology Progress Series 348: 273–284.SJ PittmanC. CaldowSD HileME Monaco2007Using seascape types to explain the spatial patterns of fish in the mangroves of SW Puerto Rico.Marine Ecology Progress Series348273284
122. Jennings S, Kaiser MJ, Reynolds JD (2001) Marine fisheries ecology. Oxford, United Kingdom: Blackwell Science. S. JenningsMJ KaiserJD Reynolds2001Marine fisheries ecologyOxford, United KingdomBlackwell Science
123. Reynolds JD, Dulvy NK, Goodwin NB, Hutchings JA (2005) Review. Biology of extinction risk in marine fishes. Proceedings of the Royal Society B: Biological Sciences 272: 2337–2344.JD ReynoldsNK DulvyNB GoodwinJA Hutchings2005Review. Biology of extinction risk in marine fishes.Proceedings of the Royal Society B: Biological Sciences27223372344
124. Grigg RW (1994) Effects of sewage discharge, fishing pressure and habitat complexity on coral reef ecosystems and reef fishes in Hawaii. Marine Ecology Progress Series 103: 25–34.RW Grigg1994Effects of sewage discharge, fishing pressure and habitat complexity on coral reef ecosystems and reef fishes in Hawaii.Marine Ecology Progress Series10325-34

# CHAPTER 6

# Diet Composition and Foraging Ecology of U.S. Pacific Coast Groundfishes with Applications for Fisheries Management

**Joseph J. Bizzarro, Mary M. Yoklavich & W. Waldo Wakefield**

*Institute of Marine Sciences, University of California, Santa Cruz, CA 95060, USA*

*Fisheries Ecology Division, Southwest Fisheries Science Center, National Marine Fisheries Service, 110 Shaffer Rd, Santa Cruz, CA 95060, USA*

*Fishery Resource Analysis and Monitoring Division, Northwest Fisheries Science Center, National Marine Fisheries Service, 2032 SE OSU Dr, Newport, OR 97365, USA*

## ABSTRACT

Determining the prey composition and foraging habitats of U.S. Pacific Coast groundfishes are specified management directives that have not received much scientific attention. To address this knowledge gap, we conducted a meta-analysis of the feeding ecology of 18 commercially important species and their life stages during a recent review of Pacific Coast groundfish essential fish habitat. A Major Prey Index was developed to evaluate relative importance among 47 prey taxa. Based on this metric, unidentified teleosts, euphausiids, and brachyuran crabs were the most important prey groups. When 14 generalized prey categories were used, fishes represented the dominant taxon (mean% weight or volume = 32.3) followed by shrimps

(11.5), crabs (10.0), and euphausiids (9.5). PERMANOVA results indicated that species-specific differences were the primary source of dietary variability among tested variables (life stage, functional group, taxonomic group). Pacific Coast groundfishes mainly were characterized as mesopredators with estimated trophic levels ranging from 3.4 to 4.2. Foraging habitats differed significantly among functional (benthic, demersal, pelagic) and taxonomic (elasmobranch, roundfish, rockfish, flatfish) groups. Using hierarchical agglomerative cluster analysis, we identified a significantly distinct trophic guild that consumes mainly polychaetes and hard-shelled molluscs (juvenile, juvenile–adult Dover Sole; juvenile–adult English Sole) and another that specializes on euphausiids (juvenile Pacific Hake; juvenile– adult Darkblotched Rockfish). Our findings filled substantial data gaps in the trophic ecology and habitatbased management of commercially important species and can be used to inform future reviews of Pacific Coast groundfish essential fish habitat.

**Keywords:** Diet composition . Dietary variability. Essential fish habitat . Foraging habitat . Major Prey Index . Trophic level . Ecosystem-based fishery management

## INTRODUCTION

Ecosystem approaches to marine fisheries management require an ecological understanding of exploited species and their communities. One emerging technique is the use of food web models, which incorporate diet composition information, to examine a variety of trophic dynamics. Because food web models can be used to determine and evaluate sources of variability in mortality and production that are not typically included in single-species stock assessments, their application may improve the management of exploited stocks (Gaichas et al. 2010, 2011; Link et al. 2011). One such application is the incorporation of steady-state mass-balance model outputs of the Gulf ofAlaska food web for consideration in groundfish stock assessments (Zador 2014). However, a lack of ecological data, such as diet composition data needed for food web models, is a common limitation that has hampered our ability to implement ecosystem-based fishery management (EBFM; Essington and Punt 2011).

Determining the prey composition and foraging habitats of federally managed aquatic species in the U.S., such as Pacific Coast groundfishes, is a specified component of the Magnuson-Stevens Fishery Conservation and Management Act (United States Department of Commerce 1996). Food habits information has been used to assess consumption impacts (Overholz and Link 2007; Link and Sosebee 2008) and fishing effects (Garrison and Link 2000; Link and Garrison 2002) on the trophic dynamics Atlantic Coast groundfish stocks. This type of information is, however, limited for many regions, including the Pacific Coast. Ecological, multi-species approaches to Pacific Coast groundfish management currently focus on determining habitat associations and establishing restricted-take zones or marine protected areas (Pacific Fishery Management Council 2016). Ecosystem approaches to management that incorporate food web interactions are, however, currently being developed (Kaplan and Levin 2009; Kaplan et al. 2013).

The need for higher levels of taxonomic specificity in prey categories was a primary recommendation in a recent review of essential fish habitat (EFH) for Pacific Coast groundfishes (Pacific Fishery Management Council 2014), and can improve the accuracy of food web models. Federal regulations concerning EFH state that Bfishery management plans should list the major prey species^ of managed groundfishes, and that Bactions that reduce the availability of a major prey species^ may be considered adverse effects to EFH[1] Once major prey species have been identified, potentially negative impacts (e.g., harvesting, habitat alteration) to them can be determined and monitored. However, no definition of what constitutes a Bmajor^ prey species is provided in the EFH regulations, and syntheses of groundfish food habits literature are limited and highly generalized (e.g., DuFault et al. 2009; Szoboszlai et al. 2015).

In order to address these data limitations, we investigated aspects of trophic ecology for a variety of species and their life stages (species and/or life stages), and for the overall assemblage of Pacific Coast groundfishes. Five primary research questions were posed: 1) What are the diet compositions, trophic levels, and foraging habitats of Pacific Coast groundfish species and/or life stages? 2) What are the major prey taxa and foraging habitats of Pacific Coast groundfishes? 3) What are the main sources of dietary variation for Pacific Coast groundfish species and/or life stages? 4) Does diet composition, trophic level, and foraging habitat differ among functional or taxonomic groups? and 5) Can trophic guilds of Pacific Coast groundfish species and/or life stages be distinguished?

# MATERIALS

## Data collection

We conducted a meta-analysis of the quantitative diet composition data of 18 federally managed Pacific Coast groundfish species (Table 1) from available food habits literature. Species were selected based on a combination of factors, including management priority, the relative amount of diet composition information, ecological and taxonomic diversity, and the relative number of species among established management categories. Selected groundfishes represent a small portion of the 117 species that are included in the fisheries management plan (Pacific Fishery Management Council 2014). Therefore, results should not be applied beyond the taxonomic or functional groups designated in this study.

**Table 1** Common name, scientific name, taxonomic group, and functional group (based on general occurrence in the water column) of 18 Pacific Coast groundfish species that were included in this study

| Common name | Scientific name | Taxonomic group | Functional group |
|---|---|---|---|
| Black Rockfish | *Sebastes melanops* | Rockfishes | Demersal |
| Brown Rockfish | *Sebastes auriculatus* | Rockfishes | Demersal |
| Copper Rockfish | *Sebastes caurinus* | Rockfishes | Demersal |
| Darkblotched Rockfish | *Sebastes crameri* | Rockfishes | Demersal |
| Dover Sole | *Microstomus pacificus* | Flatfishes | Benthic |
| English Sole | *Parophrys vetulus* | Flatfishes | Benthic |
| Greenstriped Rockfish | *Sebastes elongatus* | Rockfishes | Demersal |
| Lingcod | *Ophiodon elongatus* | Roundfishes | Demersal |
| Longspine Thornyhead | *Sebastolobus altivelis* | Rockfishes | Benthic |
| Pacific Hake | *Merluccius productus* | Roundfishes | Pelagic |
| Pacific Spiny Dogfish | *Squalus suckleyi* | Elasmobranchs | Pelagic |
| Petrale Sole | *Eopsetta jordani* | Flatfishes | Benthic |
| Rosethorn Rockfish | *Sebastes helvomaculatus* | Rockfishes | Demersal |
| Sablefish | *Anoplopoma fimbria* | Roundfishes | Demersal |
| Sand Sole | *Psettichthys melanostictus* | Flatfishes | Benthic |
| Sharpchin Rockfish | *Sebastes zacentrus* | Rockfishes | Demersal |
| Starry Skate | *Raja stellulata* | Elasmobranchs | Benthic |
| Yelloweye Rockfish | *Sebastes ruberrimus* | Rockfishes | Demersal |

Relevant literature (e.g., peer-reviewed publications, student theses, technical reports) was identified from summarized life history information in three publications (McCain et al. 2005; Love 2011; Pacific Fishery Management Council 2012), from a search of bibliographic databases (i.e., Aquatic Science and Fisheries Abstracts, BIOSIS, Web of Science, Zoological Record), and by cross-referencing citations in found literature. The geographic range of this analysis was restricted to the waters off the continental U.S. Pacific Coast; literature on groundfish diets specifically from other regions was not considered. However, studies that included some fish sampled in Canada orMexico were included when results from U.S. waters could not be differentiated.

The following information was extracted from each study and used as a basis for diet characterization and subsequent analyses: collection year, collection month, study region, sample size, collection method, collection depth, length range and maturity stage of fish, and diet metric. Only studies that reported quantitative estimates of prey weight or volume were included in our analysis because these metrics generally track energetic importance of prey taxa (Hyslop 1980; Cailliet et al. 1986). By contrast, frequency metrics (e.g., percent frequency of occurrence, %FO = number stomach samples in which a prey taxon occurs/total number of stomach samples * 100) and relative number of a prey taxon typically are a proxy for feeding behavior, but do not necessarily represent the relative energetic contribution of each prey type in a fish's diet (Hyslop 1980; Cailliet et al. 1986).Weight or volume data, however, underestimate the relative contribution of easily digestible, soft-bodied prey (e.g., gelatinous zooplankton such as jellyfishes, ctenophores, and salps; polychaetes; small crustaceans or those with thin carapaces). Soft-bodied prey taxa therefore will be underrepresented in diet composition calculations, as compared to fishes and decapod crustaceans, which typically are more abundant in groundfish diets (Love 2011). Compound measures that incorporate weight or volume (such as Index of Relative Importance [IRI], Pinkas et al. 1971) were considered only if volume or weight were not individually reported.

## Diet composition

Weight or volume of each prey category, as originally designated in a study, was converted to a percentage and then reclassified among 47 prey categories (Table 2; National Marine Fisheries Service 2013) to standardize diet composition estimates. Insufficient sampling can result in poor estimates of diet composition (Ferry and Cailliet 1996) and standard transformations can overgeneralize relative variability among various sample sizes. A weighting scheme therefore was calculated to synthesize diet composition data when more than one study was available for a species and/or life stage This scheme was developed by evaluating prey accumulation curves using individual-based diet data for five federally managed Pacific Coast skate species (Big, Longnose, California, Starry, Sandpaper; [Bizzarro 2015]), because such data were unavailable for the teleost and shark species included in this study. Prey accumulation curves were generated for a variety of sample sizes using the 47 described prey categories (Table 2; Ferry and Cailliet 1996) and the Mao Tao expected richness function in Estimate-S (Colwell 2013). Linear regression was applied to calculate the slope ($b$) of the curve terminus and determine if the curve approached an asymptote ($b < 0.05$) (Bizzarro et al. 2007, 2009). The relative slopes of prey accumulation curves were compared at different sample sizes and among species. Based on these comparisons, the following weighting scheme was created and applied: if the number of samples =1, data were not weighted; 2–10 samples (weighted 2×); 11–25 samples (4×); 26–50 samples (8×); 51–100 samples (16×); 101–500 samples (32×); > 500 samples (64×). Sample size sufficiency typically was achieved at 101–500 samples; therefore, any studies with >500 samples were weighted equally.

**Table 2** Abbreviations associated with high and low levels of taxonomic distinction among prey categories. Definitions are specific to high levels of taxonomic distinction.

| Prey category | High | Low |
|---|---|---|
| Agnathan fishes | AGNATH | FISH |
| Poachers | AGON | FISH |
| Ammodytidae | AMMO | FISH |
| Amphipods | AMPH | AMPH |
| Axiidae | AXIID | CRUST |
| Bivalves | BIVAL | MOLL |
| Cephalopods, unidentified | CEPH | CEPH |
| Chondrichthyan fishes | CHOND | FISH |
| Clupeidae | CLUP | FISH |
| Gadiformes | CODS | FISH |
| Copepods | COPE | COPE |
| Crabs, unidentified | CRAB | CRAB |
| Anomuran crabs | CRAB A | CRAB |
| Brachyuran crabs | CRAB B | CRAB |
| Other and unidentified crustaceans | CRUST | CRUST |
| Cuttlefishes | CUTT | CEPH |
| Other Decapods | DECA | CRUST |
| Echinoderms | ECHINO | ECHINO |
| Engraulidae | ENGR | FISH |
| Euphausiids | EUPH | EUPH |
| Pleuronectiformes | FLAT | FISH |
| Gastropods | GAST | MOLL |
| Herrings | HERR | FISH |
| Hexagramidae | HEX | FISH |

| | | |
|---|---|---|
| Invertebrates, unidentified | INV | INV |
| Isopods | ISO | AMPH |
| Jellyfishes and other unid. Gelatinous zooplankton | JELL | JELL |
| Loligonidae | LOLI | CEPH |
| Bivalves or Gastropods, unidentified | MOLL | MOLL |
| Myctophidae | MYCT | FISH |
| Mysids | MYSID | MYSID |
| Octopi | OCTO | CEPH |
| Squids (Oegopsina) | OEGO | CEPH |
| Osmeriformes | OSMER | FISH |
| Polychaetes | POLY | POLY |
| Rockfishes | ROCK | FISH |
| Sardines | SARD | FISH |
| Scorpaeniformes, other and identified | SCORP | FISH |
| Sculpins | SCULP | FISH |
| Shrimps, unidentified | SHRIMP | SHRIMP |
| Caridean shrimps | SHRIMP C | SHRIMP |
| Penaeid and Sergestid shrimps | SHRIMP PS | SHRIMP |
| Squids, unidentified | SQUID | CEPH |
| Other and unidentified Teleosts | TELE | FISH |
| Tunicates | TUN | INV |
| Other marine worms (e.g., Nematoda, Sipuncula) | WORM | POLY |
| Zoarcidae | ZOAR | FISH |

When possible, diet composition data were calculated by life stage (i.e., juvenile or adult). Juveniles ranged from post-settlement individuals to subadults (i.e., immature). Mature individuals were considered adults. Life stages were directly assigned from the literature if maturity information was reported. In the absence of such information, life stage was assigned based on published size-atmaturity information for a species of interest (Appendix 1). When multiple size-at-maturity estimates were available, the cut-off value was chosen from the reproductive study that was conducted in closest proximity to the relevant food habits study. When assigned from the literature, juveniles were < length at first maturity and adults were ≥ size at 100 % maturity. If size composition or life stage was not provided, or when reported sizes spanned the estimated range from first to 100 % maturity, diet was analyzed for combined juvenile-adult life stages.

## Trophic level and foraging habitat

Trophic level and foraging habitat were calculated for each species and/or life stage using the synthesized diet composition data. Trophic levels were estimated using a formula provided by Cortés (1999):

$$TL_k = 1 + \left( \sum_{j=1}^{n} P_j * TL_j \right),$$

where $TL_k$ = trophic level of species and/or life stage k, $P_j$ = proportion of prey category j in the diet of species k, $n$ = total number of prey categories, and $TL_j$ = trophic level of prey category j. Trophic level estimates for prey categories were taken from Ebert and Bizzarro (2007). Foraging habitat was calculated using a modified version of the trophic level formula:

$$FH_k = \left( \sum_{j=1}^{n} P_j * H_j \right),$$

where $FH_k$ = foraging habitat of species and/or life stage k, $P_j$ = proportion of prey category j in the diet of species k, $n$ = total number of prey categories, and $H_j$ = habitat of prey category. Three categories were used to assign prey to a general region of occurrence (i.e., functional group): 1 = benthic (on or in the seafloor), 2 = demersal (≤ ~5 m from the seafloor), 3 = pelagic (> ~5 m from the seafloor). Prey taxa were assigned a primary and, when appropriate, secondary habitat designation based on a review of pertinent literature. Primary habitats were weighted twice as much as secondary habitats. These same classifications also were used to assign groundfishes to general region of occurrence, or functional groups.

## The Major Prey Index and its calculation

Five metrics were used to assess prey importance for our selected groundfish species, including: 1) Mean % Diet Composition (Mean), 2) Median % Diet Composition (Median), 3) Prey-Specific Abundance (PSA; Amundsen et al. 1996), 4) Minimum Diet Contribution (Minimum; Minimum ≥20 %), and 5) % Frequency of Occurrence (FO). These metrics were chosen because they are commonly used to summarize diet composition data and complement each other in collectively estimating prey importance.

To account for potential inaccuracies in diet composition estimates, the following data quality metrics were established: 1) Scientific Coverage, 2) Sample Size, 3) Spatial Coverage, and 4) Temporal Coverage. Scientific Coverage was the number of studies (n) conducted on a species and/or life stage, weighted as follows: $n = 1$ (unweighted), $n = 2–3$ (weighted 2×), $n =$ 4–6 (3×),\ $n = 7–9$ (4×), $n \geq 10$ (5×) Sample Size (N, total number of stomach samples with prey items among all studies) was weighted as previously described. The distribution of each groundfish species was determined (Love et al. 2005), four regions were defined (Southern California, from the Mexican border to Monterey; Northern California, north of Monterey to the Oregon border; Oregon; Washington), and Spatial Coverage was calculated as number of regions with diet information/number of regions in which a species occurs. Temporal Coverage was calculated in the same manner as Spatial Coverage, among the following categories: 1) ≤ 1970, 2) 1971– 1985, 3) 1986–2000, and 4) 2001–2015. Ranks of data quality for each category were standardized (0.00–1.00 scale) and summed. Overall data quality for a species (i) was calculated as: value of summed data quality ranks for species *i*/greatest value of summed data quality ranks among all species.Overall data quality was presented on a relative scale, with the most robust data set assigned a value of 1.00. Relative abundance metrics (i.e., Mean, Median, PSA) for each species were adjusted for data quality using the overall data quality ranks and summed to produce overall Mean, Median, and PSA metrics for all groundfishes.

The five diet metrics were standardized (0.00–1.00) and evaluated for redundancy before being summed to create a Major Prey Index (MPI). Pearson's Correlation Coefficient ($r$) was used to determine the degree of correlation, and redundant metrics ($r > 0.70$ and $P < 0.05$ with two or more other metrics) were removed from analyses. Raw values were ranked for each nonredundant metric, summed across metrics for each prey taxon, and scaled to a range of 0.00–1.00 using the maximum aggregate value ([aggregate value of a prey category – minimum aggregate value among all prey categories]/maximum aggregate value among all prey categories). This process precluded any overwhelming contributions by particular prey taxa and enabled the contribution of each taxon to be compared on relative terms. Although EFH legislation refers to major prey "species", the MPI is used here to describe major prey "taxa" because available data did not lend itself to more specific taxonomic designations.

A randomization test (Manly 2006), iterated 9999 times, was used to determine the probability of generating the observed MPI values for each prey taxon. The ranked values among diet metrics constituted the set of observed values from which to draw, because ties limited the possible number of ranks among metrics to fewer than the number of prey categories. The expected values generated from the randomization test could fall outside the distribution of observed values; therefore, expected values were scaled by observed values. Prey taxa with MPI values significantly greater than expected by chance ($P < 0.05$) were considered to be major prey. To facilitate accurate use of the MPI, all metric calculations and the determination ofmajor prey taxa were automated through the use of an R script and a reference data set that are provided (see Online Resource).

## Sources of dietary variability

PERMANOVA was used to determine the best combination of response variables to explain the observed dietary variability in each assemblage. We summarized diet composition data among 14 generalized categories (Table 2), used a Bray-Curtis dissimilarity index (Bray and Curtis 1957) as the basis formatrix calculations, and permuted the model 9999 times. Variables (and their factors) included: SPECIES, LIFE STAGE (Juvenile, Juvenile–Adult, Adult), TAXONOMIC GROUP (elasmobranch, rockfish, roundfish, flatfish), and FUNCTIONAL GROUP (benthic, demersal, pelagic). PERMANOVA models were constructed independently for each variable to determine significant sources of dietary variability and their relative explanatory power. A final model then was calculated using forward, step-wise model selection to determine the best combination of variables to explain the observed variability in the data set. Interaction effects between variables and sample metrics also were tested to determine the influence of number of studies or number of samples among studies. Regional and temporal variables were not included in PERMANOVA models because data were insufficient to analyze the effect of their interactions with the selected variables. Including regional and temporal variables independently could yield misleading results because of dissimilar underlying species compositions.

Permutation tests of multivariate group dispersions were conducted to determine if variance differed significantly among categories for any response variable (Anderson 2006). Significant differences indicate high among-group variability that can bias P-value interpretations and increase the chance of Type-1 error. These, and all multivariate analyses, were conducted using the vegan package in R (v. 3.1.2) (Oaksanen et al. 2013).

## Trophic level and foraging habitat comparisons

Trophic level and foraging habitat indices of taxonomic and functional groups were compared using ANOVA. Index values for all groupings were homoscedastic and exhibited no extreme departures from normality.

Differences in mean trophic level and foraging habitat values were compared among species groups using one-way ANOVA on raw data because not all functional groups were represented for each taxonomic group.

## Characterization of trophic guilds

We used hierarchical agglomerative cluster analysis to determine trophic guilds among species and/or life stages based on similar diet compositions. Calculations were conducted on the more generalized diet composition data set, using Ward's Minimum Variance technique (Ward 1963), and a Bray-Curtis similarity matrix. Cluster solutions were evaluated using agglomerative coefficients (to measure the clustering structure of the dataset), cophenetic correlations (a measure of how similar two objects must be in order to be grouped into the same cluster, which we used to evaluate the goodness–of–fit between the input data and cluster output), and randomization tests (to determine significant clusters). Trophic guilds comprised species and/or life stages that formed statistically significant clusters.

# RESULTS

## Diet composition

Forty-eight sources of quantitative diet composition information were used in this study of trophic ecology of 18 Pacific Coast groundfish species (Appendix 2). One-third of the species (Pacific Hake, Sablefish, Pacific Spiny Dogfish, English Sole, Dover Sole, Black Rockfish) accounted for 87 % of 17,611 total stomach samples. By contrast, there were <65 stomach samples for each of five rockfish (Darkblotched, Greenstriped, Rosethorn, Sharpchin, Yelloweye) and one flatfish (Petrale Sole) species. Standardized diet compositions were calculated for 31 species and/or life stages mostly from combined juvenile and adult samples (Table 3). Diet composition for juveniles was calculated for 12 species, but diet composition for the adult life stage was available only for three species and generally was represented by low sample sizes.Diet composition, trophic level, and foraging habitat were estimated for adult Copper Rockfish, but this species and/or life stage was not included in subsequent analyses because it was represented by a single stomach sample.

| Common name | Life stage | *n* | N | TL | FH | INV | JELL | ECHINO | POLY | MOLL | CEPH | COPE | MYSID | AMPH | EUPH | CRAB | SHRIMP | CRUST | FISH |
|---|---|---|---|---|---|---|---|---|---|---|---|---|---|---|---|---|---|---|---|
| Black Rockfish | Juvenile | 2 | 793 | 3.9 | 1.5 | 0.3 | 0.0 | 0.0 | 0.8 | 0.2 | 0.0 | 12.6 | 18.2 | 61.6 | 0.0 | 2.6 | 2.2 | 1.2 | 0.3 |
| Black Rockfish | Juvenile – Adult | 6 | 341 | 3.9 | 1.9 | 3.2 | 10.3 | 0.0 | 3.2 | 0.2 | 3.2 | 0.0 | 1.0 | 0.4 | 10.8 | 6.7 | 1.1 | 9.0 | 51.0 |
| Brown Rockfish | Juvenile | 1 | 206 | 3.7 | 1.4 | 7.1 | 0.0 | 0.0 | 2.3 | 0.6 | 0.1 | 0.9 | 0.0 | 7.6 | 0.0 | 45.4 | 11.5 | 1.0 | 23.5 |
| Copper Rockfish | Juvenile | 3 | 348 | 3.7 | 1.5 | 0.6 | 0.0 | 0.0 | 0.1 | 3.2 | 0.0 | 9.2 | 2.4 | 24.8 | 0.0 | 26.8 | 19.5 | 0.0 | 13.3 |
| Copper Rockfish | Juvenile – Adult | 3 | 129 | 3.9 | 1.6 | 0.9 | 0.0 | 0.0 | 8.5 | 0.3 | 6.1 | 0.0 | 0.3 | 1.8 | 0.0 | 23.2 | 9.9 | 4.2 | 44.8 |
| Copper Rockfish | Adult | 1 | 1 | 4.2 | 2.0 | 0.0 | 0.0 | 0.0 | 0.0 | 0.0 | 0.0 | 0.0 | 0.0 | 0.0 | 0.0 | 0.0 | 0.0 | 0.0 | 100.0 |
| Darkblotched Rockfish | Juvenile | 1 | 18 | 3.6 | 2.2 | 0.0 | 31.3 | 0.0 | 0.0 | 0.0 | 0.0 | 9.3 | 0.0 | 19.7 | 15.3 | 0.0 | 0.0 | 24.3 | 0.0 |
| Darkblotched Rockfish | Juvenile – Adult | 1 | 20 | 3.4 | 2.4 | 0.0 | 0.0 | 0.0 | 0.0 | 0.0 | 0.0 | 4.1 | 0.0 | 5.8 | 75.2 | 0.0 | 3.0 | 0.0 | 11.8 |
| Dover Sole | Juvenile | 1 | 23 | 3.5 | 1.0 | 0.0 | 0.0 | 4.1 | 83.8 | 11.4 | 0.0 | 0.2 | 0.0 | 0.4 | 0.0 | 0.0 | 0.0 | 0.0 | 0.0 |
| Dover Sole | Juvenile – Adult | 5 | 1423 | 3.5 | 1.0 | 0.3 | 0.5 | 32.9 | 53.0 | 5.9 | 0.3 | 0.0 | 0.0 | 1.6 | 0.4 | 0.9 | 0.3 | 2.8 | 0.9 |
| English Sole | Juvenile | 2 | 303 | 3.6 | 2.0 | 0.0 | 0.0 | 0.0 | 26.7 | 0.3 | 0.0 | 46.0 | 0.0 | 17.6 | 0.0 | 0.0 | 0.0 | 9.4 0.0 | |
| English Sole | Juvenile – Adult | 9 | 1484 | 3.5 | 1.1 | 1.3 | 0.0 | 5.4 | 50.3 | 18.7 | 0.0 | 0.7 | 1.8 | 10.7 | 0.0 | 1.1 | 0.9 | 7.7 | 1.5 |
| Greenstriped Rockfish | Juvenile – Adult | 2 | 63 | 3.7 | 1.9 | 0.0 | 0.0 | 0.0 | 0.0 | 0.1 | 17.2 | 0.5 | 0.3 | 0.3 | 14.7 | 4.6 | 51.2 | 0.7 | 10.4 |
| Lingcod | Juvenile | 2 | 24 | 3.9 | 2.6 | 0.0 | 0.0 | 0.0 | 0.0 | 0.0 | 0.0 | 19.1 | 0.0 | 0.0 | 19.3 | 0.0 | 0.0 | 0.6 | 61.0 |
| Lingcod | Juvenile – Adult | 4 | 478 | 4.2 | 1.9 | 0.0 | 0.0 | 0.0 | 0.0 | 0.2 | 2.2 | 0.0 | 0.0 | 0.0 | 0.1 | 0.0 | 4.9 | 0.0 | 92.6 |
| Longspine Thornyhead | Juvenile – Adult | 1 | 281 | 3.8 | 1.5 | 1.2 | 0.0 | 5.5 | 6.5 | 1.1 | 9.1 | 0.0 | 0.4 | 1.9 | 0.0 | 19.5 | 10.2 | 13.1 | 31.4 |
| Pacific Hake | Juvenile | 5 | 1526 | 3.4 | 2.5 | 0.6 | 0.0 | 0.0 | 0.0 | 0.0 | 0.8 | 19.2 | 0.1 | 0.0 | 63.8 | 0.0 | 4.6 | 0.3 | 10.6 |
| Pacific Hake | Juvenile – Adult | 7 | 2830 | 3.7 | 2.3 | 0.0 | 0.1 | 0.0 | 0.0 | 0.0 | 1.8 | 0.0 | 0.0 | 0.0 | 45.1 | 0.3 | 11.2 | 1.0 | 40.4 |
| Pacific Hake | Adult | 4 | 778 | 4.0 | 2.3 | 0.0 | 0.0 | 0.0 | 0.0 | 0.0 | 1.0 | 0.0 | 0.0 | 0.0 | 19.4 | 0.0 | 5.2 | 3.3 | 71.0 |
| Pacific Spiny Dogfish | Juvenile | 1 | 128 | 3.5 | 2.6 | 0.0 | 70.4 | 0.0 | 0.0 | 0.0 | 0.0 | 0.0 | 0.0 | 0.0 | 0.0 | 0.0 | 0.0 | 29.6 | 0.0 |
| Pacific Spiny Dogfish | Juvenile – Adult | 8 | 2301 | 4.0 | 2.2 | 3.3 | 4.8 | 0.0 | 0.1 | 0.0 | 7.1 | 0.0 | 0.0 | 0.7 | 17.6 | 2.1 | 1.7 | 0.5 | 62.0 |
| Petrale Sole | Juvenile – Adult | 2 | 43 | 4.0 | 1.5 | 0.0 | 0.0 | 0.0 | 0.0 | 0.0 | 1.8 | 0.0 | 8.0 | 0.0 | 2.8 | 0.0 | 11.3 | 9.4 | 66.8 |
| Rosethorn Rockfish | Juvenile – Adult | 1 | 60 | 3.6 | 1.4 | 1.0 | 0.0 | 0.0 | 0.0 | 0.0 | 0.1 | 1.2 | 0.0 | 0.4 | 0.4 | 42.4 | 43.4 | 3.7 | 7.4 |
| Sablefish | Juvenile | 2 | 40 | 3.7 | 2.3 | 0.0 | 0.0 | 0.0 | 0.0 | 0.0 | 11.6 | 0.0 | 0.0 | 0.0 | 44.9 | 9.1 | 0.0 | 1.5 | 32.9 |
| Sablefish | Juvenile – Adult | 5 | 3332 | 3.8 | 1.5 | 1.2 | 0.6 | 3.5 | 3.6 | 15.4 | 15.9 | 0.1 | 0.3 | 6.1 | 6.5 | 2.0 | 4.1 | 9.0 | 31.7 |
| Sand Sole | Juvenile | 2 | 33 | 4.0 | 1.4 | 0.0 | 0.0 | 0.4 | 9.5 | 2.8 | 0.0 | 0.0 | 3.6 | 33.9 | 0.0 | 1.2 | 10.3 | 5.4 | 32.9 |
| Sand Sole | Juvenile – Adult | 4 | 414 | 3.4 | 1.9 | 0.0 | 0.0 | 0.0 | 0.1 | 0.0 | 0.0 | 0.0 | 52.1 | 5.4 | 0.0 | 0.1 | 9.4 | 0.1 | 32.7 |
| Sharpchin Rockfish | Juvenile – Adult | 2 | 44 | 3.7 | 2.3 | 0.8 | 0.0 | 0.0 | 0.0 | 0.0 | 1.0 | 1.7 | 0.0 | 0.0 | 34.3 | 5.0 | 10.8 | 10.2 | 36.3 |
| Starry Skate | Juvenile | 1 | 78 | 3.9 | 1.5 | 0.1 | 0.0 | 0.0 | 0.1 | 0.1 | 22.4 | 0.0 | 0.0 | 1.2 | 0.2 | 9.6 | 33.7 | 0.0 | 32.6 |
| Starry Skate | Adult | 1 | 32 | 4.1 | 1.7 | 0.0 | 0.0 | 0.0 | 0.0 | 0.0 | 19.6 | 0.0 | 0.0 | 0.0 | 0.0 | 7.9 | 7.7 | 0.6 | 64.1 |
| Yelloweye Rockfish | Juvenile – Adult | 2 | 37 | 4.0 | 1.4 | 0.0 | 0.0 | 0.0 | 0.0 | 0.0 | 1.0 | 0.0 | 0.0 | 0.0 | 0.0 | 26.7 | 9.3 | 0.0 | 63.0 |

Most diet composition data for Pacific Coast groundfishes ranged from Northern California toWashington, more than half of the species-specific data were published during 1971–1985, and diet data prior to 1971 were obtained for only four species (Appendix 1). Diet composition studies from Southern California were relatively rare, whereas all 18 species had diet composition information from Northern California and most species had food habits data from Oregon and Washington (Appendix 1).

Fishes were the most common and substantial dietary component of Pacific Coast groundfishes, with shrimps, crabs, and euphausiids also well represented (Table 3). Fishes were recorded in the diets of every species, with overall mean and median estimates of 32.3 % and 31.2 %, respectively. Primary f i s h t axa comprised uniden t i f i ed species (mean = 11.9 %), flatfishes (3.7 %), myctophids (2.1 %), sculpins (2.0 %), and rockfishes (2.0 %). Shrimps also were consumed by every species, although in lesser mass/volume than fishes (mean = 11.5 %, median = 9.4 %). Within this category, caridean shrimps (FO = 94.4 %, mean = 6.9 %) were more common and abundant in groundfish diets than penaeid and sergestid shrimps (FO = 50.0 %, mean = 3.2 %). Crabs were an important prey group, occurring in the diets of 15 groundfishes with mean and median contributions of 10.0 % and 4.2 %, respectively. The contribution of brachyuran crabs (FO = 72.2 %, mean = 6.2 %) exceeded that of anomuran crabs (FO = 50.5 %, mean = 4.2 %). Euphausiids were ingested by 12 species, but only contributed substantially to the diets of half of these (mean = 9.5 %, median = 1.3 %; Table 3). Some prey taxa (e.g., cephalopods, hard-shelled molluscs, unidentified crustaceans) were present in the diets of several groundfishes, but occurred in low overall proportions (Table 3).

Juvenile groundfishes generally consumed a greater proportion of small crustaceans and fewer fishes than later life stages. Increased piscivory from juvenile to later life stages was pronounced for Black Rockfish, Copper Rockfish, Starry Skate, Pacific Hake, and Lingcod (Table 3). The diets of a few species, such as Sablefish and Sand Sole, exhibited similar proportions of fishes between juvenile and juvenile-adult stages. For species in which amphipods or copepods were a substantial food item, consumption always was considerably greater for juveniles than later life stages (Table 3). A similar trend was evident for gelatinous zooplankton. Consumption of shrimps was highly variable among species and/or life stages.

## Major prey taxa

Quality of diet composition data varied greatly among species. Pacific Hake ranked highest in overall data quality (1.00). Only English Sole ranked greater than Pacific Hake among any evaluation category (Table 4). Brown Rockfish ranked lowest (overall data quality=0.32), and generally there was poorer quality diet data for the selected rockfishes than for the other groundfish species (Table 4). Spatial coverage of trophic studies was broad, and resulted in similar ranks among several species for this metric. Ranks of temporal data coverage also were similar, but only four species had diet composition data that spanned more than two time periods (Appendix 1). Scientific coverage and sample size were more variable and served to better distinguish data quality among species (Table 4).

After data quality ranks were applied, adjustments were made for highly correlated data and a significance value for major prey taxa was determined. Mean Diet Composition was found to be highly correlated with three other metrics ($r$ = 0.855 Minimum, 0.849 PSA, 0.701 FO; $P < 0.001$) and was removed from analysis. Nine of the 47 prey taxa were considered major prey, as all observed values >0.707 were significantly greater than expected ($P < 0.05$).

Major Prey Index calculations indicated that unidentified teleosts, euphausiids, and brachyuran crabs were the most important prey taxa for the 18 species of groundfish (Table 5). Unidentified teleosts were noted among 17 species to varying degrees (mean = 0.1–39.8 %). Euphausiids were well represented in the diets of Darkblotched Rockfish (mean = 45.3 %), Pacific Hake (mean = 42.8 %), and Sharpchin Rockfish (mean = 34.3 %). Brachyuran crabs were substantial dietary components for Yelloweye Rockfish, Brown Rockfish, Copper Rockfish, and Longspine Thornyhead (mean = 19.3–25.5 %).

**Table 4** Ranking of data quality used as a basis to weight the contribution of species-specific diet composition data for Major Prey Index calculations for 18 species of Pacific Coast groundfish

| Common name | *n* | N | Spatial | Temporal | Overall |
|---|---|---|---|---|---|
| Pacific Hake | 0.80 | 1.00 | 1.00 | 1.00 | 1.00 |
| Sablefish | 0.60 | 1.00 | 1.00 | 1.00 | 0.95 |
| Pacific Spiny Dogfish | 0.80 | 1.00 | 0.75 | 1.00 | 0.93 |
| English Sole | 1.00 | 1.00 | 1.00 | 0.50 | 0.92 |
| Black Rockfish | 0.80 | 1.00 | 0.75 | 0.75 | 0.87 |
| Dover Sole | 0.60 | 1.00 | 1.00 | 0.50 | 0.82 |
| Copper Rockfish | 0.80 | 1.00 | 0.75 | 0.50 | 0.80 |
| Lingcod | 0.60 | 1.00 | 0.75 | 0.50 | 0.75 |
| Greenstriped Rockfish | 0.40 | 0.25 | 1.00 | 0.50 | 0.57 |
| Sand Sole | 0.60 | 0.50 | 0.75 | 0.25 | 0.55 |
| Darkblotched Rockfish | 0.40 | 0.13 | 0.75 | 0.50 | 0.47 |
| Yelloweye Rockfish | 0.40 | 0.13 | 0.75 | 0.50 | 0.47 |
| Sharpchin Rockfish | 0.40 | 0.13 | 0.75 | 0.50 | 0.47 |
| Longspine Thornyhead | 0.20 | 0.50 | 0.75 | 0.25 | 0.45 |
| Petrale Sole | 0.40 | 0.13 | 0.75 | 0.25 | 0.40 |
| Rosethorn Rockfish | 0.20 | 0.25 | 0.75 | 0.25 | 0.38 |
| Starry Skate | 0.20 | 0.50 | 0.33 | 0.25 | 0.34 |
| Brown Rockfish | 0.20 | 0.50 | 0.25 | 0.25 | 0.32 |

Species are listed in order of descending data quality. Metric definitions and ranking criteria are provided in Methods

*n* scientific coverage, *N* sample size, *Spatial* spatial coverage, *Temporal* temporal coverage

The next series of major prey (MPI = 0.82–0.84) included caridean shrimps, polychaetes, amphipods, and unidentified crustaceans. Caridean shrimps were eaten by all species except Darkbotched Rockfish but were not consumed in abundance by any species (maximum mean = 16.4 %, Starry Skate), moderating their relative importance among major prey. Polychaetes were especially important in the diets of two groundfish species with relatively robust data sets (Dover Sole, mean = 55.7 %; English Sole, mean = 37.9 %). Amphipods were a substantial dietary component (mean = 11.4–29.1 %) for Black Rockfish, English Sole, Darkblotched Rockfish, and Copper Rockfish. Other and unidentified crustaceans were present in 17 groundfish diets but contributed <15 % to the mean diet composition of any species.

The final two major prey items were copepods and mysids, with MPI values of 0.77 and 0.71, respectively. Copepods only contributed substantially to the diet of English Sole (mean = 23.4 %) but were observed in 2/3 of the groundfish species. Similarly, mysids only were ingested in large quantities by SandSole (mean= 46.7%) but were present in the diets of half of the studied species.

Major prey taxa generally were associated with benthic and demersal habitats (Table 5). Only one (copepods) of nine major prey taxa occupied pelagic habitats, and euphausiids were associated with pelagic and demersal habitats. By contrast, five major prey taxa were strongly associated with benthic or bentho-demersal habitats (Table 5).

**Table 5** Major Prey Index (MPI) values for 47 (high level) prey taxa as determined using standardized diet composition data for 18 Pacific Coast groundfish species

| High level | Foraging habitat | MPI |
|---|---|---|
| **TELE** | **Demersal** | **0.968** |
| **EUPH** | **Pelagic-Demersal** | **0.911** |
| **CRAB B** | **Benthic** | **0.887** |
| **SHRIMP C** | **Benthic-Demersal** | **0.839** |
| **POLY** | **Benthic** | **0.831** |
| **AMPH** | **Benthic** | **0.823** |
| **CRUST** | **Benthic-Demersal** | **0.823** |
| **COPE** | **Pelagic** | **0.774** |
| **MYSID** | **Demersal** | **0.710** |
| FLAT | Benthic | 0.702 |
| JELL | Pelagic | 0.637 |
| CRAB A | Benthic | 0.629 |
| SHRIMP PS | Benthic-Demersal | 0.605 |
| SHRIMP | Benthic-Demersal | 0.573 |
| ECHINO | Benthic | 0.524 |
| ROCK | Benthic-Demersal | 0.524 |
| SCULP | Benthic | 0.508 |
| MYCT | Pelagic | 0.500 |
| BIVAL | Benthic | 0.468 |
| ENGR | Pelagic | 0.419 |
| ISO | Benthic | 0.419 |
| LOLI | Demersal | 0.403 |
| CLUP | Pelagic | 0.395 |
| GAST | Benthic | 0.379 |
| CODS | Demersal | 0.371 |
| OSMER | Demersal | 0.371 |

| | | |
|---|---|---|
| OSMER | Demersal | 0.371 |
| SCORP | Demersal | 0.355 |
| OCTO | Benthic | 0.347 |
| DECA | Benthic-Demersal | 0.339 |
| HERR | Pelagic-Demersal | 0.339 |
| ZOAR | Benthic | 0.339 |
| HEX | Demersal | 0.331 |
| AXIID | Benthic | 0.306 |
| AMMO | Pelagic-Demersal | 0.274 |
| CEPH | Demersal | 0.266 |
| TUN | Benthic | 0.234 |
| SQUID | Pelagic-Demersal | 0.218 |
| INV | Benthic-Demersal | 0.194 |
| MOLL | Benthic | 0.185 |
| OEGO | Pelagic-Demersal | 0.177 |
| CHOND | Demersal | 0.169 |
| CUTT | Benthic-Demersal | 0.113 |
| WORM | Benthic | 0.113 |
| SARD | Pelagic | 0.105 |
| AGON | Benthic | 0.065 |
| CRAB | Benthic | 0.048 |
| AGNATH | Benthic | 0.032 |

## Dietary variability

Species differences were the primary source of dietary variability for Pacific Coast groundfishes (Table 6). Independently, SPECIES explained 73.4 % of the variability in the overall data set. Two other variables, TAXONOMIC GROUP and FUNCTIONAL GROUP, exhibited significant differences among categories but explained far less overall dietary variability than SPECIES (Table 6). Although LIFE STAGE was not significant when analyzed independently, the final, best-fit model included SPECIES ($r^2$ = 0.735) and LIFE STAGE ($r^2$ = 0.090). Results of dispersion analysis were significant for both of these variables, indicating that variance was heterogeneous among constituent groups. Neither the number of studies nor the number of samples exhibited a significant interaction effect with SPECIES,

TAXONOMIC GROUP, or LIFE STAGE. Both of these factors, however, significantly influenced the results for FUNCTIONAL GROUP ($P < 0.05$).

**Table 6** PERMANOVA results of diet composition variability by species, taxonomic group, functional group, and life stage for 18 species of Pacific Coast groundfish, as calculated using low level prey categories (Table 2). Best-fit PERMANOVA model is shown first followed by individually-analyzed variables

| Model(s) | Variable(s) | df | $F$ | $r^2$ | $P$ |
|---|---|---|---|---|---|
| Best-fit | Species | 17 | 2.47 | 0.735 | 0.001* |
| | Life stage | 2 | 2.57 | 0.090 | 0.009* |
| | Residuals | 10 | | 0.175 | |
| Individual | Species | 17 | 1.95 | 0.734 | 0.003* |
| | Taxonomic group | 3 | 2.79 | 0.244 | 0.001* |
| | Functional group | 2 | 2.00 | 0.129 | 0.025* |
| | Life stage | 2 | 1.59 | 0.105 | 0.094+ |

Degrees of freedom (df), $F$-statistic, amount of variability explained. ($r^2$), and $P$-value are presented. Statistically significant variables (*) and heterogeneity of multivariate group dispersions (+) indicated. Data treatments and variable descriptions are provided in Methods

## Trophic level and foraging habitat

Estimated mean trophic levels did not vary significantly among functional ($F = 0.194$, $P = 0.825$) or taxonomic ($F = 1.106$, $P = 0.364$) groups. Median and minimum trophic levels also were similar among functional groups, ranging from 3.70–3.75, but groundfishes inhabiting benthic and demersal regions had greater maximum trophic level estimates than pelagic species (Fig. 1). Variability around the median was greatest for species occupying pelagic regions and lowest for demersal species (Fig. 1). Median trophic levels differed among taxonomic groups (3.55–3.95), with considerable within-group variability. Mean trophic level estimates therefore were more similar (3.69–3.87) than median estimates. Minimum trophic level estimates also were similar among taxonomic groups, but considerably greater maximum and quartile values were determined for elasmobranchs and roundfishes (Fig. 1). Flatfishes had the lowest median trophic level.

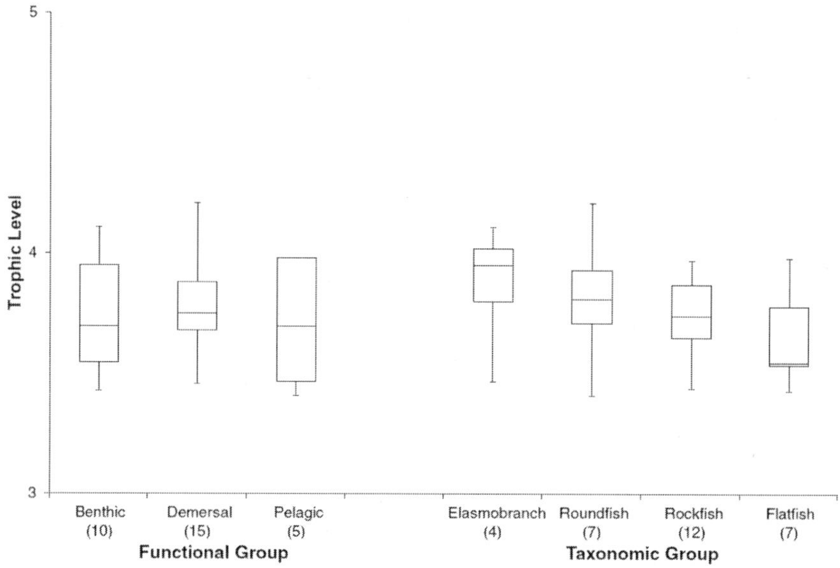


**Figure. 1** Box-and-whisker plot (median, first and third quartiles, minimum and maximum values) of trophic level for 30 Pacific Coast groundfish species and/or life stages, as calculated by Functional Group and Taxonomic Group. Numbers below labels indicate sample size

Foraging habitat estimates varied greatly among species occupying different functional and taxonomic groups (Fig. 2). Mean foraging habitat differed significantly among functional groups ($F = 10.280$, $P < 0.001$) and was highly variable for groundfishes occurring in benthic and demersal regions. Pelagic species, by contrast, foraged mainly in demersal-pelagic regions (Fig. 2). Mean (and median) foraging habitat also differed substantially by taxonomic group ($F = 5.014$, $P = 0.007$), with roundfishes foraging significantly higher in the water column than elasmobranchs and rockfishes, and all groups foraging farther off the bottom than flatfishes (Fig. 2). Flatfishes fed in benthic and demersal regions exclusively, whereas foraging habitats of elasmobranchs, roundfishes, and rockfishes ranged frombenthic-demersal to demersal-pelagic regions (Fig. 2).

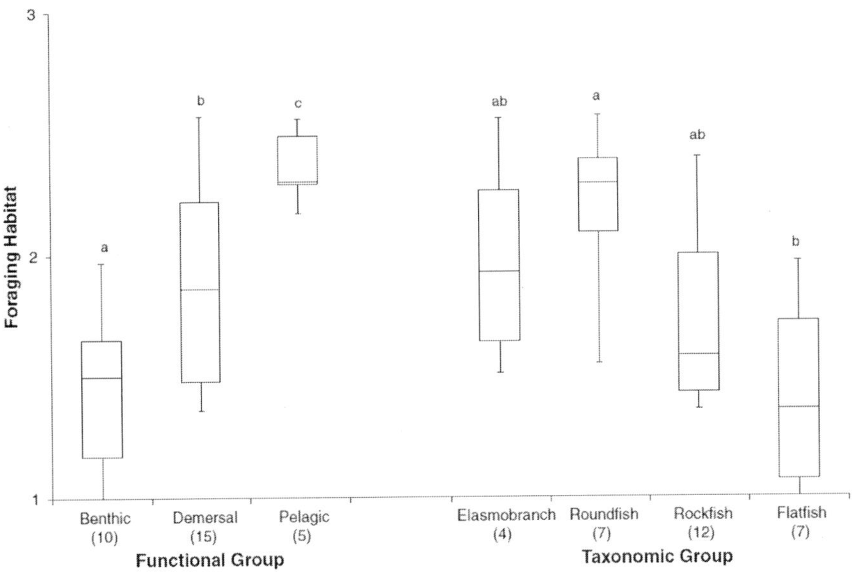


**Figure. 2** Box-and-whisker plot (median, first and third quartiles, minimum and maximum values) of foraging habitat (1 = benthic, 2 = demersal, 3 = pelagic) for 30 Pacific Coast groundfish species and/or life stages, as calculated by Functional Group and Taxonomic Group. Letters indicate significantly different groups by ANOVA (when applicable). Numbers below labels indicate sample size

## Trophic guilds

Species and/or life stages were grouped into trophic guilds based on similar diet compositions (Fig. 3). An agglomerative cluster coefficient of 0.91 indicated a high degree of clustering throughout the dissimilarity range. Additionally, pairwise species and/or life stage dissimilarities and cophenetic distances associated with the cluster dendrogram were strongly correlated (0.79), indicating that the dendrogram is an appropriate summar of diet data for Pacific Coast groundfishes.

Two small, significant trophic guilds were detected at a level of $P < 0.05$, with four larger guilds established at a level of $P < 0.10$ (Fig. 3). Dover Sole (juvenile, juvenile-adult) and English Sole (juvenile-adult) represented a significant guild with diets dominated by polychaetes (mean = > 50 %) and some hard-shelled molluscs, indicative of benthic foraging (Fig. 3). Juvenile Pacific Hake and juvenile-adult Darkblotched Rockfish formed a second significant guild, representing euphausiid predators feeding in the demersal-pelagic region (Fig. 3). Guilds formed at a level of $P < 0.10$ included groundfish species that preyed (Fig. 3, from top to bottom): 1) on gelatinous zooplankton, copepods, amphipods, and other unidentified crustaceans, 2) mainly on fishes and assorted crustacean taxa (e.g., mysids, shrimps, amphipods), 3) on shrimps, crabs, and fishes, and 4) primarily on fishes and/or euphausiids.

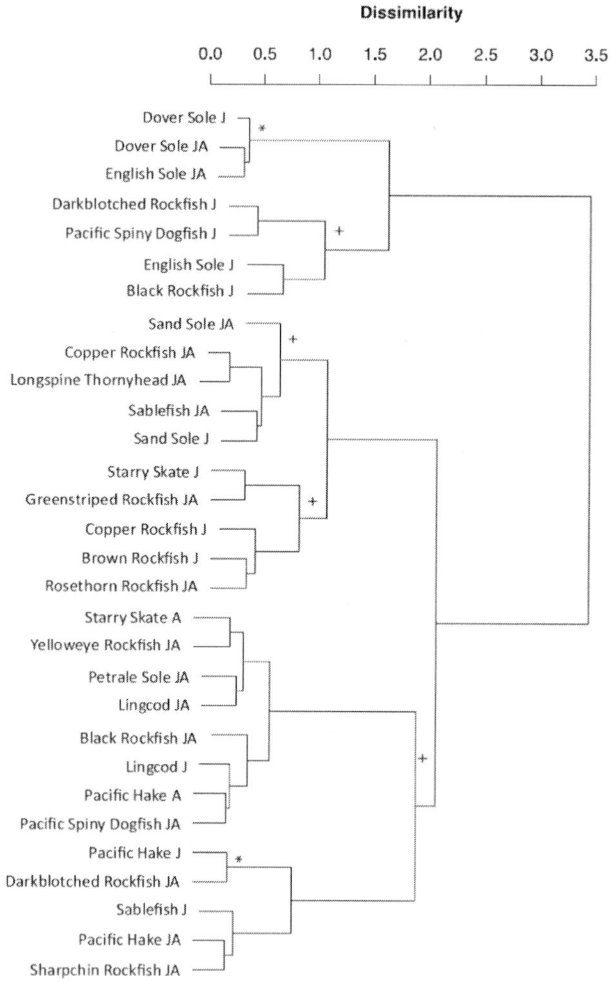


**Figure. 3** Hierarchical agglomerative cluster analysis of standardized diet composition among 30 Pacific Coast Groundfish species and/or life stages. Calculations were performed using 14 (low level) prey categories, Ward's Minimum Variance technique, and a Bray- Curtis dissimilarity matrix. J = juvenile, JA = juvenile–adult, A = adult. * = significant cluster ($P < 0.05$), + = $P < 0.10$

# DISCUSSION

Determining the composition and habitats of major prey taxa for Pacific Coast groundfishes provided new insights into the trophic ecology of this group. Epifaunal organisms that typically are common and abundant on unconsolidated seafloors (polychaetes, amphipods, brachyuran crabs, caridean shrimps) were plentiful among major prey taxa. Two major prey taxa were, however, planktonic organisms in pelagic or demersalpelagic environments (euphausiids, copepods), and mainly were ingested by juveniles of a variety of groundfish species. The designation of common, abundant taxa in pelagic, demersal, and benthic environments as major prey indicates that groundfishes are trophic generalists, regardless of foraging habitat or life stage. Consequently, widespread and abundant prey with relatively high levels of taxonomic distinction but extremely low MPI values, for example benthic poachers or pelagic sardines, are decidedly insignificant in the diets of Pacific Coast groundfishes.

Most groundfish diets were dominated by varying amounts of fishes and crustaceans, but substantial species-specific variation was evident. Because species and functional group designations lacked variability (i.e., each species and/or life stage was assigned to a single functional group), an assessment of interaction effects in the comparative analysis of diet composition was not possible. Differential foraging locations among species, however, appears to be one of the main contributing factors to dietary differences. For instance, species (e.g., Sablefish, Pacific Hake, Darkblotched Rockfish, Spiny Dogfish) that foraged on pelagic prey such as euphausiids and gelatinous zooplankton had distinctly different diet compositions than some flatfishes (Dover Sole. English Sole) that fed on benthic polychaetes and bivalve siphons (Zebold 1970; Buckley et al. 1999) or rockfishes (e.g., Brown, Copper, Rosethorn) that foraged demersally for fishes, crabs, and shrimps. The estimates of standardized diet composition of groundfishes in this study were similar to those previously calculated for 60 skate species. Like Pacific Coast groundfishes, skates ate mostly decapods and fishes, with polychaetes and amphipods of secondary importance (Ebert and Bizzarro 2007). Euphausiids were less substantial prey for skates (Ebert and Bizzarro 2007) than for Pacific Coast groundfishes, probably because of the more narrowly subscribed benthic-demersal foraging\ habitats of skates.

The far greater power of species-specific differences in explaining dietary variability relative to that of taxonomic group, functional group, or life stage suggests that each species has a distinct trophic role. Although species-specific trophic differences were well supported by the available data, even using highly generalized prey categories, the studied groundfishes do not represent a sympatric assemblage that was sampled in the same place and time. Therefore, a direct link between observed dietary differences and trophic separation among species cannot be established. Compelling evidence for trophic separation is extremely sparse in marine fishes, especially among generalist predators such as groundfishes. Even in field studies that infer resource partitioning from diet composition, sympatric fishes often exhibit dietary overlap when resources are abundant and only shift their prey spectrums when resources are scarce (Hoines and Bergstad 1999; Feyer et al. 2003).

Spatio-temporal dietary differences such as these have been reported for many of the studied groundfishes (e.g., Brodeur and Pearcy 1984; Buckley and Livingston 1997; Studebaker and Mulligan 2008; Reum and Essington 2013) and probably represent a substantial source of dietary variability. For example, the amount of rockfish consumed by lingcod was three times greater in marine reserves than in unprotected nearby waters (Beaudreau and Essington 2007). The timing and location of collection hauls represented the dominant source of dietary variability in eastern North Pacific skate assemblages, explaining an order of magnitude more variability than species or size differences (Bizzarro 2015). Brodeur and Pearcy (1984) reported seasonal, geographic, and diet variability in the diets of several groundfishes, and suggested that dietary variability seems to reflect differences in prey availability. Pooling data from different regions and time periods likely oversimplifies trophic relationships among species and/or life stages and may contribute to observed species-specific differences.

Identifying trophic guilds distinguishes fish species and/or life stages with similar predatory roles and prey taxa. Interpreting prey characteristics of the trophic guilds can provide an expanded ecological understanding of feeding relationships, foraging habitats, and improve EFH characterization for multiple life stages of exploited species. In this study, only two significant trophic guilds comprising five species and/or life stages were defined, indicating that there is broad overlap in diet composition among several prey categories for the selected groundfish species and/or life stages. These significant guilds ate large proportions of prey items (polychaetes and euphausiids) that were not abundant in the diets of most groundfish species and/or life stages. Furthermore, the members of these guilds exhibited marked spatial separation. Late juvenile and adult Dover Sole occupy much deeper habitats than comparable life stages of English Sole, though both eat large amounts of polychaetes and relatively high amounts of hard-shelled molluscs (Love 2011). Juvenile hake occur higher in the water column than Darkblotched Rockfish (Love 2011), and therefore do not use the same foraging grounds to consume euphausiids. Adding spatial associations to the specified cluster analysis would result in the assignment of ecological guilds that may differ substantially in structure and composition from the trophic guilds determined in this study.

Estimated trophic levels indicated that Pacific Coast groundfishes largely comprise meso- and upper-trophic level predators, and facilitated comparisons with trophic levels of other marine taxa. Trophic level increased with maturity for most species and reflected a shift from meso- to upper-trophic level predatory roles, typically through increased piscivory. Mean (± SE) and ranges of trophic levels of Pacific Coast groundfishes (3.77 ± 0.04; 3.41–4.24) were similar to those of skates (3.80 ± 0.02; 3.44–4.24), owing to similar diet compositions (Ebert and Bizzarro 2007). Because the Starry Skate consumed mainly fishes and cephalopods, its trophic level estimates from our study (3.91 for juveniles, 4.11 for adults) were greater than those of four other species of sympatric Pacific Coast skates (Ebert and Bizzarro 2007). Trophic level estimates of rays (3.10–4.24) also were similar to those of Pacific Coast groundfishes. Approximately 80 % of ray species were mesopredators (trophic levels 3.0–4.0; Jacobsen and Bennet 2013), as compared to 74 % of groundfish species and/or life stages in our study. Trophic levels of Pacific Coast groundfishes generally were lower than those of the sharks, but were similar to those of some carcharhiniform families (Triakidae, Sphyrhindae, Scyliorhinidae; Cortés 1999). Estimated trophic level of juvenile-adult Pacific Spiny Dogfish from our study (4.0) was comparable to that of combined Spiny Dogfish/ Pacific Spiny Dogfish (3.9; Cortés 1999); however, the trophic level estimate of juvenile Pacific Spiny Dogfish (3.5) was lower than both juvenile-adult estimates, reflecting a much greater reliance on gelatinous zooplankton instead of small, schooling fishes (Cortés 1999). Pacific Coast groundfish trophic levels estimated in our study were considerably greater than those of baleen whales or sea otters (3.2–3.4), but lower than those of pinnipeds and toothed whales (3.8–4.4) and orcas (4.5–4.6; Pauly et al. 1998).

## Management applications

The creation and application of the MPI and the determination of foraging habitats directly address current limitations in EFH characterization of Pacific Coast groundfishes. Using the MPI, fisheries scientists and managers now can consider potential impacts (e.g., harvesting, habitat destruction) to prey in the design and implementation of spatial management regulations (National Marine Fisheries Service 2013). Furthermore, differences in foraging habitats among functional and taxonomic groups and in species-specific diet compositions indicate that important foraging areas are unlikely to overlap substantially among groundfishes. Such areas should be considered on a species-specific basis, when possible.

The framework developed in this study can be expanded and applied to determine trophic relationships for evaluation and monitoring purposes, and as a precursor to the establishment of EBFM. Characterizing diet composition at high levels of taxonomic clarity can establish a comprehensive baseline condition for groundfish assemblages in a region of interest. Regular sampling in the same and nearby regions can facilitate an understanding of trophic dynamics that can be used to compare the effects of no take zones on marine communities. In addition, a food web model of the California Current (Field et al. 2006) that has been periodically updated and applied to evaluate the impacts of harvesting on food web structure (e.g., Kaplan and Levin 2009; Kaplan and Leonard 2012) can be further improved by incorporating the results of our study. Using an Atlantis model framework, it was predicted that intense fisheries exploitation would result in community replacement of relatively long-lived, kselected species, such as skates, with short-lived, rselected species, such as mackerel and anchovies (Kaplan and Levin 2009). A study by Dufault et al. (2009)modeled the food web of a diverse group of marine taxa (invertebrates, fishes, seabirds, whales) throughout thewater column off the Pacific Coast, and is considered a key input to the Atlantis model (Kaplan et al. 2013). Marked species-specific differences in groundfish diets, however, suggest that higher taxonomic and functional groupings and generalized prey categories, such as those used by Dufault et al. (2009), might not accurately characterize the complexity of trophic relationships. Our study provides more specific and robust dietary information for groundfishes that can be used to evaluate model performance by comparing outputs generated from our data with those generated using the original data. The lifestage specific diet composition information from our study also can facilitate the creation of a specific food web model for groundfishes.

## Caveats and considerations

Major, or key, ecosystem components often are distinguished through modeling exercises involving population dynamics of predators and prey, sometimes coupled with fishery interactions (Gaichas et al. 2010; Lassalle et al. 2011; Pikitch et al. 2012). Such an exercise was not possible in our study, given the nature and quality of the available diet composition data. An alternative method of prey evaluation therefore was required. Biological metrics and indices have been commonly and reliably employed to evaluate ecological data for research (Washington 1984; Krebs 1999; Abrantes et al. 2014) and fishery management (Cury et al. 2005; Jennings 2005; Samhouri et al. 2009;Kaplan and Levin 2009). Therefore, the creation and application of an index was used as the basis for the determination of major prey in this study.

The MPI is intended to address current ambiguities regarding the designation of major prey in EFH regulations; however, the available diet composition data may be too general to produce accurate estimates of prey habitat for most species of Pacific Coast groundfish. For example, the taxonomic level of the nine major prey categories ranged from infraorder (Brachyura) to subphylum (Crustacea). Broad taxonomic categories, such as unidentified teleost, may contain more accurately identified prey (e.g., flatfishes, rockfishes). In these instances, MPI values will under-represent the true importance of these prey taxa. Data at higher resolution are needed to better resolve major prey taxa (ideally to species) so that their habitats can be effectively determined. In the absence of such data, a less conservative P-value of 0.10 may be considered more appropriate than the traditional 0.05 threshold to designate major prey, depending on the objectives and needs of the user.

Quality of diet composition data varied considerably\ among species and/or life stages and influenced trophic characterizations and comparisons. Rockfishes, which comprise the great majority of federally managed Pacific Coast groundfishes (Pacific Fishery Management Council 2016), were under-represented in MPI calculations because data on diets of the selected species in this group were of relatively poor quality. Prey taxa that are common to rockfish diets, such as decapod and fish taxa, therefore contributed unequally to MPI results as compared to those of other analyses that weighed the relative contribution of each species and/or life stage equally. In addition, most diet composition data on Pacific Coast groundfishes is historic in nature. Because an environmental regime shift occurred in the California Current during the late 1970s (Hare and Mantua 2000), diet composition data collected prior to this period may not accurately reflect current trophic relationships among groundfishes.

Insufficient sample size can bias results and subsequent analyses of diet composition. The number of studies and total number of stomach samples did not influence the results of PERMANOVA analysis, and no significant differences in variance were noted among species. These results indicate that species-specific differences were robust regardless of sample size, but the accuracy of synthesized diet composition information could not be tested and represents an unknown and potentially major source of dietary variability. Additional quantitative studies with large (> 100) sample sizes, especially focused on rockfishes, are necessary to further refine estimates of major prey taxa and to facilitate more robust trophic analyses.

Differences in local prey availability may drive dietary dynamics in Pacific Coast groundfishes. Temporal and spatial factors therefore should be considered when collecting diet composition data for these generalist predators. Our study provides standardized, quantitative estimates of the diets of 18 groundfish species by life stage, which supports an improved characterization of EFH for these species. Further study that integrates spatio-temporal aspects of feeding ecology and habitat associations is a necessary step toward understanding comparative resource use by Pacific Coast groundfishes.

Because foraging habitat calculations were based on generalized prey habitats, our results should be further examined in relationship to the literature. For instance, although euphausiids were assigned to a pelagicdemersal habitat category, some marine organisms consume them when they become entrained on the benthos during vertical migrations (Isaacs and Schwartlose 1965; Genin et al. 1988; Ressler et al. 2005; Rinewalt et al. 2007; Genin 2014). Most euphausiid predators in this study, however, including some rockfishes (e.g., Greenstriped, Darkblotched, Sharpchin), are known to forage in pelagic regions (Love 2011). Therefore, the foraging habitat calculations for these species appear to be accurate. Copepods also were assigned to a pelagicdemersal functional group, and though most groundfishes foraged on pelagic copepod taxa, English Sole consumed harpacticoid copepods that are benthic (Toole 1980). In this instance, the foraging habitat for English Sole will overestimate pelagic-demersal foraging because of the generalized functional groups used in our study. Similarly, standardized diet composition should be examined in the context of fish feeding behavior. For instance, predation on bivalves by English Sole and Dover Sole is not an indication of prey excavation or durophagous feeding habits. Instead, these flatfishes nip the siphons off of bivalves as they protrude from the seafloor (Zebold 1970; Buckley et al. 1999).

## CONCLUSIONS

The findings of this study represent a necessary first step toward incorporation of trophic information into the characterization of EFH of Pacific Coast groundfishes. Establishing objective, quantitative criteria for the determination of major prey taxa addresses a need in the development of spatial management plans for these species (National Marine Fisheries Service 2013). The identification of common and abundant taxa as major prey indicates that groundfishes are trophic generalists, regardless of foraging habitat or life stage. Dietary differences were, however, highly significant among species, with varying foraging locations greatly influencing these results. These\ differences are largely a consequence of differential use of crustacean and fish taxa, as there is broad dietary overlap of these prey categories among groundfish species. Pacific Coast groundfishes were mainly mesopredators, with diet compositions and trophic levels similar to those of skates (Ebert and Bizzarro 2007).

Future work should focus on collecting quantitative diet composition information and expanding analyses to include additional species and/or life stages of groundfishes and the evaluation of spatial and temporal dietary variability. Contemporary diet composition data are limited, especially for rockfishes and for most species off Southern California. The 18 selected species constitute a small portion of the 117 species of groundfishes included in the Pacific Coast Groundfish Management Plan (Pacific Fishery Management Council 2016). Diet studies on a greater number and diversity of species, and more research on poorly studied species, will help us better define major prey taxa and trophic ecology for Pacific Coast groundfishes. Adding a spatiotemporal component to analysis will enable a more complete assessment of the relative magnitude of different sources of dietary variability for each species and/or life stage and for the overall assemblage. The incorporation of spatial and temporal diet composition data also will facilitate a better understanding of the ecological relationships among Pacific Coast groundfishes.

## ACKNOWLEDGMENTS

We thank the National Marine Fisheries Service (NMFS) Office of Habitat Conservation and NMFS Southwest Fisheries Science Center (SWFSC) for funding this project.We appreciate John Stadler's (NMFSWest Coast Region) support and insight throughout this project. Thanks also to Stephanie Flora for writing a Matlab script to automate Major Prey Index calculations and to Andrew Pike for translating the script from Matlab into R and improving its functionality. John Field (NMFS-SWFSC), two anonymous reviewers, and the scientific editor of EBF provided helpful comments and suggestions on earlier drafts of this manuscript.

## REFERENCES

1. Abrantes KG, Barnett A, Bouillon S (2014) Stable isotope-based community metrics as a tool to identify patterns in food web structure in east African estuaries. Functional Ecology 28: 270–282
2. Amundsen PA, Gabler HM, Staldvik FJ (1996) A new graphical analysis of feeding strategy from stomach contents data – modification of the Costello (1990) method. Journal of Fish Biology 48:607–614
3. Anderson MJ (2006) Distance-based test for homogeneity of multivariate dispersions. Biometrics 62:245–253
4. Beaudreau AH, Essington TE (2007) Spatial, temporal, and ontogenetic patterns of predation on rockfishes by lingcod. Transactions of the American Fisheries Society 136:1438–1452

5. Bizzarro JJ (2015) Comparative resource utilization of eastern North Pacific skates with applications for fisheries management. Ph.D. Thesis. University of Washington, School of Aquatic and Fishery Sciences, Seattle, WA, 217 pp
6. Bizzarro JJ, Robinson HJ, Rinewalt CS, Ebert DA (2007) Comparative feeding ecology of four sympatric skate species off central California, U.S.A. Environmental Biology of Fishes 80:197–220
7. Bizzarro JJ, Smith WD, Castillo-Géniz JL, Ocampo-Torres A, Márquez-Farías JF, Hueter RE (2009) The artisanal elasmobranch fishery of Sinaloa, Mexico. Pan-American Journal of Aquatic Science 4:513–531
8. Bray JR, Curtis JT (1957) An ordination of upland forest communities of southern Wisconsin. Ecological Monographs 27: 325–349
9. Brodeur RD, Pearcy WG (1984) Food habits and dietary overlap of some shelf rockfishes (genus Sebastes) from the northeastern Pacific Ocean. Fishery Bulletin 82:269–273
10. Buckley TW, Livingston PA (1997) Geographic variation in the diet of Pacific hake, with a note on cannibalism. California Cooperative Oceanic Fisheries Investigations Reports 38:53–62
11. Buckley TW, Tyler GE, Smith DM,Livingston PA (1999) Food habits of some commercially important groundfish off the coasts of California, Oregon, Washington, and British Columbia. NOAA Technical Memorandum NMFS-AFSC- 102, 173 pp
12. Cailliet GM, Love MS, Ebeling AW (1986) Fishes: a field and laboratory manual on their structure, identification, and natural history.Wadsworth Publishing Co, Belmont, CA, 194 pp
13. Colwell RK(2013) EstimateS: statistical estimation of species richness and shared species from samples. Version 9.1.0. User's Guide and application published at: http://purl.oclc.org/estimates
14. Cortés E (1999) Standardized diet compositions and trophic levels in sharks. ICES J Mar Sci 56:707–717
15. Cury PM, Shannon LJ, Roux J-P, Daskalov GM, Jarre A,Moloney CL, Pauly D (2005) Trophodynamic indicators for an ecosystem approach to fisheries. ICES J Mar Sci 62:430–442
16. Dufault AM, Marshall K, Kaplan IC (2009) A synthesis of diets and trophic overlap of marine species in the California Current. NOAA Technical Memorandum. NMFS-NWFSC- 103, 81 pp
17. Ebert DA, Bizzarro JJ (2007) Standardized diet compositions and trophic levels of skates (Chondrichthyes: Rajiformes: Rajoidei). Environmental Biology of Fishes 80:221–237
18. Essington TE, Punt AE (2011) Implementing ecosystem-based fisheries management: advances, challenges, and emerging tools. Fish and Fisheries 12:123–124
19. Ferry LA, Cailliet GM (1996) Sample size and data analysis: are we characterizing and comparing diet properly?In: MacKinlayDD, Shearer K (eds) Feeding ecology and nutrition in fish: proceedings of the symposium on the feeding ecology and nutrition in fish. International Congress on the Biology of Fishes, San Francisco, CA, pp 71–80. 14–18 July, 1996
20. Feyer F, Herbold B, Matern SA, Boyle PB (2003) Dietary shifts in a stressed fish assemblage: consequences of a bivalve invasion in the San Francisco estuary. Environmental Biology of Fishes 67:277–288
21. Field JC, Francis RC, Aydin K (2006) Top-down modeling and bottom-up dynamics: linking a fisheries-based ecosystem model with climate hypotheses in the Northern California Current. Progress in Oceanography 68:238–270
22. Gaichas SK, Aydin KY, Francis RC (2010) Using food web model results to inform stock assessment estimates of mortality and production for ecosystem–based fisheries management. Canadian Journal of Fisheries and Aquatic Science 67: 1490–1506

23. Gaichas SK, Aydin KY, Francis RC (2011)What drives dynamics in the Gulf of Alaska? Integrating hypotheses of species, fishing, and climate relationships using ecosystem modeling. Canadian Journal of Fisheries and Aquatic Science 68:1553–1578
24. Garrison LP, Link JS (2000) Fishing effects on spatial distribution and trophic guild structure in the Georges Bank fish community. Symposium proceedings on the ecosystem effects of fishing. ICES Journal of Marine Science 57:723–730
25. Genin A (2014) Bio-physical coupling in the formation of zooplankton and fish aggregations over abrupt topologies. J Mar Syst 50:3–20
26. Genin A, Haury L, Greenblatt P (1988) Interactions of migrating zooplankton with shallow topography: Predation by rockfishes and intensification of patchiness. Deep-Sea Research 35:151–175
27. Hare SR, Mantua NJ (2000) Empirical evidence for North Pacific regime shifts in 1977 and 1989. Progress in Oceanography 47:103–145
28. Hoines AS, Bergstad OA (1999) Resource sharing among cod, haddock, saithe, and pollock on a herring spawning ground. Journal of Fish Biology 55:1233–1257
29. Hyslop EJ (1980) Stomach contents analysis: a review of methods and their application. J Fish Biol 17:411–429
30. Isaacs JD, Schwartlose RA (1965) Migrant sound scatters: interactions with the sea floor. Science 150:1810–1813
31. Jacobsen IP, Bennet MB (2013) A comparative analysis of feeding and trophic level ecology in stingrays (Rajiformes: Myliobatoidei) and electric rays (Rajiformes: Torpedinoidei). PLOS ONE DOI. doi:10.1371/journal.pone.0071348
32. Jennings S (2005) Indicators to support an ecosystem approach to fisheries. Fish and Fisheries 6:212–232
33. Kaplan IC, Leonard J (2012) From krill to convenience stores: forecasting the economic effects of fisheries management on the U.S. West Coast. Marine Policy 36:947–954
34. Kaplan IC, Levin P (2009) Ecosystem-based management of what? An emerging approach for balancing conflicting objectives in marine resource management. pp 77–95. In: Beamish RJ, Rothschild BJ (eds) The future of fisheries science in North America. Fish and Fisheries Series, Springer Science. Netherlands
35. Kaplan IC, Brown CJ, Fulton EA, Gray IA, Field JC, Smith ADM (2013) Impacts of depleting forage species in the California Current. Environmental Conservation 40:380–393
36. Krebs CJ (1999) Ecological methodology. Benjamin Cummings, Menlo Park, CA, 620 pp
37. Lassalle G, Lobry J, Le Loc'h F, Bustamante P, CertainG,Delmas D, Dupuy C, Hily C, Labry C, Le Pape O,Marquis E, Petigas P, Pusineri C, Ridoux V, Spitz J, Niquil N (2011) Lower trophic levels and detrital biomass control the Bay of Biscay continental shelf food web: Implications for ecosystem management. Progress in Oceanography 91:561–575
38. Link JS, Garrison LP (2002) Changes in piscivory associated with fishing induced changes to the finfish community onGeorges Bank. Fisheries Research 55:71–86
39. Link JS, Sosebee K (2008) Estimates and implications of skate consumption in the northeastern US continental shelf. North American Journal of Fisheries Management. 28:649–662
40. Link JS, Bundy A, Overholtz WJ, Shackell N, Manderson J, Duplisea D (2011) Ecosystem-based fisheries management in the Northwest Atlantic. Fish and Fisheries 12:152–172
41. Love MS (2011) Certainly more than you want to know about the fishes of the Pacific Coast – a postmodern experience. Really Big Press, Santa Barbara, California, 650 pp
42. Love MS, Mecklenburg CW, Mecklenburg TA, Thorsteinson LK (2005) Resource inventory of marine and estuarine fishes of the west coast and Alaska: a checklist of north Pacific and Arctic Ocean species from Baja California to the Alaska– Yukon

border. United States Geological Survey, Biological Resources Division, Seattle, WA, 276 pp

43. Manly BFJ (2006) Randomization, bootstrap, and Monte Carlo methods in biology. In: Chapman and Hall/CRC Press. New York, NY, Inc., 480 pp
44. McCain, B.B., S.D. Miller, W.W. Wakefield. 2005. Pacific Coast Groundfish Fishery Management Plan. Appendix B2. Life history, geographical distribution, and habitat associations of 82 West Coast groundfish species: a literature review. National Marine Fisheries Service, Northwest Fisheries Science Center. Seattle, WA. 266 pp.
45. National Marine Fisheries Service (2013) Groundfish Essential Fish Habitat Synthesis: a report to the Pacific Fishery Management Council. NOAA NMFS Northwest Fisheries Science Center. Seattle, WA, 378 pp
46. Oaksanen J, Blanchet FG, Kindt R, Legendre P, Minchin PR, O'Hara RB, Simpson GA, Solymos P, Henry M, Stevens H,Wagner H (2013) Vegan: community ecology package. R package version 2 :–9http://CRAN.R-project.org/package=vegan
47. Overholz WJ, Link JS (2007) Consumption impacts by marine mammals, fish, and seabirds on the Gulf of Maine-Georges Bank herring complex during the years 1977–2002. ICES Journal of Marine Science 64:83–96
48. Pacific Fishery Management Council (2012) Pacific coast groundfish 5-year review of essential fish habitat. Report to Pacific Fishery Management Council. Phase 1: New information, September 2012. Portland, OR, 427 pp
49. Pacific Fishery Management Council (2014) Pacific Coast Groundfish Fishery Management Plan. May 2014. OR, Portland, 146 pp
50. Pacific Fishery Management Council (2016) Pacific Coast Groundfish Fishery Management Plan. In: March 2016. OR, Portland, 148 pp
51. Pauly D, Trites AW, Capuli E, Christensen V (1998) Diet composition and trophic levels ofmarinemammals. ICES Journal of Marine Science 55:467–488
52. Pikitch E, Boersma PD, Boyd IL, Conover DO, Cury P, Essington TE, Heppell SS, Hopude ED, Mangel M, Pauly D, Plagányi É, Sainsbury K, Steneck RS (2012) Little fish, big impact: managing a crucial link in ocean food webs. Lenfest Ocean Program, Washington, DC, 108 pp
53. Pinkas LM, Oliphant S, Iverson ILK (1971) Food habits of albacore, bluefin tuna and bonito in Californian waters. California Fish and Game 152:1–105
54. Ressler PH, Brodeur RD, Peterson WT, Pierce SD, Vance PM, Røstad A, Barth JA (2005) The spatial distribution of euphausiid aggregations in the northern California Current during August 2000. Deep-Sea Research II 52:89–108
55. Reum JCP, Essington TE (2013) Spatial and seasonal variation in ∂N15 and ∂13C values in a mesopredator shark, Squalus suckleyi, revealed through multitissue analyses. Marine Biology 160:399–411
56. Rinewalt CS, Ebert DA, Cailliet GM (2007) The feeding habits of the sandpaper skate, Bathyraja kincaidii (Garman, 1908) in central California: seasonal variation in the diet linked to oceanographic conditions. Environmental Biology of Fishes 80:147–163
57. Samhouri JF, Levin PS, Harvey CJ (2009) Quantitative evaluation of marine ecosystem indicator performance using food web models. Ecosystems 12:1283–1298
58. Studebaker RS, Mulligan TJ (2008) Temporal variation and feeding ecology of juvenile Sebastes in rocky intertidal tidepools of northern California, with emphasis on Sebastes melanops Girard. Journal of Fish Biology 72:1393–1405
59. Szoboszlai A, Thayer JA, Wood SA, Sydeman WJ, Koehn LE (2015) Forage species in predator diets: synthesis of data from the California Current. Ecological Informatics 29:45– 56
60. Toole CL (1980) Intertidal recruitment and feeding in relation to optimal utilization of nursery areas by juvenile English sole (Parophrys vetulus: Pleuronectidae. Environmental Biology of Fishes 5:383–390

61. United States Department of Commerce (1996) Magnuson- Stevens Fishery Conservation and Management Act. National Oceanic and Atmospheric Administration. National Marine Fisheries Service. Public Law:94–265
62. Ward J (1963) Hierarchical grouping to optimize an objective function. Journal of the American Statistical Association 58:236–244
63. Washington HG (1984) Diversity, biotic and similarity indices: a review with special relevance to aquatic systems. Water Research 18:653–694
64. Zador S (2014) Ecosystem considerations - 2014. Appendix C of 2014 SAFE report for Alaska. North Pacific Fishery Management Council, Anchorage, AK, 263 pp
65. Zebold SL (1970) Inter- and intraspecific comparisons of the diel distributions and the food and feeding habits of five species of demersal fishes from Duwamish Head, Puget Sound, Washington. M.S. Thesis. University of Washington, Spokane, Washington, 76 pp

# CHAPTER 7

# Fish Species Composition, Distribution and Abundance in Shadegan Wetland

**Seyed Ahmadreza Hashemi[1*], Rasoul Ghorbani[1], Farhad Kymaram[2], Seyed Abass Hossini[1], Gholareza Eskandari[3] and Aliakbar Hedayati[1]**

[1]*Department of Fishery, Faculty of Fisheries and Environment, University of Agricultural Science and Natural Resources, Gorgan, Iran*

[2]*Fisheries Research Organization, Tehran, Iran*

[3]*South of Iran aquaculture fishery research center, Ahvaz, Iran*

## ABSTRACT

Composition, distribution and abundance of fish species was carried out from April 2012 to Mars 2013 in the Shadegan wetland. Samples were collected from five stations; Mahshar, Rogbe, Khorosy, Salmane and Ateish in this local in Khuzestan provinces (Iran). A total of 3312 fish individuals comprising 26 species from 6 families were sampled throughout the entire study period. The most abundant species was Carasobarbus luteus (Cyprinidae) comprising 28.20% of the total fish caught. The mean CPUE (catch per unit effort) for all species was 29 nu/day and maximum and minimum CPUE values were 47.60 ± 11.61 nu /day (spring) and 18.40 ± 5.94 nu /day (winter), respectively. Maximum and minimum Shannon-Weinner values were 2.21 (spring) and 2 (autumn), respectively. The Shannon-Weinner value differ significantly between different station ($P<0.05$) and Margalef's, Evenness, Simpson's index didn't differ significantly between different station ($P>0.05$). A comparison of value salinity is significant between different station ($P<0.05$) and other parameters is no significant different station ($P<0.05$). The Wetland Fish Index (WFI) calculated for the entire wetland using total abundance was

2.99. The WFI should prove to be an effective tool to aid in the management and protection of important fish habitat in Shadegan wetlands.

**Keywords:** Shannon-weinner values; Wetland fish index; Shadegan wetland

# INTRODUCTION

It is important to know the fish species diversity, abundance and distribution in the reservoirs in order to develop management and conservation programs. Studies of spatial and temporal patterns of diversity, distribution and species composition of freshwater fishes are useful to examine factors influencing the structure of the fish community [1]. Biodiversity is the quantity, variety and distribution across biological scales ranging through **GENETICS** and life forms of populations, species, communities and ecosystems [2]. Biodiversity is important for the future sustainability of natural resources that include commercial fisheries. While it is axiomatic that biodiversity is essential for sustainable productive fisheries there is surprisingly little supporting evidence. Fish species are also an important indicator of ecological health. The abundance and health of fish will show the health of water bodies [3]. This is supported by Zainudin [4] who claimed that fish species diversity can be used as a biological indicator to show the level of aquatic pollution contributing to environmental quality. Biodiversity affects the capacity of living systems to respond to changes in the environment, underpins ecosystem function and provides the ecosystem goods and services that support human well-being [5,6]. This has led to the development of some management programs to conserve and to increase needs for the freshwater fish population.

Wetlands are invaluable ecological resources that serve many human needs and therefore, enhance our lives by providing a lot of opportunities. Many depend heavily on the resources of such water bodies as their main source of animal protein and family income [7]. Wetlands in the world was included about 7 to 9 million $km^2$ (4-6 percent of Earth surface). Iran wetlands are approximately 1853762 ha and between Middle East wetland was contained 25% [8]. Shadegan Wetland in Khuzestan province is one of the 18 international wetlands registered on UNESCO's Natural **HERITAGE** List. Located 52 km from Abadan and 105 km from Ahvaz, it is Iran's largest wetland and by Linking Jarahi River connect with Persian Gulf waters, the wetland is considered one of the most wonderful natural landscape of the world because of it is unique biodiversity [9]. In 2006, fish landings in Shadegan wetland reached a peak of 3700 tons, which corresponds to more than 10 billion$ [10]. The wetland is a unique natural asset for the province of Khuzestan, supporting a rich biodiversity, including threatened and **ENDEMIC** species. water resources development for agricultural intensification in the upstream catchment which would reduce the volume of the inflows, and change the seasonality of flooding development of heavy industries and infrastructures around and within the wetland pollution from industrial, urban and agricultural (including aquaculture) wastes oil spills [9].

In this study, we examined the fish species composition, distribution and abundance in shadegan wetland. Our main aims were to: (i) Estimation of fish species composition, distribution and abundance, (ii) different distribution and abundance in stations and seasons. Maramazi [11], Ansari and Mohamadi [12,13] and Ansari et al., Hashemi et al., were searched fish survey, **STOCK ASSESSMENT** and capture conditions of shadegan wetland. Lotfi et al., [9] were considered human activity and effect on shadegan wetland and also diversity and capture situation of shadegan wetland. However, one study so far has been made on fish distribution and abundance in shadegan wetland. For rational and sustainable management of this exploited resource, information on the species composition, abundance, distribution, and growth are provided. These data can be used to better fisheries stock management for these valuable fish.

## MATERIALS AND METHODS

### STUDY AREA

Composition, distribution and abundance of fish species was carried out from April 2012 to Mars 2013 in the Shadegan wetland. Samples were collected from five stations; Mahshar (48°,45′ E,30°,33′ N ), Rogbe (48°,33′ E, 30°,41′ N ), Khorosy (48°,40′ E, 30°,39′ N ), Salmane (48°,28′ E, 30°,40′ N ) and Ateish (48°,40′ E, 30°,54′ N ) in the Shadegan wetland in Khuzestan provinces (**FIGURE 1**). In each month, 5 stations were selected for sampling. Fish species was carried out by using gill net with 45 mm mesh and then Sampling transported to lab with dry ice. The gill nets were set at dusk and hauled the next morning. All fish caught were identified to species using standard taxonomic keys following Nelson and Coad [14,15].

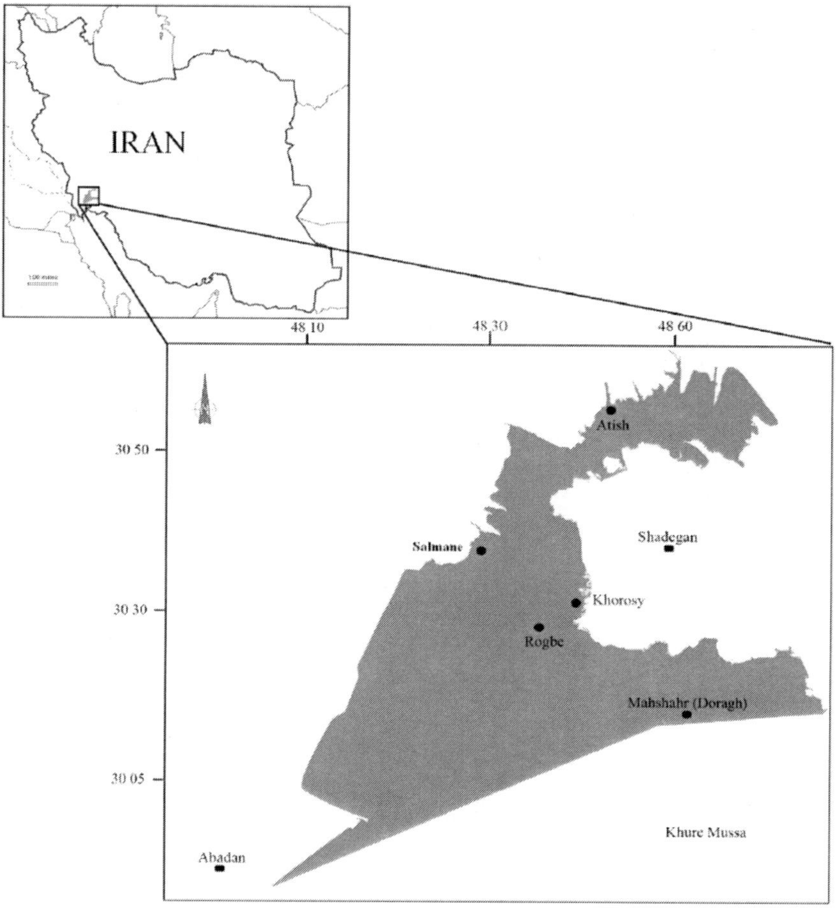


**FIGURE 1:** The map of Iran, Location of five Capture sites was sampled in Shadegan wetland (Khuzestan province, South West of Iran).

## DATA COLLECTION

Habitat monitoring and fish sampling were carried out seasonally from September 2012 to September 2013. Fishing gears set for sampling at study sites were sometimes destroyed by other fisherman in this wetland. In such cases, data collection was repeated in the next year at the same place and the same time period. Sampling started at a random point and then proceeded at evenly distributed intervals along the shoreline of the water body. Seasonally water samples for analysis of environmental parameters were collected from each station using a Nansen bottle sampler and analyses as per standard analytical procedures [16]. Eighteen environmental parameters (**TABLE 1**) were considered in this research and included water temperature (WT), water depth (WD), water salinity (WS) phosphorus (TP), nitrate (TN), PH, biological oxygen demand (BOD) and dissolved oxygen (DO).

**TABLE 1:** Table showing the list of fish species caught and the percentage composition for fish population in shadegan wetland and Water-quality optimum (U) and tolerance (T) values for 26 fish species from abundance (AB) data. (*Species occurring in <5% of wetlands were automatically assigned a T value of 1).

| Family | Species | status | N | %N | W (Kg) | % W | Y | T |
|---|---|---|---|---|---|---|---|---|
| **CYPRINIDAE** | *Carasobarbusluteus* | native | 934 | 28.20 | 73.78 | 15.74 | 2.97 | 4 |
| **CYPRINIDAE** | *Cyprinuscarpio* | native | 613 | 18.50 | 58.23 | 12.42 | 2.78 | 4 |
| **CYPRINIDAE** | *Carasiuscarasius* | native | 437 | 13.19 | 44.57 | 9.51 | 2.64 | 4 |
| **MUGILIDAE** | *Liza abu* | native | 399 | 12.04 | 13.56 | 2.89 | 2.60 | 2 |
| **CYPRINIDAE** | *Mesopotamichthyssharpeyii* | native | 336 | 10.14 | 56.78 | 12.11 | 2.52 | 3 |
| **CYPRINIDAE** | *Aspiusvorax* | native | 191 | 5.76 | 29.98 | 6.39 | 2.28 | 3 |
| **CYPRINIDAE** | *Hypophthalmichthysmolitrix* | introduce | 75 | 2.26 | 6.01 | 1.28 | 1.80 | 3 |
| **CYPRINIDAE** | *Barbusgrypus* | native | 74 | 2.23 | 11.32 | 2.41 | 1.87 | 3 |
| **SILURIDAE** | *Silurstriostegus* | native | 61 | 1.84 | 154.49 | 32.97 | 1.78 | 4 |
| **CYPRINIDAE** | *Hypophthalmichthysnobilis* | introduce | 50 | 1.50 | 5.7 | 1.21 | 1.69 | 2 |
| **SPARIDAE** | *Acantupagruslutus* | exotic | 47 | 1.41 | 2.25 | 0.48 | 1.67 | 2 |
| **CYPRINIDAE** | *Ctenopharingodonidella* | native | 25 | 0.75 | 2.6 | 0.55 | 1.39 | 3 |
| **MASTACEMBELIDAE** | *Mastacembulsmastacembuls* | native | 20 | 0.60 | 4.09 | 0.87 | 0.84 | 3 |
| **CYPRINIDAE** | *Luciobarbuspectoralis* | native | 13 | 0.39 | 1.59 | 0.33 | 1.11 | 2 |

| | | | | | | | | |
|---|---|---|---|---|---|---|---|---|
| **MUGILIDAE** | *Ellonchelonvaigiensis* | exotic* | 6 | 0.18 | 1.21 | 0.25 | 0.77 | 1 |
| **MUGILIDAE** | *Chelonsubviridis* | exotic* | 6 | 0.18 | 0.81 | 0.17 | 0.77 | 1 |
| **ENGRAULIDAE** | *Thrssahemiltoni* | exotic* | 5 | 0.15 | 0.2 | 0.04 | 0.69 | 1 |
| **SILURIDAE** | *Heteropenustifossili* | native* | 3 | 0.09 | 0.6 | 0.12 | 0.47 | 1 |
| **CYPRINIDAE** | *Luciobarbusxantropetrous* | native* | 3 | 0.09 | 0.4 | 0.08 | 0.47 | 1 |
| **CYPRINIDAE** | *Acanthobramamarmid* | native* | 2 | 0.06 | 0.2 | 0.04 | 0.3 | 1 |
| **CLUPEIDAE** | *Sardinellasindensis* | exotic* | 2 | 0.06 | 0.05 | 0.01 | 0.3 | 1 |
| **CYPRINIDAE** | *Alburnoidesbipunctatus* | native* | 2 | 0.06 | 0.04 | 0.008 | 0.3 | 1 |
| **CYPRINIDAE** | *Cyprinionmacrostumus* | native* | 2 | 0.06 | 0.04 | 0.004 | 0.3 | 1 |
| **CYPRINIDAE** | *Cyprinion kais* | native* | 2 | 0.06 | 0.02 | 0.004 | 0.3 | 1 |
| **CYPRINIDAE** | *Barbusluciobarbus* | native* | 2 | 0.06 | 0.01 | 0.002 | 0.3 | 1 |
| **CLUPEIDAE** | *Tenualosailisha* | exotic* | 2 | 0.06 | 0.01 | 0.002 | 0.3 | 1 |

## DIVERSITY OF FISH SPECIES

The diversity of fish species was estimated in terms of species **EVENNESS**, using Margalef's D index, Shannon Wiener, Evenness ratio and Simpson's. Species diversity, abundance and richness are determined by following May (1993) and used following diversity indices:

- Margalef's Index

$MD = S-1/LogN$

Where S= total number of species, N=total density and Ni=density of individual species.

- Shannon-Weinner Index

$H = -\{ni/N \log_2 ni/N\}$

- Evenness ratio

The species evenness ratio is the ratio of the observed species diversity (H') to the maximum possible for the same number of species in the sample (log S). It is expressed as:

J'=H'/log S.

If J'=1, the biomasses of the individual species are evenly distributed among all of the species in the sample [17].

- Simpson's Index

D=1-Ni(n) – I/N(N-1)

In order to test for differences in diversity among fishes in different seasons of the year, pair wise randomization tests were carried out, abundance data following May (1993). The analyses were performed as per the method of May [18] using Species Diversity and Richness software and Microsoft Excel sheet.

## WETLAND FISH INDEX (WFI)

To develop the WFI, each species was assigned U and T values according to the following equation [19].

$$WFI = \frac{\sum YiTiUi}{\sum YiTi}$$

Where Yi is the presence or log10 abundance (log[x + 1]) of species i, Ti is the value from one to three (indicating niche breadth), and Ui is the value from one to five (indicating tolerance of degradation).

CANOCO 4.5 [19] was used to run canonical correspondence analysis (CCA) as in Seilheimer and Chow-Fraser [19,20]. Canonical correspondence analysis (CCA) was used to explore the distribution of the fish communities in relation to the environmental variables. If a species occurred only once, it was excluded from the CCA analysis. Environmental variables selected for the CCA analysis included continuous variables, such as water quality data (e.g., pH and COD). All continuous environmental variables were log (1 + x) transformed and standardized to have a zero mean and unit variance. Comparison of *environmental parameters* and species diversity, abundance, richness during different spatial and temporal carried out by analysis of variance (ANOVA). **STATISTICAL ANALYSES** were performed with SPSS 21 software package and a significance level of 0.05 was adopted.

# RESULTS

A total of 3312 fish individuals comprising 26 species from 6 families were sampled from Shadegan wetland throughout the entire study period (**TABLE 1**). The most abundant species was *Carasobarbus luteus* (Cyprinidae) comprising 28.20% of the total fish caught, followed by *Cyprinus carpio* (Cyprinidae) (18.50%) and *Carasius carasius* (Cyprinidae) (13.19%). The most total weight fish was Silurs triostegus (*Siluridae* ) and *Carasobarbus luteus* (Cyprinidae) comprising 32.97% and 15.74% respectively. Maximum and minimum abundant family were Cyprinidae (82.91%) and Engralidae (0.5%) respectively (**FIGURE 2**).

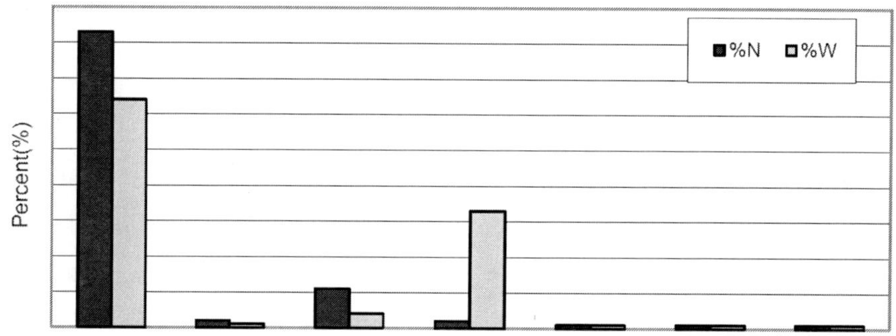


**FIGURE 2:** Fish abundant family was sampled in Shadegan wetland (%N=percent of fish number and %W= percent of fish weight).

## CPUE (CATCH PER UNIT EFFORT)

The mean CPUE (catch per unit effort) for all species was 29 nu/ day and maximum and minimum CPUE values were 47.60 ± 11.61 nu /day (spring) and 18.40 ± 5.94 nu /day (winter), respectively. The mean depth of fishing was 1 m and ranged from 0.5 m to 2.5 m. Value CPUE in each Station (**FIGURE 3**) and this value in different season was showed (**FIGURE 4**). A comparison of value CPUE is significant between different season (ANOVA, $F=16.09$, $P<0.05$) and there is significant different station them (ANOVA, $F=73.49$, $P<0.05$).

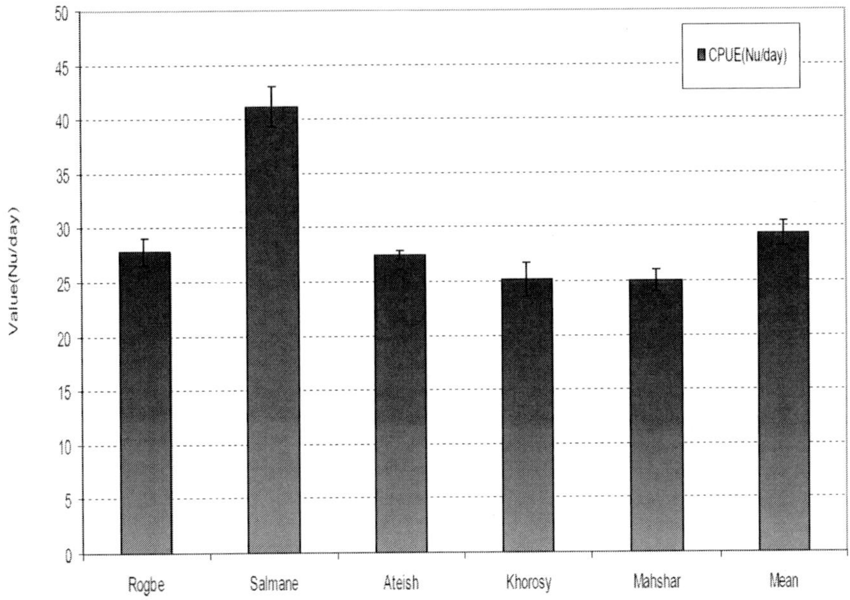


**FIGURE 3:** Value CPUE in different station from Shadegan Wetland (2012-13).

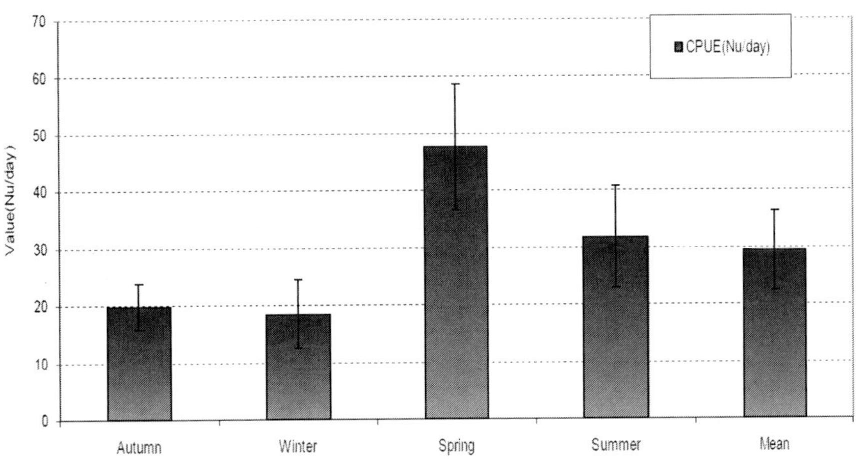


**FIGURE 4:** Value CPUE in different season from Shadegan Wetland (2012-13).

## ENVIRONMENTAL PARAMETERS AND DIVERSITY OF FISH SPECIES

Physic-chemical parameters and diversity of fish species are shown in **TABLES 2-4**. A comparison of value salinity is significant between different station (ANOVA, F=18.21, P<0.05) and other parameters is no significant different station (P<0.05).

**TABLE 2:** Physic-chemical parameters in different station from Shadegan Wetland (2012-13).

| Varible | Value | Kanal | Atish | Rogbe | Mahshar | Khorosy | P-values |
|---|---|---|---|---|---|---|---|
| DO | Max | 7.5 | 8 | 6 | 8 | 10.5 | >0.05 |
| | Min | 4.5 | 4.5 | 2 | 2 | 3 | |
| | Mean | 6.38 ± 1.31 | 6.88 ± 1.68 | 3.25 ± 1.85 | 5.13 ± 3.23 | 6.95 ± 3.35 | |
| BOD | Max | 3.5 | 4.5 | 2.5 | 3 | 6.5 | >0.05 |
| | Min | 2.5 | 2.5 | 1 | 2 | 4 | |
| | Mean | 2.88 ± 0.48 | 3.63 ± 0.85 | 1.75 ± 0.65 | 2.38 ± 0.47 | 4.74 ± 0.95 | |
| PH | Max | 8.2 | 8.3 | 8.3 | 7.5 | 8.7 | >0.05 |
| | Min | 7.5 | 8 | 8 | 7.2 | 7.2 | |
| | Mean | 8.03 ± 0.03 | 8.13 ± 0.13 | 8.10 ± 0.10 | 7.28 ± 0.12 | 7.98 ± 0.61 | |
| TEM. | Max | 22 | 22 | 25 | 24 | 21 | >0.05 |
| | Min | 10 | 8 | 10 | 9 | 15 | |
| | Mean | 16.50 ± 5.20 | 16 ± 6.20 | 17.50 ± 6.24 | 16.75 ± 6.34 | 18.20 ± 2.64 | |
| SAL. | Max | 40 | 6 | 17 | 50 | 8 | < 0.05 |
| | Min | 6 | 5 | 6 | 10 | 2 | |
| | Mean | 23 ± 14 | 5.25 ± 0.5 | 10.5 ± 4.8 | 18.75 ± 10.24 | 3.68 ± 2.97 | |
| NO3 | Max | 4.5 | 8.5 | 5 | 5.5 | 8 | >0.05 |
| | Min | 3 | 4 | 4.5 | 4 | 4.5 | |
| | Mean | 4 ± 0.71 | 5.75 ± 1.43 | 4.63 ± 0.48 | 4.88 ± 1.15 | 6.79 ± 1.32 | |

| | | | | | | | |
|---|---|---|---|---|---|---|---|
| **PO4** | Max | 0.7 | 0.6 | 0.6 | 0.8 | 1.2 | >0.05 |
| | Min | 0.2 | 0.3 | 0.1 | 0.2 | 0.2 | |
| | Mean | 0.5 ± 0.22 | 0.45 ± 0.13 | 0.5 ± 0.39 | 0.58 ± 0.26 | 0.65 ± 0.24 | |
| **DEEP** | Max | 2.2 | 2.8 | 2.4 | 2.4 | 2 | >0.05 |
| | Min | 2 | 2.1 | 1.4 | 1.4 | 1.4 | |
| | Mean | 2.08 ± 0.1 | 2.45 ± 0.29 | 1.9 ± 0.44 | 1.93 ± 0.5 | 1.73 ± 0.32 | |

**TABLE 3:** Diversity of fish species in different station from Shadegan Wetland.

| **Variable** | **Kanal** | **Atish** | **Rogbe** | **Mahshar** | **Khorosy** | ***P-values*** |
|---|---|---|---|---|---|---|
| **MARGALEF'S** | 2.03 | 2.71 | 2.31 | 2.63 | 2 | >0.05 |
| **SHANNON-WEINNER** | 1.7 | 2.04 | 1.61 | 2.39 | 1 .95 | < 0.05 |
| **EVENNESS** | 0.82 | 0.93 | 0.85 | 0.93 | 0.94 | >0.05 |
| **SIMPSON'S** | 0.25 | 0.15 | 0.16 | 0.1 | 0.4 | >0.05 |

**TABLE 4:** Diversity of fish species in different season from Shadegan Wetland.

| **Variable** | **Spring** | **Summer** | **Autumn** | **Winter** | ***P-values*** |
|---|---|---|---|---|---|
| **MARGALEF'S** | 2.62 | 2.09 | 2.11 | 2.39 | >0.05 |
| **SHANNON-WEINNER** | 2.21 | 2.06 | 2 | 2.02 | >0.05 |
| **EVENNESS** | 0.86 | 0.94 | 0.96 | 0.97 | >0.05 |
| **SIMPSON'S** | 0.14 | 0.15 | 0.14 | 0.13 | >0.05 |

In different station, Maximum and minimum Shannon-Weinner values were 2.04 (Atish) and 1.61 (Rogbe), respectively. The Shannon- Weinner value differ significantly between different station (ANOVA, $P<0.05$) and Margalef's, Evenness, Simpson's index didn't differ significantly between different station (ANOVA, $P>0.05$). Maximum and minimum Shannon-Weinner values were 2.21 (spring) and 2 (autumn), respectively. A comparison of value Shannon-Weinner, Margalef's, Evenness and Simpson's didn't differ significantly between different season (ANOVA, $P>0.05$). The correlation analysis showed that mean water temperature value was related significantly to deep water parameters used in this study; it related strongly and negatively ($r2= -0.89$, $P<0.05$).

**WFI AND CCA**

The WFI calculated for the entire wetland using total abundance was 2.99. These values fell between a possible range of 1-5, where 1 is most degraded and 5 is least degraded. An ordination of the 26 fish species with eight environmental variables produced significant correlations between species and variables associated with environmental degradation for AB data (**TABLE 1**). The position of a species on the CCA biplot is a reflection of the environmental conditions where it was found. The first axis of the CCA was strongly correlated with environmental conditions, where the positive end of CCA axis 1 was associated with species normally found in degraded conditions (e.g., BOD, DO) while the negative end was associated with species that are intolerant of water-quality impairment. The position of a species on the CCA biplot is a reflection of the environmental conditions where it was found. This location can be interpreted as representing the species' affinity for degraded vs. unimpacted habitat. It seems, distribution of *C. carasus, L. abu, C. carpio, C. luteus, H. molitrix, M. sharpeyi, species are associated with temperature, deep,* $NO_3$, PH and also *S. triostegus, A. vorax, A lutus; M. mastacembuls, C. idella* species with salinity and $PO_4$ respectively (**FIGURE 5**).

# DISCUSSION

The native marshland fish populations were originally dominated by Cyprinid fish of the genus Barbus. Hashemi et al. [21] reported that *C. carpio* was the most abundant species in their study in shadegan wetland. Overall, *B. luteus, M. sharpeyii, C. carpio, C. carasus, A. vorax* and *L. abu* are included over 70% biomass and fish main species of Shadegan wetland species [21]. The dominance of cyprinids in tropical reservoirs has been observed in Sri Lankan reservoirs, where the family formed over 50% of the species present [22]. Abundance of fish populations in river, lake with river source and reservoirs widely changed from year to year and the relative frequency of different species is different in population. This change is affected by rainfall fluctuation and floods. The increasing area and flood flow time is improved spawning, growth and survived rate [23]. Bias associated with fishing gear types can greatly influence comparisons of aquatic habitats, especially when meaningful community information is desired for habitat restoration research [24].

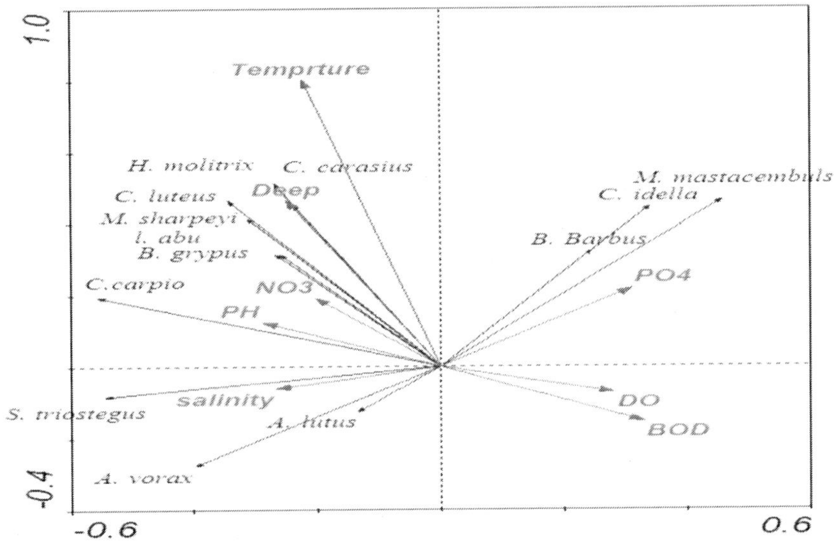


**FIGURE 5:** Ordination biplot of 26 fish species (see Table 1 for species numbers) from canonical Correspondence analysis (CCA) with 8 environmental variables in shadegan wetland.

The east stations in different seasons have high amount of Shannon-Weinner. It seems that entering the Jarahi River for east side of the wetland and entering of nutrition element was the reason of phytoplankton increasing and phytobentoz production [25] that caused Shannon-Weinner values increase in these areas. The south station have high amount of Shannon-Weinner that caused entering marine fish species in different seasons to wetland (especial summer and autumn seasons [13]. Six species of Cyprinidae (*C. carpio, H. molitrix, H. nobilis, C. idella, B, grypus, M. sharpeyii*) were introduced to Shadegan wetland in 2009 to 2012 by the Department of Fisheries Iranian as a potential for fishing in this wetland. This was done under one of the regulation and conservation programs towards maintaining and enhancing the fishery resources in this local. It seems, with induces six species of Cyprinid, abundance fish has increased. This clearly showed that the species diversity indices values using Shannon-Weaver index in this study is rather higher when compared to the values obtained from past study in this local (H=1.7-2, [26] Zakaria et al. found that species richness, species diversity and species survival in aquatic habitats were affected by several environmental factors, such as the physic-chemistry of the water, topographical, hydrological characteristics and habitat destruction.

The maximum CPUE obtained was in spring (inversely in Autumn season), it seems related to wetland climate status [13] and nutrients entering to river flow may be due to the season and also maximum phytoplankton production, wetland phytobentoz which was showed in spring time [25]. Managers commonly use catch per unit of effort (CPUE) methods for estimating population size and catchability coefficients in fish fisheries, because these parameters are key in the exploitation [27].

WFI scores was reported 2.64 in Shadegan wetland that near our finding (WFI=2.99). Interannual variation in WFI scores was not large and implies that the index will be a powerful tool for monitoring and for comparing scores calculated for wetlands sampled in different years. Fish are a suitable group of organisms for monitoring aquatic condition because they have a documented relationship with environmental perturbations (i.e., increased trophic status and degraded fish community). Variation in environmental condition cause changes in the fish community and those changes can then be extracted into ecological indicators to quantify environmental impacts [19]. For best results, we recommend consistent sampling gear and fishing effort be used when collecting fish data for use with the WFI. When WFI compared with other diversity indices (Shannon–Wiener's H′, Simpson's, evenness, and species richness), only the WFI varied as a linear function of WQI (Water Quality Index), demonstrating that the WFI is more useful than the others as an indicator of anthropogenic impact [20].

In the CCA ordination, axes 1 and 2 together explained a high percentage of variance of the species-environment biplot, with temperature, salinity, $NO_3$ and $PO_4$, as the most important variables influencing the variation of fish composition in the Shadegan wetland (**FIGURE 5**). It seems in multivariate indices; Do and BOD have low affect associated with species distribution. Thus, the fish assemblage of the freshwater-influenced habitat was characterized by the presence of numerous species that are tolerant to low-salinity conditions, and enter the system mainly for food and protection. The fish assemblage of the marine-influenced habitat was characterized by the presence of occasional and seasonal species. Multimetric indicators are especially vulnerable to bias because many of the metrics use community factors, such as percent of total catch from feeding guilds or percent of total biomass from specific species [28].

Finally, Fish are only one of many bioindicators in use today, but it is only a single tool and should be used in accord with other indices for greatest knowledge of wetland condition. Nevertheless, the Wetland Fish Index should prove to be an effective tool to aid in the management and protection of important fish habitat in Shadegan wetlands. This study provides basic information for fishery biologists in this local [29-35].

## ACKNOWLEDGEMENTS

We thank Dr.Maramazi, the manager of the South of Iran aquaculture fishery research center, Ahvaz. We are also very grateful the experts of the South of Iran aquaculture fishery research center, Ahvaz for helping the project work.

## REFERENCES

1. Galactos K, Barriga-Salazar R, Stewart DJ (2004) Seasonal and habitat influences on fish communities within the lower Yasuni River basin of the Ecuadorian Amazon. Environmental Biology of Fishes. 71: 33-51
2. MaceG, Masundire H, Baillie J, Ricketts T, Brooks T (2005) Biodiversity. In: Hassan R, Scholes R, Ash N (eds.), Ecosystems and Human Well- Being: Current State and Trends (Findings of the Condition and Trends Working Group). Island: 77-122.
3. Hamzah N (2007) Assessment on water quality and biodiversity within Sungai Batu Pahat. Master of thesis. Universiti Teknologi Malaysia: 124.
4. Zainudin MRY (2005) Assessment of fish community distribution and composition in the Perak River in order to determine biological indicators for freshwater health. Master of thesis. Universiti Sains Malaysia.
5. Hooper DU, Chapin FS, Ewel JJ, Hector A, Inchausti, P et al.,(2005)Effects of biodiversity on ecosystem functioning: a consensus of current knowledge. Ecol. Monogr. 75: 3-36.
6. Diaz S, Fargione J, Chapin III FS, Tilman D (2006) Biodiversity loss threatens human well-being. PLoS Biol4:1300-1305.
7. Haruna AB, Abubakar KA, Ladu BMB (2006) An assessment ofphysico-chemical parameters andproductivity status of Lake Geriyo,Yola, Adamawa State, Nigeria.Biological and EnvironmentalSciences Journal for the Tropics 3:18 23.
8. Mitsch WJ,JGGosselink (2000) The value of wetlands: Importance of scale and landscape setting. Ecological Economics 35:25-35.
9. Lotfi A, Ghafari H, BehroziradB, ASavari, K Kawosi (2003) Human activity and their affect in shadegan Wetland. ConselorEngining publisher: 382.
10. Ansari H,HashemiSAR, Eskandari GH (2009) Survey fishing status and Biomass fishin shadegan Wetland. The 1th Scientific Conference of Iranian Wetland, 3-4 March 2009, Ahwaz: 43
11. MaramaziGH (1997) Fish stock assessment in shadegan Wetland, South of Iran aquaculture fishery research center, Ahwaz.Iran: 57

12. Ansari H,Mohammadi GH (2001)Capture fishing status in shadegan Wetland. South of Iran aquaculture fishery research center,Ahwaz.Iran: 60.
13. HashemiS, Eskandary Gh,Ansary H, Yooneszadeh M (2011)Stock assessment and production of fish species in the Shadegan Wetland, Iran. World Journal of Fish and Marine Sciences 3:502-508.
14. Nelson JS (2006) Fishes of theWorld, 4th edition. John Wiley and Sons, Inc. Hoboken, New Jersey 601.
15. Coad BW (2012) The freshwater fishes of Iran. www.Briancoad.com.
16. Clesceri SL,AEGreenberg, RR Trussel (1989)Standard Method for the Examination of Water and Wastewater. 17th, Ecln., APHA Press.
17. May RM (1993) Patterns of species abundance and diversity. In: Cody JM (ed) Ecology and Evolution of CommunitiesCambridge, Mass.: Harvard University Press. Solow AR.. A simple test for change in community structure. J Journal of Animal Ecol, 62: 191-193.
18. Pielou EC (1966) The measurement of diversity in different types of biological collections. J. Theoret. Biol. 13: 131-144
19. Seilheimer TS, Chow-Fraser P (2006) Development and use of the wetland fish index to assess the quality of coastal wetlands in the Laurentian Great Lakes. Can J Fish Aquat Sci 63:354-366.
20. Chow-Fraser P, Lougheed VL, Crosbie B, LeThiec V, Simser L (1998) Long-term response of the biotic community to fluctuating water levels and changes in water quality in Cootes Paradise Marsh, a degraded coastal wetland of Lake Ontario. Wetlands Ecol Manag 6: 19-42.
21. Amarasinghe US (1992) Recent trends in the inland fishery of Sri Lanka. In: Baluyut EA (ed.) Indo-Pacific Fishery Commission, FAO Fishery Report No. 458 Supplement, Rome: 84-105
22. WelcommeR (2001) Inland fisheries ecology and management.Food and Agriculture Organization and Fishing News Books, Blackwell Science Ltd: 345
23. Hashemi S, Eskandary Gh,AnsaryH (2012) Biomass of Fish Species in the Shadegan Wetland, IRAN. Research Journal of Recent Sciences,1: 66-68.
24. Kholfenilsaz M (2009) Survey frequency and diversity planktonic in shadegan Wetland. Scientific Journal of Marine Biology, Islamic Azad university Ahvaz Branch, 1-12
25. Jackson DA, Harvey HH (1997) Qualitative and quantitative sampling of lake fish communities. Can. J Fish Aquat Sci 54:2807-2813.
26. Zakaria R, Mansor M, AliAB (1999) Swamp-riverine tropical fish population: A comparative study of two spatially isolated freshwater ecosystems in Peninsular Malaysia. Wetlands and Ecology Management 6: 261-268.
27. Simon TP (1999) Assessing the suitability and biological integrity of water resources using fish communities. Boca Raton, Florida: CRC Press.
28. ZhouSH, DichmontC, BurridgeCY, VenablesW, et al. (2007) Is catchability density-dependent for schooling prawns?. Fisheries Research:23-36.
29. Karr JR (2000) Health, integrity, and biological assessment: The importance of whole things. In: Pimentel D, Westra L, Noss RF (eds.) Ecological integrity: integrating environment, conservation, and health, 209-226.
30. Karr JR, Chu EW (1999) Restoring life in running waters: better biological monitoring. Washington, District of Columbia: Island Press.
31. Niemi GJ, McDonald ME (2004) Application of ecological indicators. Annu Rev Ecol Evol Syst 35:89-111.
32. Simon TP(1991) Development of index of biotic integrity expectations for the ecoregions of Indiana. I. Central Corn Belt Plain. Chicago, IL. U.S. Environmental Protection Agency.
33. Ter braakCJF, SmilauerP (1998) CANOCO reference manual and user's guide to CANOCO for windows: software for canonical community ordination. Version 4. Ithaca, New York: Microcomputer Power.

34. Verdonschot PFM (1995) Canonical correspondence analysis and related multivariate methods in aquatic ecology. Aquat Sci 57:255-289.

# CHAPTER 8

# Effect of Developmental Temperature on Swimming Performance of Zebrafish (Danio Rerio) Juveniles

**Dimitris G. Sfakianakis & Ioannis Leris & Maroudio Kentouri**

*Biology Department, University of Crete, P.O. Box 2208, Vasilika Vouton, 71409 Heraklion, Crete, Greece*

## ABSTRACT

It is widely known that water temperature affects the swimming capacity of fish. But the effect of the rearing temperature on the swimming ability of the fish at later stages, has not had similar attention. In this study, four populations of zebrafish, were reared in different water temperatures (22, 25, 28 and 31°C) and after being acclimatized in a common temperature (26.5°C) for over a month, they were subjected to swimming trials in order to evaluate the maximum relative critical velocity (RUcrit) in each case. Fish that were reared in 22°C showed statistically significant lower performance than the ones reared in 31°C (7.72±0.17 vs. 8.79±0.28, means ± S.E.). Possible explanations for the observed differentiation could be the effect of early life temperature on fish muscle ontogeny or on body shape.

**Keywords:** Critical velocity · Ucrit · RUcrit · Rearing · Plasticity

## INTRODUCTION

Swimming performance is a biological characteristic that has a very important role on fish survival (Jones et al. 1974), as it can strongly affect food capture, mating success and the avoidance of unfavourable situations (Drücker 1996). Moreover, it has been stated that the ability of locomotion is a main trait that defines to a great extent the Darwinian fitness (Reidy et al. 2000). Most of the environmental factors that affect swimming capacity have already been studied thoroughly and in a wide range of species. Among them, temperature (Fuiman and Batty 1997; Beamish 1978; Wieser and Kaufmann 1998; Ojanguren and Brana 2000; Dickson et al. 2002; Koumoundouros et al. 2002a, b; Wilson et al. 2002; Franklin et al. 2003), oxygen (Beamish 1978; Hammer 1995; Korsmeyer et al. 1996; Steffensen and Farrell 1998) and carbon dioxide concentration (Dahlberg et al. 1968; Beamish 1978), pollutants (Howard 1975; Beamish 1978; Beaumont et al. 2000; Randall and Tsui 2002; Wicks et al. 2002), salinity (Beamish 1978), light and photoperiod (Beamish 1978; Hammer 1995; Young et al. 2004) seem to play a particularly important role.

Temperature itself seems to be the most important environmental factor that affects the life of fish. Besides swimming performance, it has been shown to affect growth, metabolism, time of hatching (Herzig and Winkler 1986), development (Herzig and Winkler 1986; Fukuhara 1990; Polo et al. 1991; Gibson and Johnston 1995; Lein et al. 1997), yolk absorption (Fukuhara 1990), muscle ontogeny and development (Johnston 1981, 1993, 2006; Wilkes et al. 2001; Johnston et al. 2009), ontogeny of internal organs (Fukuhara 1990; Gibson and Johnston 1995), external morphology (Lindsey 1988; Wimberger 1992; Tudela 1999; Koumoundouros et al. 2001a; Pakkasmaa and Piironen 2001; Cabral et al. 2003; Silva 2003; Turan 2004), meristic characters (Turan 2004), appearance of skeletal deformities (Polo et al. 1991; Koumoundouros et al. 2001b), sex determination (Baroiller et al. 1999; Pavlidis et al. 2000; Koumoundouros et al. 2002a) and the overall survival (Fukuhara 1990; Lein et al. 1997) and lifespan of fish (Malek et al. 2004). Although the effect of swimming temperature on the performance of fish has been studied thoroughly and in a wide range of species, there are only a few studies focusing on the impact of rearing temperature on the swimming performance of adult fish (i.e. Koumoundouros et al. 2009).

Zebrafish is a small, colourful, tropical fish inhabiting small ponds and slow moving streams in Southeast Asia (Talwar and Jhingran 1991). For the past few decades, it has been a very important model-organism in research fields such as Genetics, Neurophysiology, Developmental Biology and Biomedicine (Amsterdam and Hopkins 2006). Numerous laboratories worldwide are using zebrafish for fundamental or applied research and there is an increasing interest for its use in new research fields (Gerlai 2003). In spite of its popularity as a research tool, only recently did some integrated studies about its biological and ecological characteristics emerge (Engeszer et al. 2007; Spence et al. 2008).

Concerning the temperature range in which zebrafish live and reproduce in the environment, more than one suggestion exists. Froese and Pauly (2010) argue that zebrafish's natural habitat temperature ranges from 18°C to 24°C, whereas Engeszer et al. (2007) report that the observed temperature range for zebrafish in the wild lies between 24.6°C and 38.6°C. Spence et al. (2008) on the other hand, mention that temperature in D. rerio's habitats ranges from 6°C in the winter to 38°C in the summer. Our personal observations throughout many rearing trials showed that the rearing temperatures of 22°C and 31°C are the lower and upper limit values for successful rearing in the laboratory. At lower or higher – respectively – temperatures, the risk of unsuccessful hatching, mass larvae mortality and severe skeletal malformations throughout the populations, is extremely high.

The purpose of this article is to study the effect of four different rearing temperatures (ranging from 22°C to 31°C) on the swimming performance of adult zebrafish.

## MATERIALS AND METHODS

A single batch of eggs of D. rerio was obtained from wild type broodstock (ZF WT2 F5, Wageningen Agricultural University, The Netherlands) that was kept in a 30 l tank at 28±0.5°C and fed three times per day with industrial dry food in flakes (Sera Vipan, Germany) and three times per week with Artemia sp. nauplii (Instar I). The eggs were collected approximately 2 h after fertilization, examined under the stereoscope (Olympus, SZX9), selected in terms of quality and submerged for 12–15 min in hydrogen peroxide (3 ml $l^{-1}$) for antifungal protection. Afterwards (approximately 3 h after fertilization), batches of 200 eggs were transferred into hatching devices that were placed in four rearing tanks of 130 l of different water temperatures (22, 25, 28 and 31°C). The regulation of temperature was achieved with the use of electrical thermostat heaters (Aquarium Systems, Visi-Therm, 100–150 W) and coolant device, when needed. The rearing was performed in duplicate (A and B).

Hatching occurred at 2 (28 and 31°C), 3 (25°C), or 6 days (22°C) after spawning. Larvae were fed four times per day, initially (at 2 days post hatching, dph) with Paramecium sp. (Blades Biological CO, UK) and later (at 33 dph) with newly hatched Artemia sp. nauplii (Instar I) (Westerfield 1995). Approximately at 73 dph, industrial dry food in flakes was introduced, which consisted the main diet of the fish throughout the acclimation and experimental periods. With every food-type change, a transitional period of three days was applied, at which both food types were provided. At approximately 143 dph, the acclimation – to the common temperature of 26.5°C – period started and lasted 1.5 month, until the start of the swimming exercises. Swimming performance trials were carried out in an apparatus described by Koumoundouros et al. (2002b). Water velocity was calibrated by means of an electromagnetic flow-meter (Valeport, Model 801). Fish forward escape was prevented using a mesh screen located at the upstream end of the swimming channel. A homogenous current velocity throughout the whole depth and length of the channel was verified by means of dye injected into the swimming channel. Temperature was maintained constant at 26.5°C using a thermostat heater, while the oxygen saturation was >99%.

For the critical velocity ($U_{crit}$) measurements to be valid, the speed tests should be conducted based on the Total Length (TL) of the specimens used (Brett 1964). Therefore, the swimming performance trials were carried out not in one but at 3 different swimming speeds in order to accurately cover different fish sizes. The swimming speeds (and the velocity increments) used to test the fish were - according to their size – the following: 30, 32.5 and 35 mm $sec^{-1}$. The TL of the fish was assessed visually before the trial to avoid excess stress from the unnecessary fish handling (Ramsay et al. 2009). If the actual TL (as measured at the end of the experiments on the digital photographs of the anaesthetized fish) differed from the estimated one and therefore from the applied speed and the velocity increment more than 1.25 mm, then the data concerning that fish were discarded. Only male fish were used for the trials, so as to avoid possible sex-dependent differences (Williams and Brett 1987). Fish were starved for 24 h prior to the swimming trials. Fish were also macroscopically examined prior to the trials for any visible morpho-anatomical malformations. Before the trial, fish were acclimated for 15 min to the holding tank's conditions and then one fish was placed in the swimming tunnel for 10 min at a water velocity of 2 TL $sec^{-1}$. Water velocity increased every 10 min at a rate of 1 TL $sec^{-1}$ until the fish fatigued. Fatigue was reached when a fish was carried away by the current, without been able to hold its position in the swimming tunnel or react to visual and acoustic stimuli.

Since body mass can affect metabolism (Hochachka 1987) and body length can affect swimming power output, we assessed both in all groups after treatments (McClelland et al. 2006). To do so, fish were anaesthetized (ethylenglycol-monophenylether, Merck, 0.2–0.3 ml $l^{-1}$), individually photographed using a digital camera (Olympus, Camedia C3030 Zoom) and weighed (M, 0.001 g), while the maximum body width (BW) was measured under a stereoscope (0.01 mm). Landmarks were also placed to the obtained pictures using an image processing software (tpsDig2, Rohlf, version 5.0.3.32), for the accurate measurement of the TL, maximum body depth ($BD_{max}$), body depth at the anus (BD) and caudal peduncle depth (CpD). Moreover, selective double staining of cartilage and bone tissue was performed on the specimens after the trials and the stained samples were examined under a stereoscope for skeletal deformities (Divanach et al. 1997; Koumoundouros et al. 1997a, b, 2001b). The abnormal fish were discarded from the analysis.

Critical swimming speed was calculated using the equation introduced by Brett (1964): $U_{crit} = U_i + U_{ii}\left(\frac{T_i}{T_{ii}}\right)$

where $U_i$ is the highest velocity maintained for the whole 10 min (mm $sec^{-1}$), $U_{ii}$ is the velocity increment (30, 32.5 and 35 mm $sec^{-1}$), Ti is the time elapsed at fatigue velocity and $T_{ii}$ is the time between velocity increments (10 min). To compare the swimming performance of many individuals that varied in length, the $U_{crit}$ had to be transformed to a new value that would be independent of the factor TL. So, after calculating the $U_{crit}$, the relative critical swimming speed ($RU_{crit}$) for every fish was calculated using this equation: $RU_{crit} = \frac{U_{crit}}{TL}$ TL (Beamish 1978)

As the assumption of homogeneity of variances was not fulfilled, the Mann-Whitney U-test was used for the comparison of swimming performance between the four different temperatures and the 2 replicates, after the appropriate correction for tied measurements (Sokal and Rohlf 1995). The non parametric Kruskal-Wallis was used to test the effect of temperature on RUcrit (Sokal and Rohlf 1995). All experimental procedures were in accordance with the European Communities Council directive (86/609/EEC) for the care and use of laboratory animals and approved by the University of Crete research committee.

# RESULTS

The relative critical velocities measured for the individuals of each duplicate of the four temperature groups are presented in Table 1. Statistical analysis showed a significant influence of the factor, rearing temperature, on the swimming performance of the juveniles ($p<0.05$, Kruskal-Wallis). At all temperature regimes, by means of Mann-Whitney U-test, no significant differences were revealed between the two duplicates.

Fish raised at 31°C have achieved the higher swimming performances in terms of $RU_{crit}$, while the ones reared in 22°C had the worst performance ($p< 0.05$, Mann-Whitney). Fish cultured in 25 and 28°C achieved intermediate critical swimming speeds although not significantly different from the others (Table 1).

**Table 1** RUcrit (SE, standard error) for each duplicate (A and B) and for each temperature group (22, 25, 28 and 31°C)

| Temperature | Dup. | $RU_{crit}$ (TL $sec^{-1}$) | SE | N | Average $RU_{crit}$ | SE | $p$ |
|---|---|---|---|---|---|---|---|
| 22°C | A | 7.68 | 0.17 | 10 | 7.72 | 0.17 | a |
| | B | 7.76 | 0.30 | 11 | | | |
| 25°C | A | 8.05 | 0.42 | 8 | 8.13 | 0.27 | ..... |
| | B | 8.20 | 0.36 | 9 | | | |
| 28°C | A | 8.23 | 0.33 | 9 | 8.36 | 0.28 | ..... |
| | B | 8.46 | 0.44 | 11 | | | |
| 31°C | A | 8.69 | 0.37 | 10 | 8.79 | 0.28 | a |
| | B | 8.89 | 0.44 | 10 | | | |

a $p<0.05$ (Mann-Whitney U-test)

The measurements of the morphometric characters performed after the swimming trials as well as the ones that derived from the morphometric analysis of the landmarks placed on the specimen photographs are presented in Table 2.

For those characters, the performed Mann-Whitney U-tests revealed no statistical significant differences between the fish raised at the different temperature regimes ($p>0.05$).

## DISCUSSION

Fish are known to react to their environment's fluctuations in all possible ways and it is an evolutionary imperative to try and adjust to these changes. Zebrafish, although considered to be a tropical fish, inhabits ecosystems with "monsoon climate", that present an extraordinary seasonal variation of temperatures that can range from as low as 6°C in winter to over 38°C in summer (Spence et al. 2008). With the phenomenon of global warming being more and more imminent, the effects that such a wide range of temperatures can have on many important processes and aspects of the fish life are certainly worth of extensive study.

Developmental temperature plays an important role on the subsequent life of fish. On the present study the role of early rearing temperature on the swimming capacity of zebrafish juveniles was examined. The results showed that fish reared at 31°C have a higher swimming ability (8.79 TL $sec^{-1}$) than fish reared at 22°C (7.72 TL $sec^{-1}$). Fish raised at the two intermediate temperatures (25 and 28°C), although not statistically different from either one of the extreme temperatures, they seem to perform better than the fish of 22°C and worse than the fish of 31°C (8.13 and 8.36 TL $sec^{-1}$ respectively).

Swimming ability of zebrafish has been studied before in the case of larval (i.e. Budick and O'Malley 2000; Levin et al. 2004) and adult specimens (Plaut and Gordon 1994; Plaut 2000; McClelland et al. 2006; Widmer et al. 2006) but never in respect to developmental temperature. Plaut (2000) studied the effect of the fin size on the swimming performance of zebrafish, Widmer et al. (2006) studied the effect of hypoxia versus normoxia and Plaut and Gordon (1994) compared wild-type and cloned zebrafish. In the present study, the maximum relative critical swimming speed was estimated from 7.7 to 8.8 TL $sec^{-1}$ which is in agreement with the results of Widmer et al. (2006), who reported it between 8.5 and 9.5 SL $sec^{-1}$, as this value is lowered if converted to TL $sec^{-1}$. Plaut (2000) measured RUcrit values around 12.6 TL $sec^{-1}$ (after the appropriate adjustment from SL to TL), and Plaut and Gordon (1994) were unable to exhaust their wild-type specimens, but conclude that they swam more than 13 SL $sec^{-1}$. Similar results were described also by McClelland et al. (2006). The presented differences in $RU_{crit}$ values between these studies and also the present one, can be explained by either the different measuring techniques and apparatus used, or the higher swimming temperature maintained by the authors during the trials (28°C).

**Table 2** Morphometric characters (SD standard deviation). TL total length, W weight, BW body width, BDmax maximum body depth, BD body depth (at anus), CpD caudal peduncle depth

| Temp. | TL (mm) | SD | W (g) | SD | BW (mm) | SD | BDmax (mm) | SD | BD (mm) | SD | Cpd (mm) | SD |
|---|---|---|---|---|---|---|---|---|---|---|---|---|
| 22°C | 32.66 | 1.60 | 0.349 | 0.04 | 4.773 | 0.32 | 6.736 | 0.36 | 5.695 | 0.37 | 3.011 | 0.19 |
| 25°C | 32.29 | 2.51 | 0.335 | 0.07 | 4.744 | 0.43 | 6.493 | 0.57 | 5.554 | 0.58 | 3.004 | 0.30 |
| 28°C | 32.71 | 1.82 | 0.347 | 0.05 | 4.882 | 0.24 | 6.770 | 0.45 | 5.789 | 0.44 | 2.961 | 0.23 |
| 31°C | 33.11 | 2.06 | 0.357 | 0.07 | 4.816 | 0.48 | 6.663 | 0.54 | 5.666 | 0.47 | 2.933 | 0.24 |

There have been of course other studies on the effect of rearing temperature on swimming performance of other – mostly farmed marine – species. For example, Koumoundouros et al. (2009) studied the effect of two rearing temperatures on the swimming performance of European sea bass juveniles. In this study, the authors argued that fish raised in lower temperature (15°C) performed better than the fish raised in higher temperature (20°C). This is in contrast with our present results (where fish raised in higher temperature ultimately performed better than the others) but both studies suggest that the rearing temperature had a profound effect on the swimming capacity of fish at later stages.

There are two possible explanations for the observed effect. The influence of developmental temperature on the muscle ontogeny of the fish which has long been established (Johnston 1981, 1993, 2006; Johnston et al. 2009; Koumoundouros et al. 2009) appears to have a substantial effect on swimming performance. Different temperatures lead to variation in body muscle mass (in terms of size and number of muscle fibres) that directly affect swimming ability. Koumoundouros et al. (2009), showed that in the case of European sea bass, the best swimmers had relatively higher red muscle to white muscle ratio and more red muscle fibres than the others. In the case of zebrafish, Johnston et al. (2009) showed that the optimal embryonic temperature for fast muscle fibre recruitment (hyperplasia) is 26°C, as in that temperature there were 18.8% more fast fibers than at 22°C and 13.7% more fibers than at 31°C. After this phase, myotube formation stops and a new phase begins (hypertrophy) that consists of nuclear accretion and increase in fibre length and diameter (Johnston et al. 2004). Even though higher developmental temperature does not lead to increased number of fast muscle fibres, it is still possible that it affects the second phase of muscle formation, resulting in longer and thicker muscle fibres.

The second is that temperature in early life affects body shape (Georgakopoulou et al. 2007; personal unpublished results) and meristic count (Blaxter 1991; Georgakopoulou et al. 2007; personal unpublished results). Georgakopoulou et al. (2007), concluded that European sea bass juveniles reared in 15°C had more slender bodies than the ones reared in 20°C, a fact that could explain the higher swimming performance of the former as observed by Koumoundouros et al. (2009).

In the present study, we tried to investigate whether there is a direct correlation between a single morphometric character and $RU_{crit}$. The fact that the examined morphometric characters do not significantly vary among the four temperature groups leads us to conclude that there is no such correlation between these size-related values and the observed differentiation of swimming performance. On the other hand, unpublished results of our team indicate that ebrafish grown in different temperatures tend to differ in both body shape and meristic character count, which is in agreement with the results of Georgakopoulou et al. (2007). It is logical to assume therefore that it is a combination of those characters (body shape in general) and it's variation among the different temperature regimes rather than the effect of a single one that eventually influences the swimming ability of a fish.

## ACKNOWLEDGEMENTS

The authors thank Maria Kampitaki for her valuable assistance during the experimental fish rearing and swimming trials. The present study was financed by the European Social Fund and National Resources (EPEAEK II-PYTHAGORAS I) to M.K.


## REFERENCES


1. Amsterdam A, Hopkins N (2006) Mutagenesis strategies in zebrafish for identifying genes involved in development and disease. Trends Genet 22:473–478
2. Baroiller JF, Guiguen Y, Fostier A (1999) Endocrine and environmental aspects of sex differentiation in fish. Cell Mol Life Sci 55:910–931
3. Beamish FWH (1978) Swimming capacity. In: Hoar WS, Randall DJ (eds) Fish physiology, vol 7. Academic, New York, pp 101–187
4. Beaumont MW, Butler PJ, Taylor EW (2000) Exposure of brown trout, Salmo trutta, to a sub-lethal concentration of copper in soft acidic water: Effects upon muscle metabolism and membrane potential. Aquat Toxicol 51(2):259–272
5. Blaxter JHS (1991) The effect of temperature on larval fishes. Neth J Zool 42:336–357 Brett JR (1964) The respiratory metabolism and swimming performance of young sockeye salmon. J Fish Res Board Can 21:1183–1226
6. Budick SA, O'Malley DM (2000) Locomotor repertoire of the larval zebrafish: swimming, turning and prey capture. J Exp Biol 203:2565
7. Cabral HN, Marques JF, Rego AL, Catarino AI, Figueiredo J, Garcia J (2003) Genetic and morphological variation of Synaptura lusitanica Capello, 1868, along the Portuguese coast. J Sea Res 50:167–175
8. Dahlberg ML, Shumway DL, Doudoroff P (1968) Influence of dissolved oxygen and carbon dioxide on swimming performance of largemouth bass and Coho salmon. J Fish Res Board Can 25:49–70
9. Dickson KA, Donley JM, Sepulveda C, Bhoopat L (2002) Effects of temperature on sustained swimming performance and swimming kinematics of the chub mackerel Scomber japonicus. J Exp Biol 205:969 Divanach P, Papandroulakis N, Anastasiadis P, Koumoundouros G, Kentouri M (1997) Effect of water currents on the development of skeletal deformities in sea bass (Dicentrarchus labrax L.) with functional swimbladder during postlarval and nursery phase. Aquaculture 156:145–155
10. Drücker EG (1996) The use of gait transition speed in comparative studies of fish locomotion. Integr Comp Biol 36:555
11. Engeszer RE, Patterson LB, Rao AA, Parichy DM (2007) Zebrafish in the wild: a review of natural history and new notes from the field. Zebrafish 4:21–40
12. Franklin CE, Wilson RS, Davison W (2003) Locomotion at -1.01 c: Burst swimming performance of five species of Antarctic fish. J Therm Biol 28:59–65
13. Froese R, Pauly D (eds) (2010) FishBase: World Wide Web electronic publication. http://fishbase.org/Summary/Species Summary.php?ID=4653. Accessed at 16 June 2010
14. Fuiman L, Batty R (1997) What a drag it is getting cold: partitioning the physical and physiological effects of temperature on fish swimming. J Exp Biol 200:1745
15. Fukuhara O (1990) Effects of temperature on yolk utilization, initial growth, and behaviour of unfed marine fish-larvae. Mar Biol 106:169–174

16. Georgakopoulou E, Sfakianakis DG, Kouttouki S, Divanach P, Kentouri M, Koumoundouros G (2007) The influence of temperature during early life on phenotypic expression at later ontogenetic stages in sea bass. J Fish Biol 70:278–291
17. Gerlai R (2003) Zebra fish: an uncharted behavior genetic model. Behav Genet 33:461–468
18. Gibson S, Johnston IA (1995) Temperature and development in larvae of the turbot Scophthalmus maximus. Mar Biol 124:17–25
19. Hammer C (1995) Fatigue and exercise tests with fish. Comp Biochem Physiol A 112:1–20
20. Herzig A, Winkler H (1986) The influence of temperature on the embryonic development of three cyprinid fishes, Abramis brama, Chalcalburnus chalcoides mento and Vimba vimba. J Fish Biol 28:171–181
21. Hochachka PW (1987) Limits: how fast and how slow muscle metabolism can go. Adv Myochem 1:3–12
22. Howard TE (1975) Swimming performance of juvenile Coho salmon (Oncorhynchus kisutch) exposed to bleached kraft pulpmill effluent. J Fish Res Board Can 32:789–793
23. Johnston IA (1981) Structure and function of fish muscles. Symp Zool Soc Lond 48:71–113
24. Johnston IA (1993) Temperature influences muscle differentiation and the relative timing of organogenesis in herring (Clupea harengus) larvae. Mar Biol 116:363–379
25. Johnston IA (2006) Environment and plasticity of myogenesis in teleost fish. J Exp Biol 209:2249
26. Johnston I, Abercromby M, Vieira V, Sigursteindottir R, Kristjansson B, Sibthorpe D, Skulason S (2004) Rapid evolution of muscle fibre number in post-glacial populations of arctic charr Salvelinus alpinus. J Exp Biol 207:4343
27. Johnston IA, Lee HT,Macqueen DJ, Paranthaman K, Kawashima C, Anwar A, Kinghorn JR, Dalmay T (2009) Embryonic temperature affects muscle fibre recruitment in adult zebrafish: Genome-wide changes in gene and microRNA expression associated with the transition from hyperplastic to hypertrophic growth phenotypes. J Exp Biol 212:1781
28. Jones DR, Kiceniuk JW, Bamford OS (1974) Evaluation of the swimming performance of several fish species from the Mackenzie river. J Fish Res Board Can 31:1641–1647
29. Korsmeyer KE, Dewar H, Lai NC, Graham JB (1996) Tuna aerobic swimming performance: physiological and environmental limits based on oxygen supply and demand. Comp Biochem Physiol B 113:45–56
30. Koumoundouros G, Gagliardi F, Divanach P, Boglione C, Cataudella S, Kentouri M (1997a) Normal and abnormal osteological development of caudal fin in Sparus aurata L. fry. Aquaculture 149:215–226
31. Koumoundouros G, Oran G, Divanach P, Stefanakis S, Kentouri M (1997b) The opercular complex deformity in intensive gilthead sea bream (Sparus aurata l.) larviculture. Moment of apparition and description. Aquaculture 156:165–177
32. Koumoundouros G, Divanach P, Anezaki L, Kentouri M (2001a) Temperature-induced ontogenetic plasticity in sea bass (Dicentrarchus labrax). Mar Biol 139:817–830
33. Koumoundouros G, Divanach P, Kentouri M (2001b) The effect of rearing conditions on development of saddleback syndrome and caudal fin deformities in Dentex dentex (L.). Aquaculture 200:285–304
34. Koumoundouros G, Pavlidis M, Anezaki L, Kokkari C, Sterioti A, Divanach P, Kentouri M (2002a) Temperature sex determination in the European sea bass, Dicentrarchus labrax (L., 1758) (Teleostei, Perciformes, Moronidae): critical sensitive ontogenetic phase. J Exp Zool 292:573–579
35. Koumoundouros G, Sfakianakis D, Divanach P, Kentouri M (2002b) Effect of temperature on swimming performance of sea bass juveniles. J Fish Biol 60:923–932

36. Koumoundouros G, Ashton C, Sfakianakis D, Divanach P, Kentouri M, Anthwal N, Stickland N (2009) Thermally induced phenotypic plasticity of swimming performance in European sea bass Dicentrarchus labrax juveniles. J Fish Biol 74:1309–1322
37. Lein I, Holmefjord I, Rye M (1997) Effects of temperature on yolk sac larvae of Atlantic halibut (Hippoglossus hippoglossus L.). Aquaculture 157:123–135
38. Levin ED, Swain HA, Donerly S, Linney E (2004) Developmental chlorpyrifos effects on hatchling zebrafish swimming behavior. Neurotoxicol Teratol 26:719–723
39. Lindsey CC (1988) Factors controlling meristic variation. In: Hoar WS, Randall DJ (eds) Fish physiology, vol 11. Academic, New York, pp 197–274
40. Malek RL, Sajadi H, Abraham J, Grundy MA, Gerhard GS (2004) The effects of temperature reduction on gene expression and oxidative stress in skeletal muscle from adult zebrafish. Comp Biochem Physiol C 138:363–373
41. McClelland GB, Craig PM, Dhekney K, Dipardo S (2006) Temperature- and exercise-induced gene expression and metabolic enzyme changes in skeletal muscle of adult zebrafish (Danio rerio). J Phys 577:739
42. Ojanguren AF, Brana F (2000) Thermal dependence of swimming endurance in juvenile brown trout. J Fish Biol 56:1342–1347
43. Pakkasmaa S, Piironen J (2001) Morphological differentiation among local trout (Salmo trutta) populations. Biol J Linn Soc 72:231–239
44. Pavlidis M, Koumoundouros G, Sterioti A, Somarakis S, Divanach P, Kentouri M (2000) Evidence of temperaturedependent sex determination in the European sea bass (Dicentrarchus labrax L.). J Exp Zool 287:225–232
45. Plaut I (2000) Effects of fin size on swimming performance, swimming behaviour and routine activity of zebrafish Danio rerio. J Exp Biol 203:813
46. Plaut I, Gordon M (1994) Swimming metabolism of wild-type and cloned zebrafish Brachydanio rerio. J Exp Biol 194:209
47. Polo A, Yufera M, Pascual E (1991) Effects of temperature on egg and larval development of Sparus aurata L. Aquaculture 92:367–375
48. Ramsay J, Feist G, Varga Z, Westerfield M, Kent M, Schreck C (2009) Whole-body cortisol response of zebrafish to acute net handling stress. Aquaculture 297:157–162
49. Randall DJ, Tsui TKN (2002) Ammonia toxicity in fish. Mar Pollut Bull 45:17–23
50. Reidy SP, Kerr SR, Nelson JA (2000) Aerobic and anaerobic swimming performance of individual Atlantic cod. J Exp Biol 203:347–357
51. Silva A (2003) Morphometric variation among sardine (Sardina pilchardus) populations from the northeastern Atlantic and the western Mediterranean. ICES J Mar Sci 60:1352
52. Sokal RS, Rohlf FJ (1995) Biometry: the principles and practice of statistics in biological research. WH Freeman and Co., New York Spence R, Gerlach G, Lawrence C, Smith C (2008) The behaviour and ecology of the zebrafish, Danio rerio. Biol Rev Camb Philos Soc 83(1):13–34
53. Steffensen J, Farrell AP (1998) Swimming performance, venous oxygen tension and cardiac performance of coronary-ligated rainbow trout, Oncorhynchus mykiss, exposed to progressive hypoxia. Comp Biochem Physiol A 119:585–592
54. Talwar PK, Jhingran AG (1991) Inland fishes of India and adjacent countries, vol 1. Oxford & IBH Pub. Co., Rotterdam Tudela S (1999) Morphological variability in a Mediterranean, genetically homogeneous population of the European anchovy, Engraulis encrasicolus. Fish Res 42:229–243
55. Turan C (2004) Stock identification of Mediterranean horse mackerel (Trachurus mediterraneus) using morphometric and meristic characters. ICES J Mar Sci 61:774
56. Westerfield M (1995) The zebrafish book: a guide for the laboratory use of zebrafish (Danio rerio). University of Oregon Press, Eugene Wicks BJ, Joensen R, Tang Q, Randall DJ (2002) Swimming and ammonia toxicity in salmonids: The effect of sub lethal ammonia exposure on the swimming performance of Coho salmon and the

acute toxicity of ammonia in swimming and resting rainbow trout. Aquat Toxicol 59:55–69
57. Widmer S, Moore FBG, Bagatto B (2006) The effects of chronic developmental hypoxia on swimming performance in zebrafish. J Fish Biol 69:1885–1891
58. Wieser W, Kaufmann R (1998) A note on interactions between temperature, viscosity, body size and swimming energetic in fish larvae. J Exp Biol 201:1369
59. Wilkes D, Xie SQ, Stickland NC, Alami-Durante H, Kentouri M, Sterioti A, Koumoundouros G, Fauconneau B, Goldspink G (2001) Temperature and myogenic factor transcript levels during early development determines muscle growth potential in rainbow trout (Oncorhynchus mykiss) and sea bass (Dicentrarchus labrax). J Exp Biol 204:2763–2771
60. Williams IV, Brett JR (1987) Critical swimming speed of Fraser and Thompson river pink salmon (Oncorhynchus gorbuscha). Can J Fish Aquat Sci 44:348–356
61. Wilson RS, Kuchel LJ, Franklin CE, Davison W (2002) Turning up the heat on subzero fish: thermal dependence of sustained swimming in an Antarctic notothenioid. J Therm Biol 27:381–386
62. Wimberger PH (1992) Plasticity of fish body shape. The effects of diet, development, family and age in two species of Geophagus (Pisces; Cichlidae). Biol J Linn Soc 45:197–218
63. Young PS, Swanson C, Cech JJ Jr (2004) Photophase and illumination effects on the swimming performance and behavior of five California estuarine fishes. Copeia 2004:479–487

# CHAPTER 9

# Testing the Influence of Environmental Heterogeneity on Fish Species Richness in Two Biogeographic Provinces

**Philippe Massicotte*, Rapha¨el Proulx, Gilbert Cabana and Marco A. Rodr´ıguez**

*Centre de Recherche sur les Interactions Bassins Versants-E´ cosyste`mes Aquatiques (RIVE),*

*Universit´e du Qu´ebec `a Trois-Rivi`eres, Trois-Rivi`eres, Canada*

## ABSTRACT

Environmental homogenization in coastal ecosystems impacted by human activities may be an important factor explaining the observed decline in fish species richness. We used fish community data (>200 species) from extensive surveys conducted in two biogeographic provinces (extent >1,000 km) in North America to quantify the relationship between fish species richness and local (grain <10 km$^2$) environmental heterogeneity. Our analyses are based on samples collected at nearly 800 stations over a period of five years. We demonstrate that fish species richness in coastal ecosystems is associated locally with the spatial heterogeneity of environmental variables but not with their magnitude. The observed effect of heterogeneity on species richness was substantially greater than that generated by simulations from a random placement model of community assembly, indicating that the observed relationship is unlikely to arise from veil or sampling effects. Our results suggest that restoring or actively protecting areas of high habitat heterogeneity may be of great importance for slowing current trends of decreasing biodiversity in coastal ecosystems.

**Keywords:** Aquatic community assembly, Conservation biology, Random placement model, Diversity, Heterogeneity hypothesis, Simulation

## INTRODUCTION

The environmental heterogeneity hypothesis (*MacArthur & MacArthur, 1961; MacArthur & Wilson, 1967; Ricklefs, 1977*) states that species richness increases with the number of ecological niches; that is, species coexistence is facilitated in more heterogeneous environment because different taxa can capitalize on different environmental conditions. The hypothesis has been tested using many taxonomic groups across different spatial grains (average distance among observations) and extents (size of the whole study area) ranging from meters to thousands of kilometers. An extensive meta-analysis by *Field et al. (2009*) found that environmental heterogeneity was the primary factor driving species richness for 63 of the 273 cases (23%) assessing the relative importance of environmental heterogeneity versus other environmental factors. Environmental heterogeneity, however, had a stronger effect on species richness in studies conducted at small grain sizes (39% of the cases), suggesting that the relationship is contingent on the spatial scale. Furthermore, only 4 of the 393 relationships (1%) were from surveys of aquatic ecosystems having small grain size (<10 $km^2$) and large geographical extent (>1,000 km). Thus, there appears to be no consensus on the effects of small-grain environmental heterogeneity on species richness when investigated over large geographical areas.

This paucity of broad-scale studies may be related to the difficulties faced by aquatic ecologists in quantifying heterogeneity across different temporal and spatial scales (*Kovalenko, Thomaz & Warfe, 2011; Yeager, Layman & Allgeier, 2011; Tisseuil et al., 2013*) possibly reflecting the difficulties of obtaining the data needs to quantify such relationship. As a consequence, the term 'heterogeneity' has been used rather loosely, as it could refer to habitat complexity, habitat diversity or environmental variability in both space and time (*Palmer, Menninger & Bernhardt, 2010*). For example, *Oberdorff et al. (2011*) assessed habitat heterogeneity at the continental scale using the proportion of different biomes found within river drainage basins, whereas *Guégan, Lek & Oberdorff (1998*) used the mean annual flow discharge as a proxy for environmental heterogeneity in 183 rivers throughout the world. Although these two studies found a positive relationship between heterogeneity and fish species richness, their measures of environmental heterogeneity were confounded with biogeographic factors, such as the size of the drainage area and with other global environmental descriptors including seasonality of rainfall in lotic systems. Recent meta-analyses of the relationship concluded that environmental homogenization has a consistent and negative impact on animal diversity (*Smokorowski & Pratt, 2007; Seiferling, Proulx & Wirth, 2014*).

Empirical evidence of the relationship between local (small grain) fish species richness and environmental heterogeneity remains sparse for aquatic ecosystems of broad spatial extent. Further evaluation of the relationship is needed, especially when considering that: (1) species richness is declining in both freshwater and marine ecosystems (*Ricciardi & Rasmussen, 1999; Worm et al., 2006*), and (2) aquatic ecosystems are increasingly impacted by human activities, such as systematic embankment, river damming and seafloor trawling that are causing environmental homogenization (*Lotze et al., 2005; Jackson, 2008)* as well as changes in water quality variables (*Rabalais, 2002*). Fish communities are affected by structural characteristics of the environment such as reef structure and the presence of vegetation (*Kuffner et al., 2006*) and also by water quality variables such as salinity, turbidity, and oxygen concentration (*Rabalais, 2002; Bejarano & Appeldoorn, 2013*). The objective of this study was to evaluate the effect of local environmental heterogeneity in environmental variables (spatial grain <10 km$^2$) on fish species richness at the scale of biogeographic regions (spatial extent >1,000 km). We used data on fish communities (26 orders, 73 families, 136 genera, 204 species), obtained from extensive surveys in two coastal ecosystems of North America. Using a set of environmental variables routinely measured by monitoring programs, we demonstrate that fish species richness in coastal ecosystems responds positively to the spatial heterogeneity of environmental conditions and quantify the magnitude of this effect. We implemented a random placement model of community assembly to ensure that the empirical relationship found between species richness and environmental heterogeneity did not result of a veil or sampling effect.

# MATERIAL AND METHODS

## Study site and data collection

Fish abundances and environmental measurements were obtained from two extensive surveys conducted by the U.S. Environmental Protection Agency's Environmental Monitoring and Assessment Program (EMAP). The first data set consisted of four sampling campaigns conducted in the Virginian biogeographic province between 1990 and 1993 (*Hale et al., 2002*). Stations were located along the coastline and in large river estuaries of the East Coast (Delaware, Hudson, Potomac, York; Fig. 1A). The second data set was assembled from four sampling campaigns conducted in the Louisianan biogeographic province between 1991 and 1994. Stations were located along the Gulf of Mexico from the Rio Grande, Texas, to Anclote Island, Florida (Fig. 1B). Field campaigns in the two biogeographic provinces were carried out between July and September of each year.

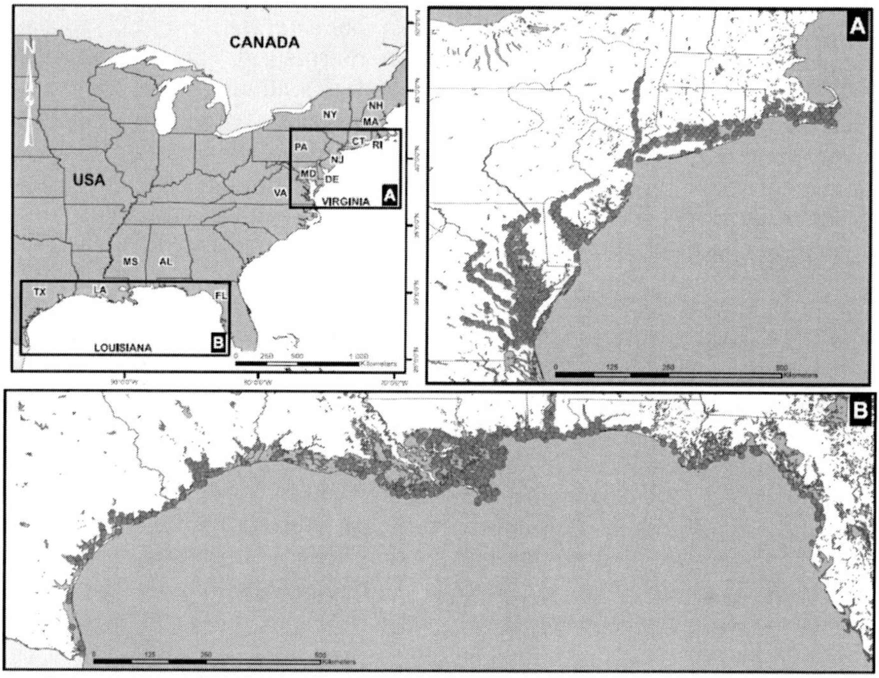


**Figure** 1 Spatial Distribution Of Sampling Sites.

Fish were sampled using balloon trawls (funnel-shaped nets, 4.9 m wide with 2.5 cm stretched mesh) deployed from a research vessel using a hydraulic-powered boom in the vicinity of the sampling stations. The duration of the trawl was 10 ± 2 (mean ± SD) minutes at a speed of 2–3 knots. This corresponds to a length of 0.77 ± 0.15 (mean ± SD) km. Following a successful trawl, the net was hauled aboard and the catch was released into a plastic trough, or a fish sorting table, where species composition and abundance were recorded (see Appendix S1). A total of

2,237 individuals (fork length: min. = 2.2 cm; max. = 91.18 cm; mean ± SD = 12.08 ± 7.33 cm) were captured from the Louisianan biogeographic province and 1,883 individuals (fork length: min. = 2.5 cm; max. = 92.6 cm; mean ± SD = 16.03 ± 10.37 cm) were captured from the Virginian biogeographic province, yielding a total of 4,120 individuals (Table 1, Appendix S1).

**Table** 1 Loadings And Summary Statistics For Environmental Variables.

| Variable | Virginian | | | | | | Louisianan | | | | | |
|---|---|---|---|---|---|---|---|---|---|---|---|---|
| | Loadings | | | | | | Loadings | | | | | |
| | Comp. 1 | Comp. 2 | Comp. 3 | Mean | Std. Dev. | Range (min–max) | Comp. 1 | Comp. 2 | Comp. 3 | Mean | Std. Dev. | Range (min–max) |
| Water density ($\sigma_t$) | −0.49 | 0.02 | 0.12 | 9.08 | 8.68 | −4.36–23.94 | | | | | | |
| Dissolved oxygen (mg $L^{-1}$) | −0.10 | −0.69 | 0.03 | 6.90 | 1.25 | 3.0–11.2 | −0.42 | 0.55 | −0.10 | 6.89 | 1.33 | 3.4–14.8 |
| Fluorescence | 0.28 | −0.34 | 0.42 | 11.82 | 7.70 | 0–30 | | | | | | |
| PAR ($\mu^{-2}s^{-1}$) | −0.05 | −0.27 | −0.85 | 545.76 | 464.29 | 9–3621 | −0.51 | −0.41 | −0.10 | 813.25 | 477.61 | 12–1,870 |
| pH | −0.28 | −0.53 | 0.16 | 7.93 | 0.48 | 6.3–9.4 | −0.40 | 0.47 | 0.41 | 8.00 | 0.46 | 5.3–9.5 |
| Salinity (ppt) | −0.49 | 0.00 | 0.11 | 16.18 | 11.05 | 0.03–32.89 | −0.06 | −0.14 | 0.84 | 13.47 | 10.70 | 0.01–37.35 |
| Temperature (°C) | 0.39 | −0.21 | −0.16 | 25.40 | 2.46 | 11.80–30.85 | −0.50 | 0.02 | −0.32 | 29.77 | 1.41 | 24.7–34.0 |
| Transmissivity (%) | −0.44 | 0.10 | −0.14 | 53.37 | 23.19 | 0–93 | −0.39 | −0.54 | 0.11 | 63.97 | 16.12 | 2–133 |

The environmental data comprised physical and chemical measurements. Dissolved oxygen concentrations (mg × $L^{-1}$) were determined using an air-calibrated oxygen meter (Yellow Springs Instruments, Yellow Springs, OH, USA) on surface water samples (625 mL) obtained with a Go-Flo bottle (General Oceanics, Miami, Florida, USA). Salinity (ppt), temperature (°C), pH, transmissivity (% of ambient light transmitted through the water column), photosynthetically active radiation ($\mu$E × $m^{-2}$ × $s^{-1}$), fluorescence (unitless) and water density ($\sigma$t, kg × $m^{-3}$−1, 000) were measured using a SeaBird CTD meter (Sea-Bird Electronics, Bellevue, Washington, USA) lowered through the water column at a rate of approximately 0.25 m × $s^{-1}$until it reached the bottom (Table 1). Fluorescence and water density data were not available for the Louisianan surveys. Implicit to our approach is that the gradient of environmental conditions captures a range of habitat and resource types. For example, temperature and dissolved oxygen may to some degree correlate with water depth or nutrient loading, whereas the photosynthetically active radiation is a more direct measure of primary production in the water column. Other variables, such as salinity, impose physiological constraints to the distribution of fish species in coastal transition zones. Detailed information about the sampling and analytical procedures can be found on the EMAP web site (http://www.epa.gov/emap/index.html). Although other environmental variables such as macrophyte cover might be important determinants of environmental heterogeneity, the selected variables are known to affect the ecology of individual fish species (*Mandrak, 1995*).

## Environmental heterogeneity

To represent the gradient of environmental conditions among stations of the same biogeographic province, we used the scores of a principal component analysis (PCA) performed on the environmental variables. The first three PCA axes (Table 1) were retained based on Kaiser's criterion (*Kaiser, 1960*) and explained nearly 75% of the environmental variability in both Virginian (PC1 = 42.28%, PC2 = 19.7%, PC3 = 12.6%) and Louisianan (PC1 = 32.5%, PC2 = 23.8%, PC3 = 19.6%) biogeographic provinces. We quantified the degree of local spatial autocorrelation in environmental conditions near each station as a reciprocal measure of environmental heterogeneity.

We calculated the local Moran *I* statistic on the scores of the first three PCA axes using the localmoranfunction of the spdep package in R (*Bivand et al., 2013; Bivand & PIras, 2015*). Only the first PCA axis was retained for further analysis because we did not find any relationship between Moran's *I* calculated for PCA axes 2 or 3 and species richness. The Moran *I* statistic identifies station neighborhoods where environmental conditions of similarly high or low values cluster spatially (high *I*), as well as neighborhoods where environmental conditions are more contrasted (low *I*). High *I* values indicate low heterogeneity (positive autocorrelation), whereas values around zero indicate high heterogeneity. Negative *I* values indicate local over-dispersion patterns (i.e., negative autocorrelation), which are rarely observed in nature (*Borcard, Gillet & Legendre, 2011*). The *I* statistic is given by *Anselin (1995)*:

$$I = (n-1)\frac{x_i - \bar{X}}{\sum_{i=1}^{n}(x_i - \bar{X})^2}\sum_{j=1}^{n} w_{ij}(x_j - \bar{X}) \qquad (1)$$

where $x_i$ is the value of the observation $i$, $\bar{X}$ is the mean of the variable, $w_{ij}$ is

the spatial weight ($1/distance^2$) between observations $i$ and $j$, and $n$ is the number of stations sampled. Each *I* (one per station) have been calculated by including all surrounding neighbours in a 75 km radius using the dnearneigh function of the spdep package. The chosen radius was large enough to include a sufficient number of neighboring stations in the calculation of Moran's *I* (average number of stations: Virginian = 54.3; Louisianan = 52.7), while small enough to prevent the inclusion of stations that are only remotely connected. Spatial weights were scaled as $1/distance^2$ in Eq. (1), thus varying the search radius had little effect on *I* values. Because we could not determine whether patterns of over-dispersion should be associated with high or low levels of environmental heterogeneity, the few stations (less than 4%) with negative *I* values were removed from subsequent statistical analyses. We did not find substantial differences between results for *I* calculated using all the data pooled at the biogeographic level (spatio-temporal *I*) and *I* calculated for each sampling year separately (spatial *I*). Consequently, we view *I* as a measure of spatial heterogeneity in local environmental conditions across space (Appendix S2, Fig. 1, Eq. (1)).

## Numerical simulations

We developed a random placement model of community assembly to determine the heterogeneity–species richness relationship in the absence of explicit habitat selection mechanisms. The model has two main components: (1) environmental heterogeneity and (2) species richness, each being simulated independently of the other on a two-dimensional surface (Fig. 2). This approach has been successfully used in various ecological studies aiming to highlight the effect of landscape structures on different aspects of animal biodiversity (McGill, 2011; Campos et al., 2013).

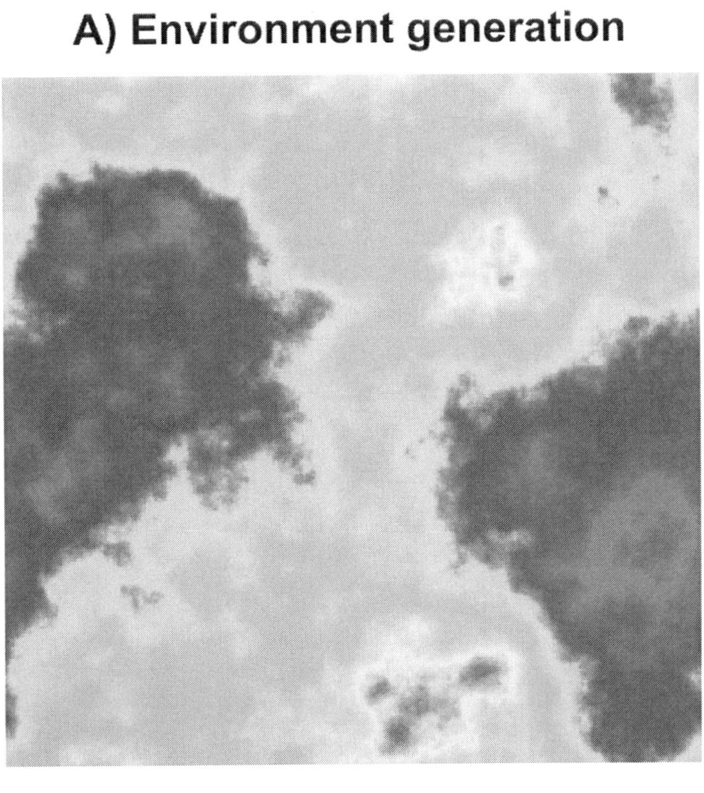

B) Random placement
C) Local Moran (I)

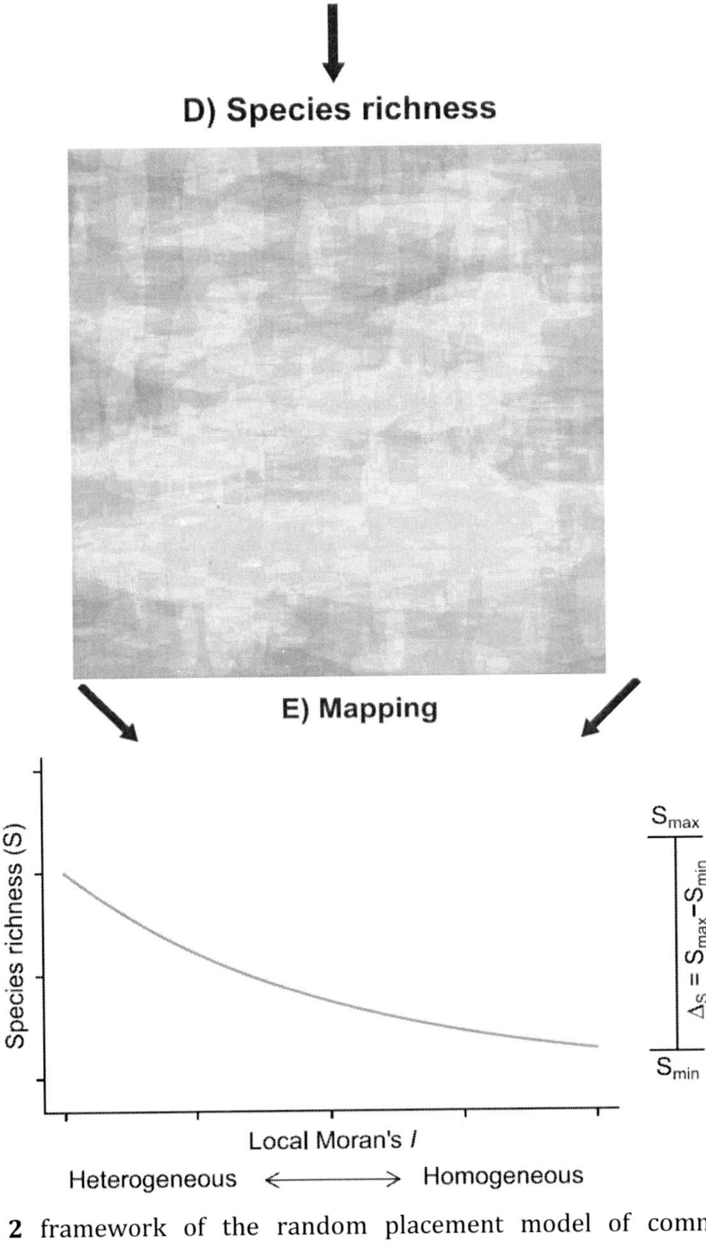


**Figure 2** framework of the random placement model of community assembly used to determine the relationship between fish species richness (*s*) and habitat heterogeneity in the absence of any particular habitat selection mechanisms.

The first model component simulates the spatial patterns of environmental conditions (Fig. 2A). Environmental spatial patterns can be modeled as a fractional Brownian function. The spectral density *S*(*f*) of a two-dimensional surface follows a power spectrum $S(f) \propto 1/f^{\beta}$ (*Keitt, 2000*), where *f* is frequency and $\beta = 1 + 2H$. The Hurst exponent (*H*) controls the degree of auto-correlation in environmental conditions; a large $H (H \longrightarrow 1)$ results in relatively homogeneous spatial patterns, whereas a lower $H (H \longrightarrow 0)$ produces more heterogeneous patterns. To generate the environmental spatial patterns in our simulations, we used the Matlab function noiseonf, which uses the inverse Fourier transformation of a power spectrum with a predetermined Hurst exponent (*Kovesi, 2000*). This procedure generates 'neutral' landscapes (e.g., *With, 1997*; *Keitt, 2000*) that share several statistical properties with environmental patterns observed in nature. The Hurst exponent of the simulated surface was parameterized using the linear slope of the log–log semi-variogram (*Gallant et al.,* 1994) computed on the scores of the first axis of the PCA of environmental conditions, yielding values of $H \approx 0.4$ in both biogeographical provinces.

The second component (Fig. 2B) of our model simulates the random placement of species with different distribution ranges. We based our random placement model of community assembly on two premises (*McGill & Collins, 2003*; *McGill, 2010*): (1) the centroid of each species range is determined by sampling from a uniform distribution over the surface and (2) the range size of species is distributed according to a power distribution. *McGill & Collins (2003)* reported that implementing either a log-normal or a power distribution did not affect the results of random placement model. Each of our simulation runs proceeded as described in Algorithm 1. Local species richness is then calculated by summing the overlap of different species ranges. On the basis of the observed regional distributions of the sampled species (Appendix S2, Fig. 3), we used the following parameters to implement the random placement model: $G = 1, 000$, $r_{min} = 10$ km and $r_{max} = 1, 000$ km.

---

Algorithm 1. Random placement of species (component 1, Fig. 2A)

1. Generate a surface of size $G \times G$.
2. Randomly pick the distribution range *r* of a new species from a power function $f(r) = r^{-a}$ where $r_{min} \leq r \leq r_{max}$ (Appendix S2, Fig. 2).
3. Choose the species centroid randomly from a uniform distribution over the surface.
4. Repeat previous steps until the surface is completely covered by species ranges (ranges are allowed to overlap).

---

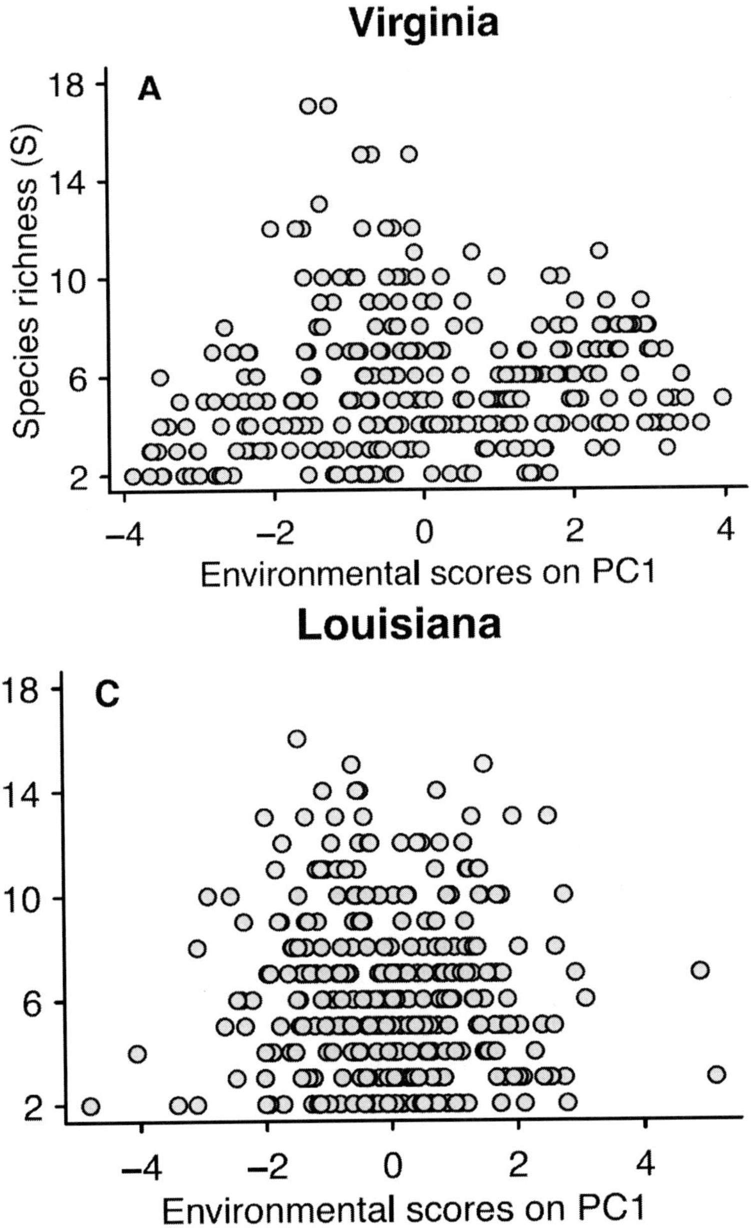
Virginia
A
Species richness (S)
18
14
10
6
2
−4
−2
0
2
4
Environmental scores on PC1
Louisiana
C
18
14
10
6
2
−4
−2
0
2
4
Environmental scores on PC1

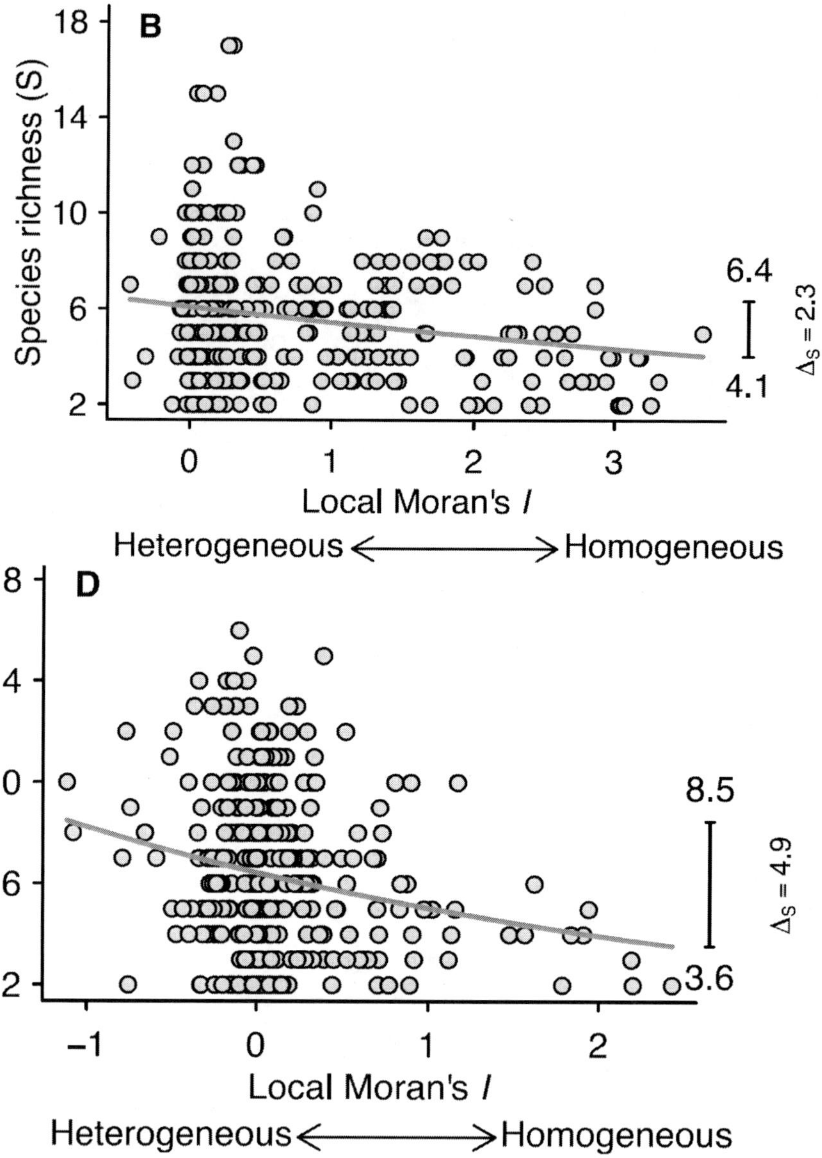


**Figure** 3 Relationships between species richness (***s***) and pca scores for the first axis (a and c) and local moran's ***i*** (b and d) for the virginian and louisianan biogeographic provinces.

To represent the range of each species on the surface, we used ellipses with major axis length $r$ and minor axis length sampled from a uniform in the interval $[r/4, r/2]$ as described in *Proulx et al. (2014)*. To simulate an anisotropic spatial process, we placed the elliptical ranges with their major axis oriented either horizontally (with probability = 0.75) or vertically (with probability = 0.25). This decision was motivated by the fact that species ranges in both biogeographical provinces are preferentially oriented along rivers and coastlines that broadly conform to the proposed alignment. Finally, to determine the parameter $\alpha$ empirically, we calculated the range of all fish species in each biogeographical province (Appendix S2, Fig. 3) and estimated the power coefficient of the frequency using the log-ratio formula (Eq. (5) in *Newman, 2005*). We obtained values of $\alpha$ = 1.214 for the Virginian province and $\alpha$ = 1.189 for the Louisianan province, and therefore used a value of 1.2 in our simulations. Using different combinations of ellipse shape ratio and orientation, we found that the species richness was robust to these changes. Most importantly, varying the shape ratio and orientation of the ellipse (species range) did not affect the general direction and relative effect size of the simulated environmental heterogeneity–species richness relationship. We generated the two model components on grids of 1, 000 × 1, 000 cells (Figs. 2 A and and 2 B).2B). A total of 10,000 simulations where performed according to Algorithm 2. It is to be noted that the model does not aim to approximate the absolute number of species at each location. Consequently, we used relative changes in species richness ($\Delta_S$) to compare modeled and observed results.

In each of the biogeographic provinces surveyed, approximately 5% of the stations yielded species richness values of zero. These zeros may partly arise from a 'veil effect' (*Preston, 1948*), and so reflect insufficient sampling effort rather than true absences. Truncation of samples at the veil may induce a spurious negative relationship between richness and predictor variables (Fig. 2E). To represent this effect in the simulated data, we set three veil lines at percentiles 0%, 5% and 15% and excluded species richness values below these thresholds (Fig. 4).

Algorithm 2. Global simulation procedure

1. Generate an environmental grid (component 1, Fig. 2A).
2. Generate a species placement grid (component 2, Fig. 2B).
3. Randomly subsample 400 grid cells (roughly corresponding to the total number of sampling stations in each biogeographic province, Appendix S2, Fig. 4).
4. Calculate the local Moran's $I$ at each subsampled cell on the environmental grid following the procedure described in the *Environmental heterogeneity* section (Eq. (1), Appendix S2, Fig. 3).
5. Pair each local $I$ value to its associated species richness value on the environmental and the species placement grid, respectively.
6. Fit a negative binomial regression between the paired values of local Moran's $I$ and species richness (Fig. 2E).
7. Calculate the relative increase in species richness ($\Delta_S$) predicted by the regression curve.

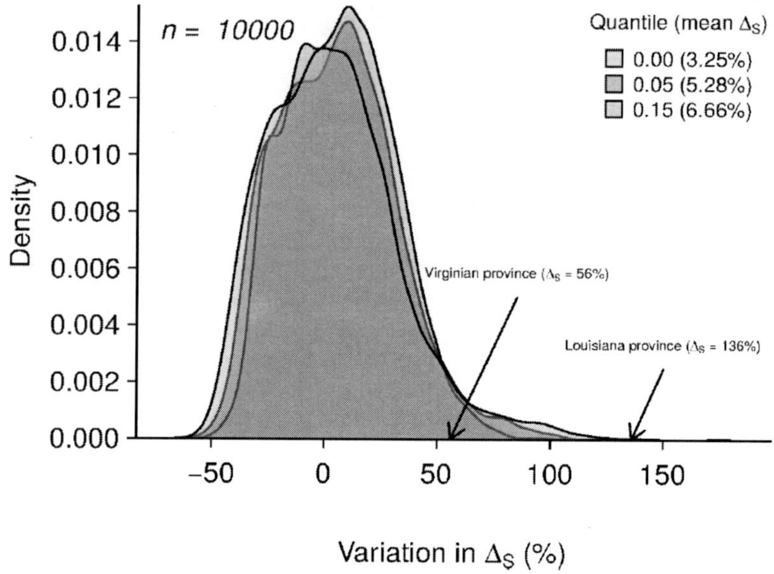


**Figure** 4 Results of 10,000 simulations showing the influence of quantile cut (veil effect) on modeled species richness.

## Statistical analyses

We used regression analyses to examine the relationships between species richness and the scores from the first PCA axis of environmental variables. To determine whether environmental heterogeneity had an influence on species diversity for both observed and simulated data, negative binomial regressions were fitted to the points above the veil effect threshold using the glm.nb function of the MASS package in R (version 3.0.1). We also checked for the presence of spatial autocorrelation in the model residuals.

# RESULTS

Fish species richness was not correlated with any of the first three principal components from the analysis of environmental variables (Table 1; Figs. 3A and and3C),3C), or with any of the individual environmental variables (results not shown). However, species richness was related to environmental heterogeneity (Figs. 3B and and3D).3D). For both biogeographic provinces, the negative binomial regressions showed that species richness was greater in more heterogeneous environments (Figs.

3B and and3D).3D). In the Virginian province (Fig. 3B), the mean species richness increased from 4.1 in the most homogeneous environments to 6.4 in the most heterogeneous environments, representing a gain of 2.3 ± 0.11 (95% confidence limits) species which correspond to 56% relative increase. A similar pattern was found for the Louisianan province (Fig. 3D) where mean species richness increased from 3.6 in the most homogeneous environments to 8.5 in the most heterogeneous environments, representing a gain of 4.9 ± 0.16 (95% confidence limits) species which correspond to 136% relative increase. We did not find spatial autocorrelation in the model residuals.

Averaging the results of 10,000 model simulations, the mean species richness relative increase ($\Delta_S$) were of 3.25%, 5.28% and 6.66% for the 0%, 5% and 15% veil effects, respectively (Fig. 4). The probabilities of observing $\Delta_S$ greater or equal to 56% (Virginian province) due to a sampling effect for different veils (0%, 5%, 15%) were of 4.68%, 3.7% and 2.12%, respectively (Table 2). Considering a $\Delta_S$ of 136% threshold (Louisianan province), these probabilities dropped to 0.05%, 0.01% and 0% (Table 2).

**Table** 2 The probabilities of observing $\delta_s$ greater or equal than 56% (virginian) or 136% (louisianan) due to sampling effect (i.e., random) under different scenarios of veil effects (0%, 5%, 15%).

| | Veil at 0% | Veil at 5% | Veil at 15% |
|---|---|---|---|
| Virginian (56%) | 4.68 | 3.70 | 2.12 |
| Louisianan (136%) | 0.05 | 0.01 | 0.00 |

# DISCUSSION

Many factors, including environmental heterogeneity, have been reported to affect the diversity of aquatic communities (*Field et al., 2009).* However, it is likely that the set of factors influencing species richness differs across spatial and temporal scales (*Fausch et al., 2002).* Moreover, environmental heterogeneity has been identified as a key factor maintaining animal biodiversity in aquatic ecosystems (*Levin et al., 2010*). Our study combines data from extensive surveys and simulations to demonstrate a strong positive influence of environmental heterogeneity on the species richness of fish communities. Interestingly, species richness was associated with the spatial heterogeneity of environmental variables but not with their magnitude. For both biogeographic provinces, mean species richness in the most heterogeneous environments was markedly greater than in the most homogeneous environments, as quantified by the negative binomial regressions. Furthermore, the observed effect of heterogeneity on species richness was substantially greater (Fig. 3) than that generated by the simulations based on a random community assembly model, so it seems unlikely that the observed relationship arose solely as a byproduct of veil or sampling effects.

## Environmental variables

Climate, energy, and primary productivity have a major influence on species richness at the regional, continental and global scales (*Guégan, Lek & Oberdorff, 1998; Hawkins, Field & Cornell, 2003; Field et al., 2009*). Studies conducted at small grain also indicate that environmental variables can influence the occurrence of species and abundance in local fish communities in both space and time (*Menge & Olson, 1990; Thiel et al., 1995; Rodríguez & Lewis, 1997*). In contrast to these findings, we did not observe any direct effect of the magnitude of individual environmental conditions (Table 1), including salinity, chlorophyll-*a* concentration and water temperature, on the species richness of local fish communities in either the Virginian (Fig. 3A) or Louisianan (Fig. 3C) biogeographic provinces.

In contrast with a simple randomization procedure, the simulation approach used in the random placement models allowed us to make a number of assumptions regarding the ecology of coastal fishes: (1) the spatial pattern of environmental conditions follows a two-dimensional power spectrum; (2) the centroid of each species on the seascape is determined by sampling from a uniform distribution and its range size by sampling from a power distribution; (3) fish species richness is independent from environmental conditions at the site of capture. We note that all three assumptions were supported by the empirical data. Another major assumption of random placement models is that the probability of finding a fish species at a particular site is independent of other species. Such ecological independence between co-occurring species has been shown to accurately reproduce a number of community patterns (*McGill, 2010; McGill, 2011*). For example, a recent study of shrubland plant communities reported that only 7–19% of all species pairs showed strong and consistent spatial associations, leading the authors to conclude that ecological processes are left no discernible spatial signature (*Perry et al., 2014*). In contrast with these findings, our results suggest that coastal fish communities may show a spatial signature, as fish species richness was not associated locally with the magnitude of environmental variables, but rather with their spatial heterogeneity.

## Environmental heterogeneity

Environmental heterogeneity influences many ecological processes such as fluxes of organisms, material and energy among riverscape elements (*Pickett & Cadenasso, 1995*). Our results demonstrate that fish species richness responded positively to increased environmental heterogeneity (Figs. 3B and and3D)3D) in both the Virginian and Louisianan biogeographic provinces. Simulations using a random placement model of community assembly showed that species richness increased only slightly in more heterogeneous environments (Fig. 4). For instance, less than 5% of the 10,000 simulations generated $\Delta_S$ greater than the conservative value of 56% observed in the Virgina biogeographic province (Figs. 3 and and4,4, Table 2). Hence, it is unlikely that the positive relationship observed between environmental heterogeneity and species richness in both biogeographic provinces is the result of a sampling effect (*sensu McGill, 2011*).

Aquatic ecologists often use the term 'heterogeneity' rather loosely to refer to habitat complexity, habitat diversity or environmental variability over time (reviewed in Palmer, Menninger & Bernhardt, 2010). For example, at small scales, heterogeneity usually refers to the variability in structural physical properties of the aquatic habitat such as riparian vegetation, channel configuration, artificial riffles and substrate granulometry (*Palmer, Menninger & Bernhardt, 2010*). Conversely, studies conducted at regional or continental scales have used large-grained variables such as percentage of different types of biome or drainage area as a proxy for habitat heterogeneity (*Guégan, Lek & Oberdorff, 1998; Field et al., 2009; Oberdorff et al., 2011*), possibly reflecting the difficulty of obtaining information at a finer resolution. Consequently, studies conducted at regional or continental scales are likely to capture broad-scale environmental heterogeneity that is coarse relative to the local heterogeneity to which individual fish respond, particularly for species having ranges smaller than the study grain size (*O'Neill et al., 1986; Turner et al., 1989; Wiens, 1989*). However, the question of which scale is optimal for quantifying the heterogeneity-diversity relationship is still open (*Chase & Leibold, 2002; Durance, Lepichon & Ormerod, 2006; Pittman et al., 2007)*.

# CONCLUSIONS

Over the last century, coastal ecosystems have become increasingly impacted by anthropogenic pressures (*Lotze et al., 2006*), including many human–driven activities that reduce the temporal and spatial heterogeneity of coastal habitats. For example, commercial fish trawlers are known to reduce the spatial heterogeneity of the sea floor structure (*Helfman, 2007)*. Similarly, the temporal variability of water flows in many of the world's largest rivers are regulated by dams (*Nilsson et al., 2005)*. This reduced variability in runoffs has been shown to increase the homogeneity of water channels, as well as to degrade fish habitats (see *Moyle & Mount, 2007 and references therein*). The current study shows that, independently of the environmental conditions prevailing locally, more homogeneous habitats can support fewer fish species. Hence, restoring or actively protecting areas of high habitat heterogeneity appears to be of great importance for slowing actual trends of decreasing biodiversity in coastal ecosystems.

## ACKNOWLEDGMENTS

We thank the U.S. Environmental Protection Agency's Environmental Monitoring and Assessment Program (EMAP) for freely providing their data. Although the data described in this article have been funded wholly or in part by the U.S. Environmental Protection Agency through its EMAP Estuaries Program, it has not been subjected to Agency review, and therefore does not necessarily reflect the views of the Agency and no official endorsement should be inferred. K Roach made helpful comments on an earlier version of the manuscript.

Support was given by the Groupe de Recherche Interuniversitaire en Limnologie et Environnement Aquatique (GRIL). P Massicotte was supported by a postdoctoral fellowship from RIVE. The funders had no role in study design, data collection and analysis, decision to publish, or preparation of the manuscript.

## REFERENCES

1. Anselin (1995) Anselin L. Local indicators of spatial association—LISA. Geographical Analysis. 1995;27(2):93–115. doi: 10.1111/j.1538-4632.1995.tb00338.x
2. Bejarano & Appeldoorn (2013) Bejarano I, Appeldoorn R. Seawater turbidity and fish communities on coral reefs of Puerto Rico. Marine Ecology Progress Series. 2013;474:217–226. doi: 10.3354/meps10051
3. Bivand et al. (2013) Bivand R, Hauke J, Kossowski T. Computing the Jacobian in Gaussian spatial autoregressive models: an illustrated comparison of available methods. Geographical Analysis. 2013;45(2):150–179. doi: 10.1111/gean.12008
4. Bivand & PIras (2015) Bivand R, PIras G. Comparing implementations of estimation methods for spatial econometrics. Journal of Statistical Software. 2015;63(18):1–36. *Available at http://www.jstatsoft.org/v63/i18/*
5. Borcard, Gillet & Legendre (2011) Borcard D, Gillet F, Legendre P. Numerical ecology with R. New York, NY: Springer; 2011

6. Campos et al. (2013) Campos PRA, Rosas A, de Oliveira VM, Gomes MAF. Effect of landscape structure on species diversity. PLoS ONE. 2013;8(6):e760 doi: 10.1371/journal.pone.0066495
7. Chase & Leibold (2002) Chase JM, Leibold MA. Spatial scale dictates the productivity-biodiversity relationship. Nature. 2002;416:427–430. doi: 10.1038/416427a
8. Durance, Lepichon & Ormerod (2006) Durance I, Lepichon C, Ormerod SJ. Recognizing the importance of scale in the ecology and management of riverine fish. River Research and Applications. 2006;22(10):1143–1152. doi: 10.1002/rra.965
9. Fausch et al. (2002) Fausch KD, Torgersen CE, Baxter CV, Li HW. Landscapes to riverscapes: bridging the gap between research and conservation of stream fishes. BioScience. 2002;52(6):483–498. doi: 10.1641/0006-3568(2002)052[0483:LTRBTG]2.0.CO;2
10. Field et al. (2009) Field R, Hawkins BA, Cornell HV, Currie DJ, Diniz-Filho JAF, Guégan J-F, Kaufman DM, Kerr JT, Mittelbach GG, Oberdorff T, O'Brien EM, Turner JRG. Spatial species-richness gradients across scales: a meta-analysis. Journal of Biogeography. 2009;36(1):132–147. doi: 10.1111/j.1365-2699.2008.01963.x
11. Gallant et al. (1994) Gallant JC, Moore ID, Hutchinson MF, Gessler P. Estimating fractal dimension of profiles: a comparison of methods. Mathematical Geology. 1994;26(4):455–481. doi: 10.1007/BF02083489
12. Guégan, Lek & Oberdorff (1998) Guégan J-F, Lek S, Oberdorff T. Energy availability and habitat heterogeneity predict global riverine fish diversity. Nature. 1998;391:382–384. doi: 10.1038/34899
13. Hale et al. (2002) Hale SS, Hughes MM, Strobel CJ, Buffum HW, Copeland JL, Paul JF. Coastal ecological data from the Virginian Biogeographic Province, 1990–1993. Ecology. 2002;83(10):2942–2942
14. Hawkins, Field & Cornell (2003) Hawkins B, Field R, Cornell H. Energy, water, and broad-scale geographic patterns of species richness. Ecology. 2003;84(12):3105–3117. doi: 10.1890/03-8006
15. Helfman (2007) Helfman GS. Fish conservation: a guide to understanding and restoring global aquatic biodiversity and fishery resources. Washington D.C.: Island Press; 2007
16. Jackson (2008) Jackson JBC. Colloquium paper: ecological extinction and evolution in the brave new ocean. Proceedings of the National Academy of Sciences of the United States of America. 2008;105(Suppl):11458–11465. doi: 10.1073/pnas.0802812105
17. Kaiser (1960) Kaiser HF. The application of electronic computers to factor analysis. Educational and Psychological Measurement. 1960;20(1):141–151. doi: 10.1177/001316446002000116
18. Keitt (2000) Keitt T. Spectral representation of neutral landscapes. Landscape Ecology. 2000;15:479–493. doi: 10.1023/A:1008193015770
19. Kovalenko, Thomaz & Warfe (2011) Kovalenko KE, Thomaz SM, Warfe DM. Habitat complexity: approaches and future directions. Hydrobiologia. 2011;685(1):1–17. doi: 10.1007/s10750-011-0974-z
20. Kovesi (2000) Kovesi PD. 2000. MATLAB and octave functions for computer vision and image processing. *Available at http://www.csse.uwa.edu.au/ ~pk/research /matlabfns/*
21. Kuffner et al. (2006) Kuffner IB, Brock JC, Grober-Dunsmore R, Bonito VE, Hickey TD, Wright CW. Relationships between reef fish communities and remotely sensed rugosity measurements in Biscayne National Park, Florida, USA. Environmental Biology of Fishes. 2006;78(1):71–82. doi: 10.1007/s10641-006-9078-4
22. Levin et al. (2010) Levin LA, Sibuet M, Gooday AJ, Smith CR, Vanreusel A. The roles of habitat heterogeneity in generating and maintaining biodiversity on continental margins: an introduction. Marine Ecology. 2010;31(1):1–5. doi: 10.1111/j.1439-0485.2009.00358.x

23. Lotze et al. (2006) Lotze HK, Lenihan HS, Bourque BJ, Bradbury RH, Cooke RG, Kay MC, Kidwell SM, Kirby MX, Peterson CH, Jackson JBC. Depletion, degradation, and recovery potential of estuaries and coastal seas. Science. 2006;312:1806–1809. doi: 10.1126/science.1128035
24. Lotze et al. (2005) Lotze HK, Reise K, Worm B, Van Beusekom J, Busch M, Ehlers A, Heinrich D, Hoffmann RC, Holm P, Jensen C, Knottnerus OS, Langhanki N, Prummel W, Vollmer M, Wolff WJ. Human transformations of the Wadden sea ecosystem through time: a synthesis. Helgoland Marine Research. 2005;59(1):84–95. doi: 10.1007/s10152-004-0209-z
25. MacArthur & MacArthur (1961) MacArthur R, MacArthur J. On bird species diversity. Ecology. 1961;42(3):594–598. doi: 10.2307/1932254
26. MacArthur & Wilson (1967) MacArthur RH, Wilson EO. The theory of island biogeography. Princeton: Princeton University Press; 1967
27. Mandrak (1995) Mandrak NE. Biogeographic patterns of fish species richness in Ontario lakes in relation to historical and environmental factors. Canadian Journal of Fisheries and Aquatic Sciences. 1995;52(7):1462–1474. doi: 10.1139/f95-141
28. McGill & Collins (2003) McGill B, Collins C. A unified theory for macroecology based on spatial patterns of abundance. Evolutionary Ecology Research. 2003;5:469–492
29. McGill (2010) McGill BJ. Towards a unification of unified theories of biodiversity. Ecology Letters. 2010;13(5):627–642. doi: 10.1111/j.1461-0248.2010.01449.x
30. McGill (2011) McGill BJ. Linking biodiversity patterns by autocorrelated random sampling. American Journal of Botany. 2011;98(3):481–502. doi: 10.3732/ajb.1000509
31. Menge & Olson (1990) Menge BA, Olson AM. Role of scale and environmental factors in regulation of community structure. Trends in Ecology & Evolution. 1990;5(2):52–57. doi: 10.1016/0169-5347(90)90048-I
32. Moyle & Mount (2007) Moyle PB, Mount JF. Homogenous rivers, homogenous faunas. Proceedings of the National Academy of Sciences of the United States of America. 2007;104(14):5711–5712. doi: 10.1073/pnas.0701457104
33. Newman (2005) Newman M. Power laws, Pareto distributions and Zipf's law. Contemporary Physics. 2005;46(5):323–351. doi: 10.1080/00107510500052444. [Cross Ref]
34. Nilsson et al. (2005) Nilsson C, Reidy C, Dynesius M, Revenga C. Fragmentation and flow regulation of the world's large river systems. Science. 2005;308:405–408. doi: 10.1126/science.1107887
35. Oberdorff et al. (2011) Oberdorff T, Tedesco PA, Hugueny B, Leprieur F, Beauchard O, Brosse S, Dürr HH. Global and regional patterns in riverine fish species richness: a review. International Journal of Ecology. 2011;2011:1–12. doi: 10.1155/2011/967631
36. O'Neill et al. (1986) O'Neill RV, Deangelis DL, Waide JB, Allen GE. A hierarchical concept of ecosystems. Princeton: Princeton University Press; 1986. p. 253
37. Palmer, Menninger & Bernhardt (2010) Palmer MA, Menninger HL, Bernhardt E. River restoration, habitat heterogeneity and biodiversity: a failure of theory or practice? Freshwater Biology. 2010;55(Suppl. 1):205–222. doi: 10.1111/j.1365-2427.2009.02372.x
38. Perry et al. (2014) Perry GLW, Miller BP, Enright NJ, Lamont BB. Stochastic geometry best explains spatial associations among species pairs and plant functional types in species-rich shrublands. Oikos. 2014;123(1):99–110. doi: 10.1111/j.1600-0706.2013.00400.x
39. Pickett & Cadenasso (1995) Pickett ST, Cadenasso ML. Landscape ecology: spatial heterogeneity in ecological systems. Science. 1995;269:331–334. doi: 10.1126/science.269.5222.331
40. Pittman et al. (2007) Pittman S, Christensen J, Caldow C, Menza C, Monaco M. Predictive mapping of fish species richness across shallow-water seascapes in the

Caribbean. Ecological Modelling. 2007;204(1–2):9–21. doi: 10.1016/j.ecolmodel.2006.12.017
41. Preston (1948) Preston FW. The commonness, and rarity, of species. Ecology. 1948;29(3):254–283. doi: 10.2307/1930989
42. Proulx et al. (2014) Proulx R, Roca IT, Cuadra FS, Seiferling I, Wirth C. A novel photographic approach for monitoring the structural heterogeneity and diversity of grassland ecosystems. Journal of Plant Ecology. 2014;7(6):518–525. doi: 10.1093/jpe/rtt065
43. Rabalais (2002) Rabalais NN. Nitrogen in aquatic ecosystems. Ambio. 2002;31:102–112
44. Ricciardi & Rasmussen (1999) Ricciardi A, Rasmussen JB. Extinction rates of north american freshwater fauna. Conservation Biology. 1999;13(5):1220–1222. doi: 10.1046/j.1523-1739.1999.98380.x
45. Ricklefs (1977) Ricklefs R. Environmental heterogeneity and plant species diversity: a hypothesis. American Naturalist. 1977;111:376–381. doi: 10.1086/283169
46. Rodríguez & Lewis (1997) Rodríguez MA, Lewis WM. Structure of fish assemblages along environmental gradients in floodplain lakes of the Orinoco River. Ecological Monographs. 1997;67(1):109–128. doi: 10.1890/0012-9615(1997)067 [0109:SOFAAE] 2.0.CO;2
47. Seiferling, Proulx & Wirth (2014) Seiferling I, Proulx RL, Wirth C. Disentangling the environmental-heterogeneity-species-diversity relationship along a gradient of human footprint. Ecology. 2014;95(8):2084–2095. doi: 10.1890/13-1344.1
48. Smokorowski & Pratt (2007) Smokorowski K, Pratt T. Effect of a change in physical structure and cover on fish and fish habitat in freshwater ecosystems—a review and meta-analysis. Environmental Reviews. 2007;15:15–41. doi: 10.1139/a06-007
49. Thiel et al. (1995) Thiel R, Sepulveda A, Kafemann R, Nellen W. Environmental factors as forces structuring the fish community of the Elbe Estuary. Journal of Fish Biology. 1995;46(1):47–69. doi: 10.1111/j.1095-8649.1995.tb05946.x
50. Tisseuil et al. (2013) Tisseuil C, Cornu J-F, Beauchard O, Brosse S, Darwall W, Holland R, Hugueny B, Tedesco PA, Oberdorff T. Global diversity patterns and cross-taxa convergence in freshwater systems. The Journal of Animal Ecology. 2013;82(2):365–376. doi: 10.1111/1365-2656.12018
51. Turner et al. (1989) Turner MG, O'Neill RV, Gardner RH, Milne BT. Effects of changing spatial scale on the analysis of landscape pattern. Landscape Ecology. 1989;3(3–4):153–162. doi: 10.1007/BF00131534
52. Wiens (1989) Wiens JA. Spatial scaling in ecology. Functional Ecology. 1989;3(4):385–397. doi: 10.2307/2389612
53. With (1997) With KA. The application of neutral landscape models in conservation biology. Conservation Biology. 1997;11(5):1069–1080. doi: 10.1046/j.1523-1739.1997.96210.x
54. Worm et al. (2006) Worm B, Barbier EB, Beaumont N, Duffy JE, Folke C, Halpern BS, Jackson JBC, Lotze HK, Micheli F, Palumbi SR, Sala E, Selkoe KA, Stachowicz JJ, Watson R. Impacts of biodiversity loss on ocean ecosystem services. Science. 2006;314:787–790. doi: 10.1126/science.1132294
55. Yeager, Layman & Allgeier (2011) Yeager LA, Layman CA, Allgeier JE. Effects of habitat heterogeneity at multiple spatial scales on fish community assembly. Oecologia. 2011;167(1):157–168. doi: 10.1007/s00442-011-1959-3

# CHAPTER 10

# Zebra Mussel (Dreissena Polymorpha) Affects the Feeding Ecology of Early Stage Striped Bass (Morone Saxatilis) in the Hudson River Estuary

**Michael G. Smircich & David L. Strayer & Eric T. Schultz**

*Department of Ecology and Evolutionary Biology, University of Connecticut, 75 North Eagleville Road, Storrs, CT 06269-3043, USA*

*Cary Institute of Ecosystem Studies, 2801 Sharon Turnpike; P.O. Box AB, Millbrook, NY 12545, USA*

## ABSTRACT

Variability in the feeding ecology of young fishes over short and long time scales in estuaries is likely to affect population dynamics. We studied 14 years of early stage Striped Bass feeding ecology in the Hudson River Estuary over a 25-year time span, including years in which invasive zebra mussels markedly altered energy flow within the estuary.We predicted that feeding success would be low and that diet composition would be altered during years of high zebra mussel impact, particularly in upriver locations where mussels occur. Feeding success in the short termwas indicated by volume of gut contents and in the long term by dry mass at length, i.e. condition; these measures were positively intercorrelated and varied significantly year to year. We tested for associations between condition and multiple biotic and abiotic environmental variables. There was a strong negative effect of zebra mussel grazing rate on condition in upriver locations and a weak positive effect in downriver locations. In upriver locations, condition was 33% higher when local salinity was high and zebra mussel grazing rates were low, whereas in downriver locations, conditionwas 35% higher when zebra mussel grazing rates and copepod abundance were high and local dissolved oxygen was low. Copepods, amphipods, mysids, and Leptodora constituted the highest prey-specific index of relative importance throughout the estuary. There was no evident effect of the zebra mussel invasion on diet composition. This long-term study corroborates the inferences of earlier studies that zebra mussels reduced early-stage striped bass growth rate.

**Keywords**: ANOSIM. Condition . Invasive species . Trophic ecology

## INTRODUCTION

Estuaries are dynamic, variable systems that serve as a critical nursery habitat for many coastal fishes (MacLusky and Elliott 2004), offering favorable feeding conditions (Blaber and Blaber 1980; Friedland et al. 1996). Food abundance as well as abiotic conditions can fluctuate greatly over space and time. Given this variability, longterm studies of estuarine ecology can be particularly informative. Long-term data are critical for quantifying ecological response to environmental changes, understanding expression of key ecological trends, providing core data for parameterizing models, influencing management of species or ecosystems, and acting as the basis of collaborative studies (Lindenmayer et al. 2012). The goal of this study is to document the long-term feeding ecology of early stage estuary-dependent Striped Bass (Morone saxatilis [Walbaum, 1792]).

Studying fish feeding ecology addresses fundamental questions including what prey was consumed and when it was consumed. Diet analyses also provide insight towards broader subjects such as population dynamics (Braga et al. 2012), habitat use (Feitosa and Ferreira 2015), evolution (Collar et al. 2009), energy flow between ecosystems (Baxter et al. 2005), and conservation (Alcaraz et al. 2015; Donadelli et al. 2015). Characterizing the diet of early-stage fishes is particularly important because this is when fish grow most rapidly and undergo ontogenetic shifts that are critical to survival. The early life stages of anadromous fish within estuaries are typically the only times these fish can easily be captured until they arrive back into freshwater as spawning adults. Generally the feeding of young fishes has not been studied as much as that of adults due to additional difficulties associated with collection and analysis of specimens (Gerking

1994), and patterns in feeding ecologies of estuarine larvae relative to other habitats remain poorly characterized (Llopiz 2013).

The Hudson River Estuary is a body of water that hosts many early stage fish species, including nearly a dozen diadromous fishes (Waldman 2006). The estuary is a complex,well-studied system in which several longterm environmental data collection programs have been in place for multiple decades (Levinton and Waldman 2006). The estuary has historically undergone major ecological changes, including the recent invasion of zebra mussels (Dreissena polymorpha [Pallas, 1771]). Zebra mussels first appeared in the Hudson River in 1991, and have been abundant throughout the freshwater portion of the estuary since then (Strayer et al. 2011). The mussels sharply reduced phytoplankton biomass and markedly altered estuarine energy flow (Pace et al. 1998; Pace et al. 2010; Strayer et al. 2014b).Within the first two years of mussel establishment, total phytoplankton biomass declined 85% (Caraco et al. 1997), zooplankton biomass declined by more than 70% (Paceet al. 1998), and zoobenthos biomass declined by 40% (Strayer and Smith 2001). Strayer et al. (2004) found that growth and abundance decreased, and distribution changed, of early stage pelagic fishes in the estuary following the invasion. Subsequently, many musselinduced ecological impacts within the river diminished towards pre-invasion levels due to a decrease in average mussel size (Strayer et al. 2014a). The effects of these changes on feeding success and diet composition of larval fish in the Hudson River has not yet been studied, excepting one finding that larval shad consume zebra mussel veligers (Nack et al. 2015).

Long-term analysis of early stage fish feeding ecology in the estuary is facilitated by extensive monitoring. In particular, The Hudson River Utilities Longitudinal River Survey began in 1973 to estimate the impact of five electric generating stations on Striped Bass, White Perch (Morone americana [Gmelin, 1789]), and Atlantic Tomcod (Microgadus tomcod [Walbaum, 1792]). Annual reports of this survey ("year-class reports" e.g., ASA Analysis and Communication 2016) provide data on distribution and abundance of early life stages of selected fish species throughout the estuary, with associated physicochemical data. Furthermore, ichthyoplankton samples from this survey have been archived and made available for research through Normandeau Associates and the New York State Museum. With well-defined objectives and sound statistical design, such long-term monitoring program offer important opportunities for scientific research on the effects of environmental change (Lindenmayer and Likens 2009).

Striped Bass is a species of particular interest in the Hudson River Estuary ichthyoplankton. Striped Bass once comprised a commercial fishery in the Hudson River until it was closed in the mid-1970's because of PCB contamination (Waldman 2006). Today they are perhaps the most popular sport fish in the river. Striped Bass are anadromous, spending most of their adult life in marine waters and entering the estuary in the spring to spawn in the fresh waters of the middle and upper estuary (O'Connor et al. 2012). They are typically described as feeding opportunists (Hurst and Conover 2001), but at least one study documented selective feeding (Howe et al. 2008). Prior studies of larval Striped Bass diet showed that they consumed mainly copepods, Bosmina, and amphipods (Brett and Groves 1979; Hurst and Conover 2001; Jordan et al. 2003; Limburg et al. 1997; Morgan et al. 1981). Based on analysis of seasonal change in size distribution of juvenile striped bass, Strayer et al. (2004) concluded that the zebra mussel invasion had reduced the feeding success and growth rate of this species in the estuary. Data on how feeding success and diet were affected would be useful and have not been provided in other studies of early stage Striped Bass or other species.

The purpose of this study is to characterize early-stage Striped Bass diet composition and feeding success over a 25-year period in the Hudson River Estuary.We focus on three main goals: 1) to describe diet composition and evaluate feeding success over a multi-decade time span; 2) to test for effects of zebra mussels on the feeding ecology of Striped Bass; and, 3) determine what ecological factors most influence Striped Bass long-term feeding success. We predicted reduced feeding success and a change in diet composition during years of mussel impact, particularly in freshwater areas of the estuary inhabited by the mussels. Given the variable nature of estuaries, we expected that changes in additional environmental factors could also affect feeding success.

# METHODS

Samples were collected and archived as part of the Hudson River Utilities Longitudinal River Survey. The ichthyoplankton survey samples the entire length of the estuary with a 1.0-$m^2$ Tucker trawl in a stratified random design; sampling is weekly from early April through June and then biweekly until September. During the season that sampleswe analyzed are collected, all samples are taken at night (20:00 to 05:00). Larval fish sampleswere preserved in 10% formalin. No consistent effect of formalin preservation on fish length and mass is apparent across multiplestudies (Johnston andMathias 1993; Parker 1963; Schultz et al. 2005; Shields and Carlson 1996; Yeh and Hodson 1975). We assumed shrinkage to be uniform throughout our samples and did not correct for it in our analyses. The years analyzed for this study included: 1988, 1991–93, 1997, 1999, 2003, 2005–07, 2009–12. Years were selected based on sample availability within our study time frame. For analysis of zebra mussel effects, we grouped these years into three periods (Pace et al. 2010): "preinvasion" (1988–1992), "invasion impact" (1993–2004), and "recovery" (2005-present).

Undamaged early-stage Striped Bass were selected for analysis based on developmental stage, time of year, and location. Sixty fish per year were selected from samples containing larval and juvenile Striped Bass, taken during the growing season (early June until mid- September). In total, 840 specimens were used for this study. When possible, 30 fish each year were analyzed from each of two estuary locations (Table 1), upriver (river km ≥ 96) and downriver (river km < 96). This partitioning of the river corresponds roughly to the average seasonal position of the salt front. Zebra mussels are generally restricted to upriver freshwater areas.

Each selected specimen was measured for length, dissected for gut contents, and then dried. All specimens were photographed and digitally measured to the nearest 0.01 mm using SigmaScan® Pro 5.0 (Systat Software Inc., San Jose, CA). Specimen length (Table 1) ranged from 6 mm to 60 mm (mean 24 mm, standard deviation 11 mm). This size range encompasses larval and juvenile stages, which can be differentiated at about 30 mm (Hardy 1978). Specimens were placed into a drying oven for 48 h at 60 °C for determination of dry mass. We preserved gut contents in 70% ethanol (Hyslop 1980; Jude 1973).We estimated volume of gut contents after they had settled for 24 h in 1.5 ml conical vials.

**Table 1** Summary of sample sizes per bass life stage (L = larvae, J = juvenile), mean length per year (mm), and standard deviation of length (SD) in upriver and downriver locations. Fish less than 30 mm were considered to be larvae

| | Upriver | | | Downriver | | |
|---|---|---|---|---|---|---|
| Year | L/J | Mean length | (SD) | L/J | Mean length | (SD) |
| 1988 | 11/28 | 22.1 | (5.4) | 10/20 | 36.0 | (11.4) |
| 1991 | 20/44 | 20.8 | (7.4) | 5/5 | 40.0 | (20.4) |
| 1992 | 17/30 | 14.2 | (2.6) | 17/13 | 28.2 | (17.2) |
| 1993 | 13/23 | 21.7 | (15.4) | 24/6 | 23.6 | (13.2) |
| 1997 | 7/17 | 17.7 | (8.1) | 36/5 | 19.4 | (7.0) |
| 1999 | 12/18 | 19.8 | (4.3) | 32/9 | 25.2 | (11.8) |
| 2003 | 16/23 | 23.9 | (8.7) | 11/19 | 40.1 | (15.7) |
| 2005 | 7/30 | 19.1 | (3.4) | 26/4 | 21.8 | (6.8) |
| 2006 | 9/29 | 20.9 | (3.9) | 17/13 | 29.1 | (9.0) |
| 2007 | 10/19 | 16.7 | (5.5) | 26/15 | 26.2 | (9.7) |
| 2009 | 6/30 | 18.2 | (2.4) | 28/2 | 19.7 | (6.2) |
| 2010 | 4/29 | 20.9 | (4.4) | 23/7 | 26.6 | (4.0) |
| 2011 | 9/18 | 21.4 | (8.1) | 23/16 | 26.2 | (8.7) |
| 2012 | 5/22 | 22.3 | (4.4) | 33/4 | 25.8 | (4.9) |

The height of the settled contents was measured to the nearest 0.01 mmand converted to volume in μl using an empirical calibration formula.

We tested for temporal and spatial effects (year and zebra mussel period, upriver and downriver locations) on two measures of feeding success. Volume of gut contents represented a measure of short-term feeding success and condition (dry mass at length as suggested by Jakob et al. 1996) represented a longer-term measure of feeding success. Only specimens that had gut contents, comprising about 75% of specimens, were included in analysis of short-term feeding success. Analysis of temporal change in both measures of feeding success (log10-transformed) was conducted in analysis of covariance (ANCOVA) in which predictors were log10(length), zebra mussel period (pre-invasion, invasion impact, recovery), year nested within period, and the interaction between length and year. Year nested within period was treated as a random effect, the mean square of which was the denominator in F-tests of the effect of period. We conducted separate analyses by location (upriver or downriver) because of significant interactions with other main effects. We also conducted mixed linearmodeling by location, in which sample tow was encoded as a random effect to account for the nonindependence of specimens collected in the same ichthyoplankton sample. The ANCOVA and mixed linear model analyses were designed to test the prediction that feeding success was reduced during the impact period relative to the pre-invasion and recovery periods. Because we expected the reduction in feeding success to be most pronounced in the upriver location, we tested for an effect of mussel period, via one-way analysis of variance, on the difference in each year between the upriver and downriver predicted values of each measure of feeding success, adjusted to a common mean size. Analyses were run in SAS for Windows 9.4 (SAS Institute, Inc., Cary NC: PROC GLM, PROC CORR, PROC MIXED, and PROC ANOVA).

We tested seven abiotic and biotic environmental factors (Table 2) via mixed linear modeling to identify those factors that were most associated with longer-term feeding success, as measured by dry mass-at-length.We conducted separate analyses by location (upriver and downriver). Abiotic data (salinity, dissolved oxygen, temperature) were collected at the same times and locations as ichthyoplankton samples. In contrast, biotic data were obtained from monitoring programs of the Cary Institute of Ecosystem Studies and were represented in our analyses as annual averages because they were not taken on the same time scale or in same locations as ichthyoplankton samples. Biotic data included measures of water column production (chlorophyll a, copepod abundance), benthic production (amphipod abundance) and zebra mussel impact (zebra mussel filtration rate; see Strayer et al. 1996 for details about how this is estimated). In pre-invasion years, zebra mussel filtration rate was zero or negligibly small (Strayer et al. 1996). In predictive regressions, we included quadratic terms for abiotic variables (centered to a mean of 0 before

**Table 2** Summary statistics of environmental variables used in mixed modeling. Two values (upriver/downriver) are presented for abiotic variables that were measured in conjunction with sampling for early-stage Striped Bass, whereas one value is presented for biotic variables that were measured at a fixed single site. The 5$^{th}$ and 95th percentiles were used to quantify the effect of selected variables on condition

| Environmental variable | Mean | 5th Percentile | 95th Percentile |
|---|---|---|---|
| Amphipods[a] (no. $m^{-2}$) | 1100 | 380 | 3300 |
| Chlorophyll *a*[a] ($\mu g\ l^{-1}$) | 9.8 | 4.1 | 29 |
| Copepods[a] ( no. $l^{-1}$) | 7 | 3.4 | 17 |
| DO[b] (mg $l^{-1}$) | 7.5/6.6 | 6.2/5.6 | 8.9/7.9 |
| Salinity[b] (ppt) | 0.21/3.3 | 0.1/0.1 | 0.9/7.8 |
| Temperature[b] | 24/25 | 21/22 | 26/28 |
| ZM filtration rate[a] ($m^3 m^{-2} d^{-1}$) | 3.4 | 0.063 | 8.3 |

[a] Cary Institute of Ecosystem Studies data; data are annual averages collected at the Kingston long-term monitoring station

[b] Hudson River Utilities data; data were collected in conjunction with ichthyoplankton tows throughout the estuary; squared values also used in modeling to test for a quadratic response

squaring to reduce collinearity between first-order and second-order terms), because we had no prior expectations that the response to abiotic predictors would be linear. In all models, the response variable was log10(dry mass), and log10(length) was included as a covariate. Sample tow was encoded as a random effect. The best models of 1023 candidate models were selected by information-theoretic criteria, i.e. AIC and related statistics (Akaike 1974; Burnham and Anderson 2002). Models were ranked as best to worst in order of increasing AIC; each model was compared to the best in terms of their difference in AIC, and the relative likelihood of each model was estimated as Akaike weight. We evaluated the predictive power of zebra mussel filtration rate (ZMFR) by computing the summed Akaike weight of models that included ZMFR. To quantify the effect of the environmental factors on feeding success, we predicted dry mass at mean length for low (5th percentile) and high (95th percentile) values of the predictors in the best model. We performed mixed linear modeling using SAS for Windows 9.4 (PROC MIXED).

For analysis of diet composition, prey items from each stomach were identified to the lowest taxonomic level possible. We aggregated some prey taxa into higher levels in order to increase the frequency of occurrence. Prey weight was determined directly or was estimated using regressions of mass versus number. The proportional frequency of occurrence and the abundance of prey in the diet in terms of numbers and weight were calculated with the following equations (Brown et al. 2012):

Proportional frequency of occurrence (pFO):

$$pFO_i = \frac{n_i}{n} \quad ,(1)$$

in which $n_i$ is the number of specimens with guts containing prey i, and n is the total number of specimens;

Proportional abundance in terms of number and weight $(pPN_i, pPW_i)$:

$$pN_i = \sum_{j=1}^{n} pN_{ij}/n, \quad pW_i = \sum_{j=1}^{n} pW_{ij}/n \quad ,(2)$$

in which $pN_{ij}$ is the proportional numerical abundance of prey category i in stomach sample j and $pW_{ij}$, is the proportional abundance in weight of prey category i in stomach sample j;

Proportional prey-specific abundance in terms of number and weight ($pPN_i$, $pPW_i$):

$$pPN_i = \sum_{j=1}^{n} pN_{ij}/n_i, \quad pPW_i = \sum_{j=1}^{n} pW_{ij}/n_i \quad .(3)$$

The prey-specific index of relative importance (%PSIRI) was calculated as (Brown et al. 2012):

$$\%\mathrm{PSIRI}_i = \frac{pFO_i * (pPN_i + pPW_i) * 100}{2} \quad (4)$$

All values were calculated by aggregating all individuals by location.

To assess differences in diet composition among years and periods, we performed a multivariate twoway analysis of similarity (ANOSIM). Analysis was conducted on a Bray-Curtis dissimilarity matrix that was constructed from the average log10 transformed numerical percentages of prey types (adding 1 to all values before log transformation). A Similarity Percentage Analysis (SIMPER) was performed to identify those prey items accounting for the most dissimilarity between years. For visual representation of mean dietary differences between years, we used non-metric multidimensional scaling ordination (NMDS) of the dissimilarity matrix produced by the

ANOSIM analysis. Kruskal's stress statistic 1 was used to determine the best spatial representation of the different years; a stress of <0.2 was considered an acceptable fit (Clarke 1993). We predicted that a multivariate grouping of diet composition would be evident, such that diet during the mussel impact period would be distinct from the period before the invasion and that during the invasion recovery. All dietary comparisons were made with PRIMER Version 6 software (Primer-E Ltd., Plymouth UK).

# RESULTS

Two measures of feeding success varied concordantly among individuals. The short-term measure of individual feeding success (volume of gut contents) and the longer-term measure (condition as dry mass-at-length) were significantly but weakly correlated with each other after size adjustment ($R = 0.11$, $N = 642$, $p = 0.005$).

The volume of gut contents varied among years in both locations. Upriver and downriver values for annual mean volume of gut contents varied concordantly among years ($R = 0.73$, $N = 14$, $p = 0.003$). There was no overall difference with location (t-test $P = 0.8$; Fig. 1a). Volume of gut contents varied among years upriver (ANCOVA: $F_{11,293} = 4.0$, $p < 0.0001$; mixed linear model: $F_{11,254} = 1.8$, $p = 0.05$) and downriver (ANCOVA: years $F_{11,321} = 7.6$, $p < 0.0001$; mixed linear model $F_{11,255} = 3.3$, $p = 0.0003$) but did not change with zebra mussel period in either river location.

Condition also varied among years in both locations. Upriver and downriver values for annual mean condition were not correlated with each other. There was no overall difference with location (t-test $P = 0.28$; Fig. 1b). Condition varied among years upriver (ANCOVA: $F_{11,374} = 5.0$, $p < 0.0001$; mixed linear model: $F_{11,332} = 2.4$, $p = 0.004$) and downriver (ANCOVA: $F_{11,425} = 3.91$; $p < 0.0001$; mixed linear model: $F_{11,354} = 2.7$, $p = 0.003$). Condition did not change with zebra mussel invasion period in either river location.

Regressions taking multiple environmental variables into account indicate that zebra mussel filtration was associated with decreased condition in upriver locations (Table 3). No single set of upriver environmental variables yielded a best-fit model; ten models were nearly as good (i.e. ΔAIC <2.0) as the model with the lowest AIC value. Zebra mussel filtration rate (ZMFR) was a strong predictor. All of the best five models included ZMFR.

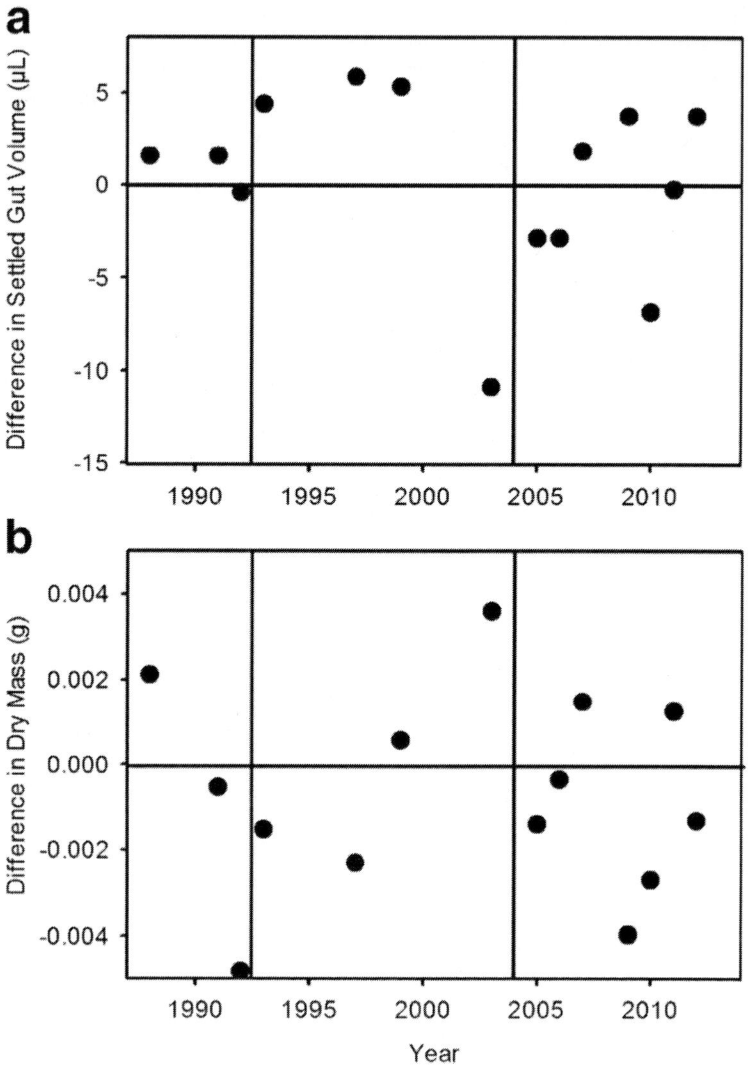


**Figure. 1** Difference in (a) volume of gut contents and (b) condition between up- and downriver samples. Each point represents 60 fish. Values greater than zero (indicated by horizontal reference line) indicate that feeding success was higher in upriver locations than downriver locations. The figure is split into thirds by vertical lines to mark the zebra mussel invasion periods: pre-invasion, invasion impact, and recovery

The sum of Akaike weights over models including ZMFR was 0.99 (hence, the summed weights over models that omitted ZMFR was 0.01). Predictors in the model with the lowest AIC included salinity as well as ZMFR (Table 3a). Condition increased with salinity and decreased with ZMFR in all five best models (Table 3b). To assess the magnitude of environmental variability on condition, we estimated dry weight at mean length under low-condition (salinity at its 5th percentile, ZMFR at its 95th percentile) and high-condition (salinity at its 95th percentile, ZMFR at its 5th percentile) values for predictors. Condition was 33% higher when environmental predictors were favorable (i.e. dry mass = 0.017 g and 0.012 g respectively).

In downriver locations, three models that included copepods, dissolved oxygen and dissolved oxygen2 yielded the best fit (Table 3a). Three of the five best models also included ZMFR. ZMFR was a weak predictor relative to copepods and dissolved oxygen (summed Akaike weights of models with ZMFR =0.67; with copepods =0.95; with dissolved oxygen and/or dissolved oxygen$^2$ = 0.92). Condition increased with higher values of copepods and with higher values of ZMFR (Table 3b). The slopes for dissolved oxygen and dissolved oxygen$^2$ were negative and positive respectively, indicating that there was a decelerating negative effect of dissolved oxygen on condition. Condition was 35% higher when environmental predictors were favorable (i.e. dry mass = 0.029 g and 0.039 g respectively).

Copepods and amphipods were predominant prey in the diet of Striped Bass in both river locations over the 14 years sampled (Table 4). In the upriver location, amphipods and copepods (prey-specific index of relative importance = 40%and 39%respectively) represented the dominant prey items. In the downriver location, amphipods (41%), copepods (37%), and mysids (19%) constituted the majority of the diet (Table 4, Fig. 2). The pN of mysids in the downriver location was over five times that of mysids in the upriver location, whereas the pN of Leptodora in the upriver location was over ten times that of Leptodora in the downriver location. Diet composition did not differ among zebra mussel periods in both river locations; however, diet composition differed among years upriver (ANOSIM, global $R$ = 0.27, $p$ = 0.001) and downriver (ANOSIM, global $R$ = 0.13, $p$ = 0.001). No temporal pattern in diet composition is evident in the upriver location (Fig. 3a), but 2009 diverged from other years, reflecting high consumption of Leptodora in that year. In the downriver location, the first four years lie on one edge of the multidimensional space opposite from one of the invasion-recovery years, 2010 (Fig. 3b). These differences are mostly explained by low consumption of mysids in the early years and high consumption of mysids in 2010. Diet of larval and juvenile Striped Bass was similar; copepods and amphipods dominated the diet of both life stages and their representation in the diet was similar in upriver and downriver locations (Table 5).

# DISCUSSION

To our knowledge, this study is the first to examine the feeding ecology of early-stage fishes over a multidecade time span. We found that zebra mussel filtration was associated with reduced feeding success of earlystage Striped Bass in upriver locations of the Hudson River Estuary where the mussels now live. The invasion of zebra mussels did not have a clear effect on diet composition.

**Table 3** Selected regression models to predict condition in upriver and down river samples. Response variable in each regression is log(dry weight). Predictors include chlorophyll *a* (Chl, μg $l^{-1}$), copepod density (Cop, $l^{-1}$), dissolved oxygen (DOmg $l^{-1}$), salinity (sal, ppt), temperature (Temp, °C). a) Statistics used in model evaluation. For each model 1$^{-5}$, listed in order of model quality, table lists predictors in the model (with the exception of log(length), which is in every model), Akaike's Information Criterion (AIC), difference in AIC between models 2–4 and model 1 (ΔAIC), Akaike weight (Wt). b) Summary of variable estimates. For each model 1–5, table lists slopes of predictors in the model, with the exception of predictors log(length) (upriver range: 3.32– 3.35; downriver range: 3.34–3.36) and the intercept (upriver range: -6.19 – -6.17; downriver range: -6.16 – -6.14)

a)

| Upriver | | | | | Downriver | | | | |
|---|---|---|---|---|---|---|---|---|---|
| Model | Predictors | AIC | ΔAIC | Wt | Model | Predictors | AIC | ΔAIC | Wt |
| 1 | Sal, ZMFR | -720.4 | 0 | 0.0440 | 1 | Cop, DO, DO$^2$, ZMFR | -1059 | 0 | 0.288 |
| 2 | DO, Sal, ZMFR | -720.3 | 0.100 | 0.0419 | 2 | Chl, Cop, DO, DO$^2$, ZMFR | -1057 | 1.80 | 0.117 |
| 3 | Cop, DO, Sal, ZMFR | -719.3 | 1.100 | 0.0254 | 3 | Cop, DO, DO$^2$ | -1057 | 1.80 | 0.117 |
| 4 | Sal, Temp$^2$, ZMFR | -718.9 | 1.500 | 0.0208 | 4 | Cop, DO, ZMFR | -1056 | 2.80 | 0.0710 |
| 5 | Cop, Sal, ZMFR | -718.9 | 1.500 | 0.0208 | 5 | Chl, Cop, DO, DO$^2$ | -1054.8 | 3.80 | 0.0431 |

b)

| Upriver | | | | | | Downriver | | | | | |
|---|---|---|---|---|---|---|---|---|---|---|---|
| Model | Chl | Cop | Sal | Temp$^2$ | ZMFR | Model | Chl | Cop | DO | DO$^2$ | ZMFR |
| 1 | | | 0.0822 | | -0.0101 | 1 | | 0.00388 | -0.0345 | 0.0138 | 0.00352 |
| 2 | | 0.0134 | 0.100 | | -0.0111 | 2 | 0.000376 | 0.00404 | -0.0346 | 0.0144 | 0.00376 |
| 3 | 0.00196 | 0.0148 | 0.106 | | -0.0109 | 3 | | 0.00324 | -0.0334 | 0.0130 | |
| 4 | | | 0.0776 | 0.000543 | -0.0100 | 4 | | 0.00366 | -0.0263 | | 0.00334 |
| 5 | 0.00141 | | 0.0848 | | -0.00992 | 5 | -0.000090 | 0.00322 | -0.0334 | 0.0129 | |

**Table 4** Prey of striped bass fromupriver and downriver locations in the Hudson River estuary, quantified by mean proportional numerical abundance (pN), mean proportional prey-specific numerical abundance (pPN), mean proportional abundance in weight (pW), mean proportional prey specific abundance in weight (pPW), mean proportional frequency of occurrence (pFO), and percent prey specific index of relative importance (%PSIRI)

| Prey Taxon | Upriver | | | | | | Downriver | | | | | |
|---|---|---|---|---|---|---|---|---|---|---|---|---|
| | pN | pPN | pW | pPW | pFO | %PSIRI | pN | pPN | pW | pPW | pFO | %PSIRI |
| Copepoda | 0.45 | 0.73 | 0.36 | 0.58 | 0.62 | 40.2 | 0.41 | 0.79 | 0.33 | 0.64 | 0.52 | 37.1 |
| Amphipoda: | | | | | | | | | | | | |
| Gammaridae | 0.20 | 0.53 | 0.26 | 0.68 | 0.38 | 22.9 | 0.17 | 0.56 | 0.19 | 0.61 | 0.30 | 17.8 |
| Corophiidae | 0.0100 | 0.78 | 0.0089 | 0.69 | 0.013 | 0.94 | 0.013 | 0.34 | 0.018 | 0.46 | 0.039 | 1.55 |
| Unidentified | 0.15 | 0.55 | 0.18 | 0.67 | 0.27 | 16.4 | 0.18 | 0.60 | 0.20 | 0.65 | 0.31 | 19.2 |
| Mysida | 0.022 | 0.74 | 0.028 | 0.95 | 0.030 | 2.52 | 0.16 | 0.71 | 0.21 | 0.93 | 0.22 | 18.3 |
| *Leptodora* | 0.085 | 0.79 | 0.088 | 0.82 | 0.11 | 8.61 | 0.0070 | 0.54 | 0.0049 | 0.38 | 0.013 | 0.59 |
| *Bosmina* | 0.025 | 0.39 | 0.030 | 0.47 | 0.064 | 2.77 | 0.0042 | 0.14 | 0.0058 | 0.19 | 0.030 | 0.50 |
| Sididae | 0.012 | 0.47 | 0.016 | 0.61 | 0.026 | 1.39 | 0.017 | 0.50 | 0.019 | 0.55 | 0.034 | 1.80 |
| Podocopa | 0.019 | 0.21 | 0.0037 | 0.042 | 0.090 | 1.13 | 0.0030 | 0.17 | 0.0004 | 0.023 | 0.017 | 0.17 |
| Chydoridae | 0.0045 | 0.35 | 0.0043 | 0.33 | 0.013 | 0.44 | 0.0079 | 0.46 | 0.0018 | 0.10 | 0.017 | 0.48 |
| *Daphnia* | 0.0011 | 0.25 | 0.0001 | 0.031 | 0.0043 | 0.0600 | 0.0033 | 0.38 | 0.0004 | 0.042 | 0.0086 | 0.18 |
| Cumacea | 0.0053 | 0.31 | 0.0046 | 0.27 | 0.017 | 0.50 | 0.0027 | 0.21 | 0.0004 | 0.028 | 0.013 | 0.15 |
| Bivalvia | 0.0056 | 0.33 | 0.0043 | 0.25 | 0.017 | 0.49 | 0.0029 | 0.67 | 0.0023 | 0.54 | 0.0043 | 0.26 |
| Chironomidae | 0.0056 | 0.44 | 0.0072 | 0.56 | 0.013 | 0.64 | 0.0099 | 0.23 | 0.016 | 0.38 | 0.043 | 1.31 |
| Clupeidae | 0.0078 | 0.61 | 0.012 | 0.94 | 0.013 | 0.99 | 0.0038 | 0.44 | 0.0079 | 0.92 | 0.0086 | 0.58 |

The diet was composed predominantly of copepods and amphipods, and was similar between locations. Previous studies (Hurst and Conover 2001; Jordan et al. 2003; Limburg et al. 1997) also found that earlystage bass fed primarily on copepods and amphipods as well as Bosmina, which was not well represented in the diets of our samples. The scarcity of Bosmina within our samples is likely a result of high seasonal variation in abundance: Bosmina displays a pattern of dramatic increase and decline in May through June (Limburg et al. 1997; Strayer et al. 1999). Most of our samples were taken after the major Bosmina bloom would be expected. In terms of overall %PSIRI, copepods were the most important prey item upriver and second-most important downriver. The importance of copepods in the diet of young-of-the-year Striped Bass has been highlighted in other studies (Gardinier and Hoff 1982; Limburg et al. 1997; Markle and Grant 1970) and may be explained by their relatively stable abundance. Even in years of zebra mussel impacts, when some types of zooplankton were greatly depleted, copepod abundances remained stable (Pace et al. 1998).

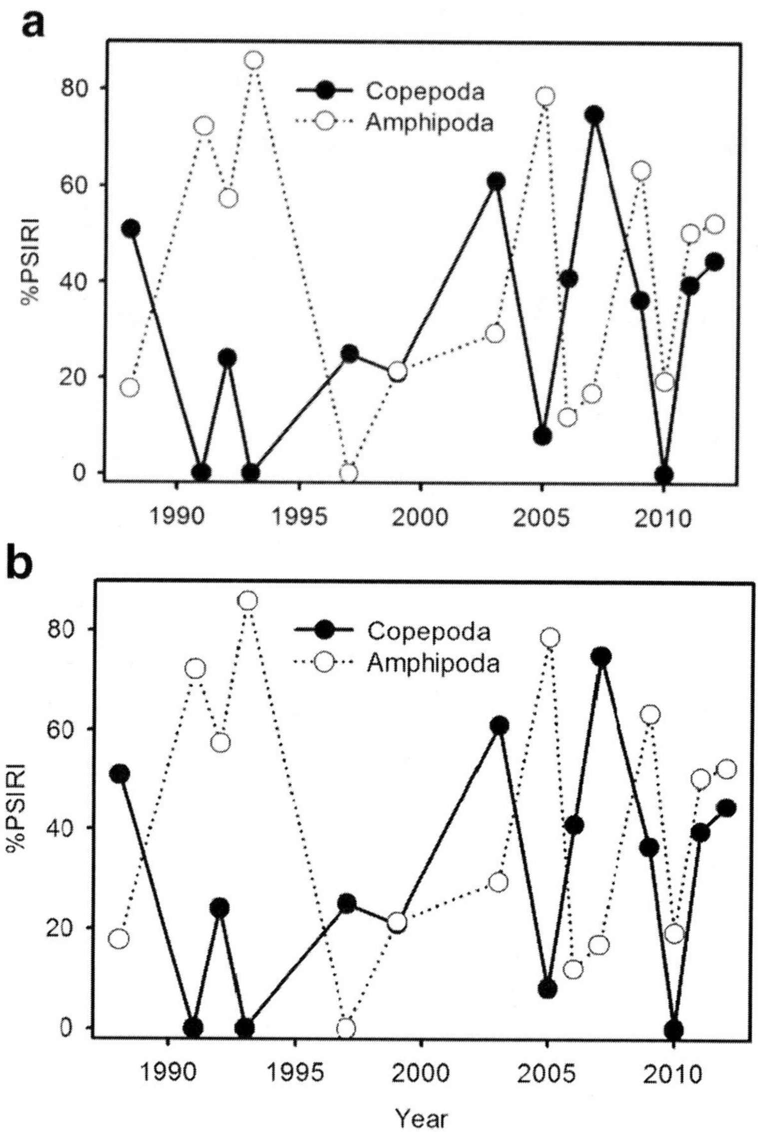


**Figure. 2** Annual mean %PSIRI of copepods and amphipods (all families combined) for (a) upriver and (b) downriver samples

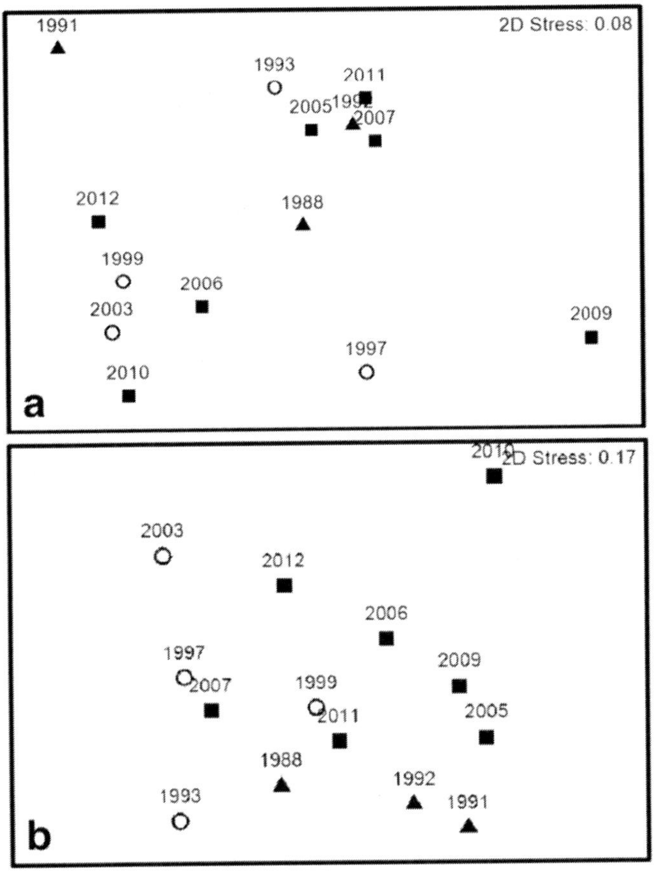


**Figure. 3** Non-metric multidimensional scaling plot of (a) upriver and (b) downriver diet composition. Distance between points indicates dissimilarity. Solid triangles represent years of the preinvasion period, open circles represent years of the invasionimpact period, and solid squares represent years of the recovery period

The prevalence of amphipods in the diet across years, particularly downriver, was similarly observed by Jordan et al. (2003) and Gardinier and Hoff (1982). Prey that shifted in importance along-river included Leptodora, which is a freshwater cladoceran (Browman et al. 1989), and mysids, most of which are found in marine and brackish waters (Smith 2001).

We analyzed two indices of feeding success in this study. Volume of gut contents provides a short-term (i.e. hours) record of recent feeding given short passage time (Govoni et al. 1986), whereas morphometric indices of condition can accurately assess feeding success over a period of several days (Ferron and Leggett 1994). The short-term measure should be more subject to stochastic variability and high-frequency influences such as diel timing of sampling. As expected, volume of gut contents was less well predicted by explanatory variables than dry mass (e.g., in regression against length, $R^2$ = 0.24 and 0.98 respectively). Nonetheless, the two measures of feeding success were intercorrelated, indicating that some of the processes affecting variability in volume of gut contents were sufficiently sustained to affect condition as well. We found a strong effect of year on both indices of feeding success.We expected to find that measures of long-term feeding success were higher downriver. In juvenile euryhaline species, higher salinities are typically associated with higher specific growth rate, efficient food conversion ratio, and energy absorption efficiency (Altinokand and Grizzle 2001). This prediction was not supported by our data.

**Table 5** Predominant prey of larval (a) and juvenile (b) striped bass from upriver and downriver locations in the Hudson River estuary, quantified by mean proportional numerical abundance (pN), mean proportional prey-specific numerical abundance (pPN), mean proportional abundance in weight (pW), mean proportional prey specific abundance in weight (pPW), mean proportional frequency of occurrence (pFO), and percent prey specific index of relative importance (%PSIRI). Table presents only values for copepod and amphipod prey types to facilitate comparisons between life stages. Each value represents a 14-year average

| | Upriver | | | | | | Downriver | | | | | |
|---|---|---|---|---|---|---|---|---|---|---|---|---|
| Prey Taxon | pN | pPN | pW | pPW | pFO | %PSIRI | pN | pPN | pW | pPW | pFO | %PSIRI |
| a) | | | | | | | | | | | | |
| Copepoda | 0.46 | 0.73 | 0.36 | 0.57 | 0.63 | 40.6 | 0.47 | 0.88 | 0.39 | 0.73 | 0.54 | 43.1 |
| Amphipoda: | | | | | | | | | | | | |
| Gammaridae | 0.18 | 0.50 | 0.24 | 0.67 | 0.36 | 20.8 | 0.20 | 0.65 | 0.21 | 0.69 | 0.31 | 20.7 |
| Unidentified | 0.16 | 0.57 | 0.19 | 0.69 | 0.28 | 17.3 | 0.15 | 0.62 | 0.16 | 0.68 | 0.24 | 15.4 |
| b) | | | | | | | | | | | | |
| Copepoda | 0.38 | 0.75 | 0.35 | 0.69 | 0.50 | 36.2 | 0.33 | 0.66 | 0.23 | 0.47 | 0.49 | 28.0 |
| Amphipoda: | | | | | | | | | | | | |
| Gammaridae | 0.40 | 0.68 | 0.42 | 0.72 | 0.58 | 40.9 | 0.13 | 0.42 | 0.15 | 0.48 | 0.30 | 13.6 |
| Unidentified | 0.08 | 0.40 | 0.09 | 0.43 | 0.21 | 8.66 | 0.24 | 0.58 | 0.26 | 0.62 | 0.42 | 25.1 |

We found that associations of environmental variables with long-term feeding success differed with location. In upriver locations, high values of salinity and low values of zebra mussel grazing rate were associated with higher condition in the best explanatory models, and some of these models indicated associations with higher values of copepod abundance, higher values of dissolved oxygen, and a quadratic effect of temperature. The association between condition and an estimate of annual mean zebra mussel filtration rate supports the inference of previous studies (Strayer et al. 2004, 2014b) that depletion of production in the water column by zebra mussels has reduced feeding opportunities for young fish. Salinity at the time and location that the fish were sampled modulated the association with zebra mussel grazing, perhaps indicating a transient inhibition of grazing or an intrusion of forage for early-stage fishes from downriver locations. In downriver locations, high values of copepod abundance and low values of dissolved oxygen were associated with higher condition in the best explanatory models, and some of these models indicated associations with higher values of zebra mussel grazing and higher or lower values of chlorophyll. This positive effect of annual mean copepod abundance on condition indicates that the availability of a predominant prey item drives feeding opportunities. The negative effect of dissolved oxygen is surprising, given that decreasing dissolved oxygen generally results in decreased growth and food intake in fish (Brett and Groves 1979; Jobling 1993), and that low dissolved oxygen levels influence pathways of energy flow in estuaries (Breitburg et al. 1997; Wu 2002). On the other hand, the samples we analyzed were not subject to hypoxic conditions (i.e. the 5th percentile of $dO_2$ was 5.6 mg $l^{-1}$ , well above a generally accepted value for hypoxia of 2 mg $l^{-1}$). It is likely that dissolved oxygen at these levels doesn't directly affect condition, but instead it reflects the influence of some other factor that we have not accounted for in our modeling, or is aliased with combinations of other measured variables such as temperature and salinity.

We emphasized information theoretic criteria for model selection in our analyses. We favored this approach over selection of variables according to pvalues,b ecause we sought the set of variables that best fit the data, rather than selecting variables that were least likely to have had no effect. Lively discussion continues on the relative merits of information-theoretic criteria versus significance testing (Murtaugh 2014), but there is general agreement that either should be supplemented by estimates of effect size.We have done so by showing that an individual in a favorable environment, according to the selected model, would be 33% - 35% heavier at length than an individual in a poor environment. We regard these as relatively dramatic differences in condition. We note also that p-values of all predictor variables in the best-selected models were ≤0.05.

This study is the first to investigate zebra mussel effects on early-stage Striped Bass feeding ecology. Given the mussel's dramatic impacts on the estuary's trophic structure (Strayer et al. 1999, 2014a), it was plausible to expect that feeding success had declined during years of high impact, particularly upriver where zebra mussels are abundant. Prior work suggesting this had motivated the present study: Strayer et al. (2004) found that apparent growth rates of juvenile Striped Bass was lower in the freshwater portion of the estuary during what we refer to here as the invasion impact period, relative to the pre-invasion period. Subsequently, during what is termed the recovery period, Striped Bass growth rate did not rebound, unlike that of other open-water feeding species such as American Shad (Strayer et al. 2014b). We find that feeding success didnot vary according to the pre-invasion, invasion impact and recovery periods, despite pronounced interannual variability. The association with zebra mussel was revealed in regressions that modelled zebra mussel effects as a continuous variable in conjunction with other predictors, thus quantitatively representing annual differences in zebra mussel grazing as well as variability in other environmental factors that could directly or indirectly influence feeding success.

Zebra mussel have affected diet composition and feeding success of fishes in other systems. In the Great Lakes, the appearance of zebra mussel depleted populations of the amphipod Diporeia, which forced Lake Whitefish and Alewives to shift their prey use; feeding success declined, as did growth rate and condition (Madenjian et al. 2015; Pothoven et al. 2001; Pothoven and Madenjian 2008). Some fishes incorporate zebra mussels into their diets (French and Bur 1996; Nack et al. 2015; Pothoven et al. 2001; Ruetz et al. 2012;Watzin et al. 2008), but we found no evidence of early stage Striped Bass feeding on zebra mussel veligers. Other molluscs occurred as rare prey in Striped Bass and other early-stage fishes in our samples; thus we regard as possible but unlikely that the absence of young bivalves in our samples might be attributable to preservation in formalin.

About 25% of early stage striped bass guts were empty in our samples. Specimens collected for this study were all taken at night so that survey catchabilityw ould be high, whereas most dietary studies that report sampling time occurred during daylight hours. Although Striped Bass larvae do feed at night (Chesney 1989; McHugh and Heidinger 1977), Duston and Astatkie (2012) found that the prey capture rate of larval Striped Bass at night depends on prey densities; visual feeding was up to four-fold more effective than nonvisual feeding at low prey densities but similar at the highest densities. Nonetheless, we found no effect of tow timeo n gut fullness (results not shown). We acknowledget hat this does not ensure that nighttime tows provide a complete picture of feeding ecology. Future studies in the Hudson might incorporate daytime samples to compliment the data we have shown.

This research underscores the value of studying longterm ecological data by enabling us to assess how Striped Bass feeding ecology has responded to complex phenomena that occurred over a prolonged period within an estuary. If we had taken a cross-sectional approach,w e might have drawn different conclusions based on isolated events or fluctuations in environmental factors that wouldn't have been accounted for in our data set. This study also sheds light on the importance of longterm environmental sampling programs and associated samples. Archived samples enabled us to characterize 14 years of Striped Bass diet composition and to demonstrate the effect on trophic ecology of early stage Striped Bass of zebra mussels within the Hudson River estuary. The value of long-term monitoring studies is closely tied to decisions about what and how to monitor (Lindenmayer and Likens 2009); we also recommend areful consideration of what samples to preserve and how to preserve them. It is unfortunate that our formalin-fixed and preserved samples could not furnish direct estimates of growth rate via otolith daily increment analysis. Future studies that examine the trophic impact of further zebra mussel expansion and that of other invasive species on early-stage fishes should combine analysis of feeding ecology as well as measures of individual vital rates such as growth.

## ACKNOWLEDGEMENTS

We thank Normandeau Associates and the New York State Museum for providing the samples for this study. Alan Bullock, Hannah Proctor, and BryanWeatherwax coordinated specimen transfers. John Young of ASA Analysis and Communications assisted with water quality datasets. John Young and Sebastian A. Klarian helpfully commented on manuscript drafts. The following University of Connecticut students helped with specimen processing and data collection: Allison Villa, Harley Pastore, Keith Berggren, Mary Ward, Kayla Delventhal, Jacob Hudobenko, Lucie Rodriguez, Mark Schuman, Allison Ballek, Hayley Morway, Sophie Van Etten, Allison Brodeur, Miranda Lin, Chloe Toso, and Brian Ruger. Funding for this work was provided by the Hudson River Foundation (Grant #008/13A) Some of the data used herein were collected with the support ofthe LTREB program (DEB-1119739) of the National Science Foundation.

## REFERENCES

1. Akaike H (1974) A new look at the statistical model identification. IEEE T Automat Contr 19:716–723
2. Alcaraz C, Gholami Z, Esmaeili HR, García-Berthou E (2015) Herbivory and seasonal changes in diet of a highly endemic cyprinodontid fish (Aphanius farsicus). Environ Biol Fish 98:1541–1554
3. Altinokand I, Grizzle JM (2001) Effects of brackish water on growth, feed conversion and energy absorption efficiency by juvenile euryhaline and freshwater stenohaline fishes. J Fish Biol 59:1142–1152
4. ASA Analysis and Communication (2016) 2014 year class report for the Hudson River estuary monitoring program. Washingtonville, NY

5. Baxter CV, Fausch KD, Saunders WC (2005) Tangled webs: reciprocal flows of invertebrate prey link streams and riparian zones. Freshw Biol 50:201–220
6. Blaber SJM, Blaber TG (1980) Factors affecting the distribution of juvenile estuarine and inshore fish. J Fish Biol 17:143–162
7. Braga RR, Bornatowski H, Vitule JRS (2012) Feeding ecology of fishes: an overview of worldwide publications. Rev Fish Biol Fisher 22:915–929
8. Breitburg DL, Loher T, Pacey CA, Gerstein A (1997) Varying effects of low dissolved oxygen on trophic interactions in an estuarine food web. Ecol Monogr 67:489–507
9. Brett J, Groves T (1979) Physiological energetics. Fish Physiology 8:279–352
10. Browman HI, Kruse S, O'Brien WJ (1989) Foraging behavior of the predaceous cladoceran, Leptodora kindti, and escape responses of their prey. J Plankton Res 11:1075–1088
11. Brown SC, Bizzarro JJ, Cailliet GM, Ebert DA (2012) Breaking with tradition: redefining measures for diet description with a case study of the Aleutian skate Bathyraja aleutica (Gilbert 1896). Environ Biol Fish 95(1):3–20
12. Burnham KP, Anderson D (2002) Model selection and multimodel inference: a practical information-theoric approach, 2nd edn. Springer, New York
13. Caraco NF, Cole JJ, Raymond PA, Strayer DL, Pace ML, Findlay SE, Fischer DT (1997) Zebra mussel invasion in a large, turbid river: phytoplankton response to increased grazing. Ecology 78:588–602
14. Chesney EJ Jr (1989) Estimating the food requirements of striped bass larvae Morone saxatilis: effects of light, turbidity and turbulence. Mar Ecol-Prog Ser 53:191–200
15. Clarke KR (1993) Non-parametric multivariate analyses of changes in community structure. Aust J Ecol 18:117–143
16. Collar DC, O'Meara BC, Wainwright PC, Near TJ (2009)Piscivory limits diversification of feeding morphology incentrarchid fishes. Evolution 63:1557–1573
17. Donadelli V, Longobardi A, FinoiaMG,Marino G (2015) Feeding hatchery-reared dusky grouper Epinephelus marginatus juveniles on live prey: implications for restocking. Environ Biol Fish 98:1757–1766
18. Duston J, Astatkie T (2012) Prey density and nonvisual feeding by larval striped bass. T Am Fish Soc 141:610–614
19. Feitosa JLL, Ferreira BP (2015) Distribution and feeding patterns of juvenile parrotfish on algal-dominated coral reefs. Mar Ecol 36:462–474
20. Ferron A, LeggettWC (1994) An appraisal of condition measures for marine fish larvae. Adv in Mar Biol 30:217–303
21. French JRP, Bur MT (1996) The effect of zebra mussel consumption on growth of freshwater drum in Lake Erie. J Freshw Ecol 11:283–289
22. Friedland KD, Ahrenholz DW, Guthrie JF (1996) Formation and seasonal evolution of Atlanticmenhaden juvenile nurseries in coastal estuaries. Estuaries 19:105–114
23. GardinierMN, Hoff TB (1982) Diet of striped bass in the Hudson River estuary (Morone saxatilis). New York Fish Game J 29: 152–165
24. Gerking SD (1994) Feeding ecology of fish. Academic Press, San Diego, California
25. Govoni JJ, Boehlert GW, Watanabe Y (1986) The physiology of digestion in fish larvae. Environ Biol Fish 16:59–77
26. Hardy JD (1978) Development of fishes of the Mid-Atlantic Bight - an atlas of egg, larval, and juvenile stages. - Volume IIIAphredoderidae through Rachycentridae. Department of the Interior Fish and Wildlife Service, Washington DC
27. Howe DV, Jordan RC, Juanes F (2008) Selective feeding in a generalist invertivore, age-0 striped bass. Ecol Freshwater Fish 17:495–501
28. Hurst TP, Conover DO (2001) Diet and consumption rates ofoverwintering YOY striped bass, Morone saxatilis, in the Hudson River. Fishery B-NOAA 99:545–553
29. Hyslop EJ (1980) Stomach content analysis- a review of methods and their applications. J Fish Biol 17:411–429
30. Jakob EM, Marshall SD, Uetz GW (1996) Estimating fitness: a comparison of body condition indices. Oikos 77(1):61–67

31. Jobling M (1993) Bioenergetics: feed intake and energy partitioning. In: Rankin JC, Jensen FB (eds) Fish ecophysiology. Chapman and Hall, London, pp. 1–44
32. Johnston TA, Mathias JA (1993) Length reduction and dry weight loss in frozen and formalin-preserved larval walleye, Stizostedion vitreum (Mitchill). Aquac Res 24:365–371
33. Jordan RC, Howe DV, Hurst TP, Juanes F (2003) Feeding habits of age-0 striped bass, Morone saxatilis, in the mid-Hudson River estuary: temporal, spatial, and ontogenetic variation. Estuaries 26:1486–1493
34. Jude DJ (1973) Food and feeding habits of gizzard Shad in pool 19 Mississippi River. T Am Fish Soc 102:378–383
35. Levinton JS,Waldman JR (eds) (2006) The Hudson River estuary.Cambridge University Press, New York
36. Limburg KE, Pace ML, Fischer D, Arend KK (1997)Consumption, selectivity, and use of zooplankton by larval striped bass and white perch in a seasonally pulsed estuary. T Am Fish Soc 126:607–621
37. Lindenmayer DB, Likens GE (2009) Adaptive monitoring: a new paradigm for long-term research and monitoring. Trends Ecology Evolution 24:482–486
38. Lindenmayer DB, Likens GE, Andersen A, Bowman D, Bull CM, Burns E, Dickman CR, Hoffmann AA, Keith DA, Liddell MJ, Lowe AJ, Metcalfe DJ, Phinn SR, Russell-Smith J,Thurgate N, Wardle GM (2012) Value of long-term ecological studies. Austral Ecol 37:745–757
39. Llopiz JK (2013) Latitudinal and taxonomic patterns in the feeding ecologies of fish larvae: a literature synthesis. J Marine Syst 109-110:69–77
40. MacLusky DS, Elliott M (2004) The estuarine ecosystem: ecology, threats, and management, 3rd edn. Oxford University Press, New York
41. Madenjian CP et al (2015) Changes in the Lake Michigan food web following dreissenid mussel invasions: a synthesis.Journal of Great Lakes research 41. Supplement 3:217–231
42. Markle DF, Grant GC (1970) The summer food habits of youngof-the year striped bass in three Virginia Rivers. Chesap Sci 11:50–54
43. McHugh JJ, Heidinger RC (1977) Effects of light on feeding and egestion time of striped bass fry. Prog Fish Cult 39:33–34
44. Morgan RP, Rasin VJ, Copp RL (1981) Temperature and salinity effects on development of striped bass eggs and larvae. Tam Fish Soc 110:95–99
45. Murtaugh PA (2014) In defense of P values. Ecology 95:611–617
46. Nack CC, Limburg KE, Schmidt RE (2015) Diet composition and feeding behavior of larval American Shad, Alosa sapidissima (Wilson), after the introduction of the invasive zebra mussel, Dreissena polymorpha (Pallas), in the Hudson River estuary, NY. Northeast Nat 22:437–450
47. O'Connor MP, Juanes F, McGarigal K, Caris J (2012) Describing juvenile American shad and striped bass habitat use in the Hudson River estuary using species distributionmodels. EcolEng 48:101–108
48. Pace ML, Findlay SEG, Fischer D (1998) Effects of an invasivebivalve on the zooplankton community of the Hudson River. Freshw Biol 39:103–116
49. PaceML, Strayer DL, Fischer D, MalcomHM(2010) Recovery of native zooplankton associated with increased mortality of an invasive mussel. Ecosphere 1:1–10
50. Parker RR (1963) Effects of formalin on length and weight of fishes. Journal Fish Board Can 20:1441–1455
51. Pothoven SA, Madenjian CP (2008) Changes in consumption by alewives and lake whitefish after dreissenid mussel invasionsin lakes Michigan and Huron. N Am J FishManage 28:308– 320
52. Pothoven SA, Nalepa TF, Schneeberger PJ, Brandt SB (2001) Changes in diet and body condition of lake whitefish insouthern LakeMichigan associated with changes in benthos. N Am J Fish Manage 21:876–883
53. Ruetz CR, Reneski MR, Uzarski DG (2012) Round goby predation on Dreissena in coastal areas of eastern LakeMichigan. J Freshw Ecol 27:171–184

54. Schultz ET, Young J, Martin JM, Lwiza KM (2005) Tracking cohorts: analysis of migration in the early life stages of an estuarine fish. Estuaries 28:394–405
55. Shields PA, Carlson SR (1996) Effects of formalin and alcohol preservation on lengths and weights of juvenile sockeye Salmon. Alaska Fish Res Bull 3:81–93
56. Smith DG (2001) Minor Malacostraca. In: Pennak's freshwater invertebrates of the United States: Porifera to Crustacea, 4th edn. John Wiley and Sons, Canada, pp. 545–551
57. Strayer DL, Smith LC (2001) The zoobenthos of the freshwater tidal Hudson River and its response to the zebra mussel (Dreissena polymorpha) invasion. Arch Hydrobiol Suppl 139:1–52
58. Strayer DL, Powell J,Ambrose P, Smith LC, PaceML, Fischer DT (1996) Arrival, spread, and early dynamics of a zebra mussel(Dreissena polymorpha) population in the Hudson River estuary. Can J Fish Aquat Sci 53:1143–1149
59. Strayer DL, Caraco NF, Cole JJ, Findlay S, Pace ML (1999) Transformation of freshwater ecosystems by bivalves: a case study of zebra mussels in the Hudson River. Bioscience 49: 19–27
60. Strayer DL, Hattala KA, Kahnle AW(2004) Effects of an invasive bivalve (Dreissena polymorpha) on fish in the Hudson River estuary. Can J Fish Aquat Sci 61:924–941
61. Strayer DL, Cid N, Malcom HM (2011) Long-term changes in a population of an invasive bivalve and its effects. Oecologia 165:1063–1072
62. Strayer DL, Cole JJ, Findlay SE, Fischer DT, Gephart JA,Malcom HM, Pace ML, Rosi-Marshall EJ (2014a) Decadal-scale change in a large-river ecosystem. Bioscience 64:496–510
63. Strayer DL, Hattala KA, Kahnle AW, Adams RD (2014b) Has the Hudson River fish community recovered from the zebra mussel invasion along with its forage base? Can J Fish Aquat Sci 71:1146–1157
64. Waldman JR (2006) The diadromous fish fauna of the Hudson River: life histories, conservation concerns, and research avenues. In: Levinton JS, Waldman JR (eds) The HudsonRiver estuary. Cambridge University Press, New York, pp. 171–188
65. WatzinMC, Joppe-Mercure K, Rowder J, Lancaster B, Bronson L( 2008) Significant fish predation on zebra mussels Dreissena polymorpha in Lake Champlain, U.S.a. J Fish Biol 73:1585– 1599
66. Wu RS (2002) Hypoxia: from molecular responses to ecosystem responses. Mar Pollut Bull 45:35–45
67. Yeh CF, Hodson RG (1975) Effects of formalin on length and weight of bluegill and white crappie from Lake Nasworthy. Texas Southwest Nat:315–321

# CHAPTER 11

# Conservation Physiology of Marine Fishes: State of The Art and Prospects for Policy

**David J. McKenzie**[1]

[1]*Centre for Marine Biodiversity Exploitation and Conservation, UMR MARBEC (CNRS, IRD, IFREMER, UM), Place E. Bataillon cc 093, 34095*

## ABST RACT

The state of the art of research on the environmental physiology of marine fishes is reviewed from the perspective of how it can contribute to conservation of biodiversity and fishery resources. A major constraint to application of physiological knowledge for conservation of marine fishes is the limited knowledge base; international collaboration is needed to study the environmental physiology of a wider range of species. Multifactorial field and laboratory studies on biomarkers hold promise to relate ecophysiology directly to habitat quality and population status. The 'Fry paradigm' could have broad applications for conservation physiology research if it provides a universal mechanism to link physiological function with ecological performance and population dynamics of fishes, through effects of abiotic conditions on aerobic metabolic scope. The available data indicate, however, that the paradigm is not universal, so further research is required on a wide diversity of species. Fish physiologists should interact closely with researchers developing ecological models, in order to investigate how integrating physiological information improves confidence in projecting effects of global change; for example, with mechanistic models that define habitat suitability based upon potential for aerobic scope or outputs of a dynamic energy budget. One major challenge to upscaling from physiology of individuals to the level of species and communities is incorporating intraspecific variation, which could be a crucial component of species' resilience to global change. Understanding what fishes do in the wild is also a challenge, but techniques of biotelemetry and biologging are providing novel information towards effective conservation. Overall, fish physiologists must strive to render research outputs more applicable to management and decision-making. There are various potential avenues for information flow, in the shorter term directly through biomarker studies and in the longer term by collaborating with modellers and fishery biologists.

**Keywords:** Biomarkers, ecological models, fisheries, Fry paradigm, individual variation, telemetry

# INTRODUCTION

Marine ecosystems provide essential resources and services, with fishes being of prime socio-economic importance. There are alarming global trends of excessive exploitation and habitat degradation of marine fishes, with most commercial stocks either overfished or nearing capacity (Pauly *et al.*, 2002; FAO, 2014; Pauly and Zeller, 2016). Global climate change is also altering patterns of marine biodiversity, with more pronounced effects expected in the future (Perry *et al.*, 2005; IPCC, 2014; McNeil and Sasse, 2016). The consequences of over-exploitation, habitat degradation and global climate change are unknown, but there is legitimate concern about irreversible loss of fisheries resources and biodiversity, leading to reduced food security and even direct threats to ecosystem integrity (Perry *et al.*, 2005; Cheung *et al.*, 2009; Sumaila *et al.*, 2011; Duarte, 2014; Elliott *et al.*, 2015; Pauly and Zeller, 2016). There is a need, therefore, to improve the scientific knowledge base underpinning advice on the sustainable management of marine fish biodiversity and fisheries resources.

There is a growing belief that physiological information, understanding how animals function, can contribute significantly to the resolution of management and conservation problems for marine fishes and to the ability accurately to project potential impacts of environmental pressures (Wang and Overgaard, 2007; Pörtner and Farrell, 2008; Rijnsdorp *et al.*, 2009; Wilson *et al.*, 2009; Pörtner and Peck, 2010; Jørgensen *et al.*, 2012; Seebacher and Franklin, 2012; Wilson, 2014). Physiologists typically take a Darwinian view (Fig. 1), whereby the abiotic factors within a given habitat can define which animals survive and reproduce there, based upon their physiology. Over the course of generations, there is natural selection of physiological adaptations to prevailing conditions (Prosser, 1950; Schmidt-Neilsen, 1982; Garland and Carter, 1994), with the evolution of a functional niche for each species (Hutchinson, 1957). In each new generation, the physiology of the individuals contributes to their performance, behaviour and fitness in a realized niche (Arnold, 1983; Garland and Carter, 1994; Feder *et al.*, 2000; Huey *et al.*, 2012). This influences the abundance and distribution of their population and species (Buckley *et al.*, 2012; Huey *et al.*, 2012; Chave, 2013; Heffernan *et al.*, 2014) and, by logical extension, the composition of communities and assemblages in the ecosystem (Buckley *et al.*, 2012; Chave, 2013; Cooke *et al.*, 2014; Heffernan *et al.*, 2014).

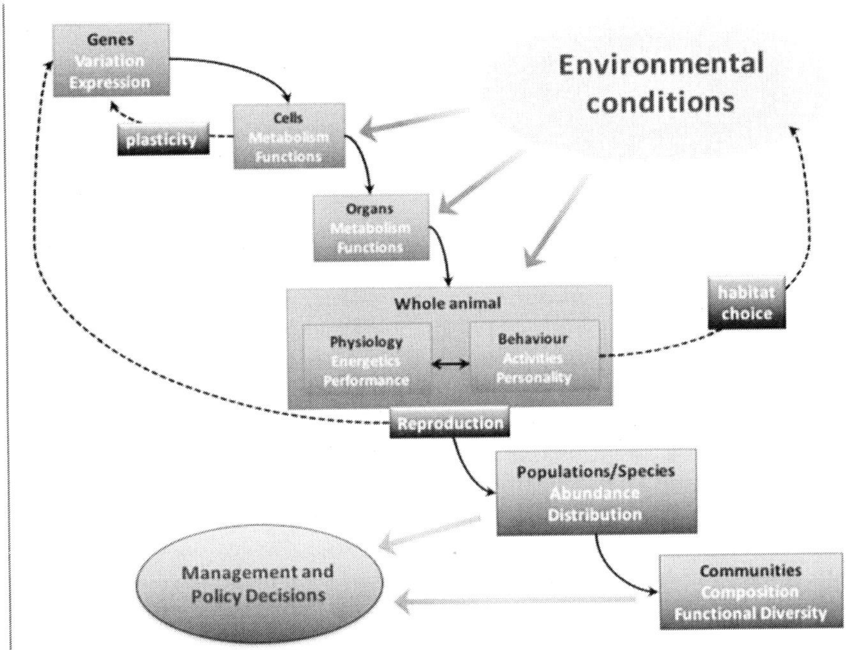


**Figure 1:** How the physiology of an individual marine fish can influence species population dynamics and community biodiversity, through hierarchical levels of biological organization (inspired by Le Quesne and Pinnegar, 2012). Environmental conditions influence whole-animal physiology (energetics and performance) by influencing gene expression and the biochemistry and physiology (metabolism and function) of cells and organs. Physiology has a complex interplay with the behaviour (activities and personality) of the individual that, together, influence its ecological performance and reproductive output (fitness). This, then, influences the abundance and distribution of the species in the environment. This, in turn, influences the composition and functional diversity of communities and assemblages in the ecosystem. This is the scale at which management decisions are required and made. The dotted black lines show feedbacks. The effects of environmental conditions on cells feed back to influence gene expression, which then influences function by phenotypic plasticity/flexibility. Behavioural habitat choice feeds back to influence

environmental conditions and, therefore, their effects on lower levels of organization. Reproduction each generation feeds back to influence genetic variation and drive the evolution of local adaptation in the species.

The vast majority of marine fishes are water-breathing ectotherms; therefore, physiological and behavioural responses to increasing temperatures, growing zones of hypoxia, ocean acidification, eutrophication and general habitat degradation are to be expected. This has obvious implications for conservation research because the prediction would be that, as environmental conditions change, so will the distribution of habitat that any given species chooses to, or is able to, occupy (Buckley *et al.*, 2012; Huey *et al.*, 2012; Le Quesne and Pinnegar, 2012; Cooke *et al.*, 2014; Martin *et al.*, 2015).

This overall premise was the impetus to establish a network of interested scientists, funded by the European Union's Cooperation in Science and Technology (COST) programme. The COST Action FA1004 'Conservation Physiology of Marine Fishes' (2011–2015) provided a forum for dialogue on this topic, through a series of international conferences and workshops (http://fish-conservation.nu). This article reviews the main issues that were raised and discussed over COST FA1004's lifetime, with general perspectives on a broad series of topics. These topics fall into four main themes: (i) the state of basic physiological knowledge for marine fishes and how this might be applied directly towards conservation goals; (ii) how physiological knowledge can be integrated into ecological models; (iii) how biotelemetry and biologging studies can contribute to conservation research; and (iv) how conservation physiology research might be rendered relevant and available to decision-makers. These perspectives set the stage for this Theme Section, which comprises reviews, perspectives and research articles that, together, provide an indication of the state of the art of thinking and research in the field.

## The physiological knowledge base is limited

The restricted knowledge base is currently a major constraint to the use of physiology for conservation of marine fishes. Although there are >30 000 species of fishes, knowledge on marine fishes is confined to tens of species, which occur in countries with developed fish ecophysiology research communities. Within these countries, there is a focus on species that are economically or ecologically important and/or are relatively easy to obtain and maintain in captivity. These include temperate species, such as Atlantic cod (*Gadus morhua*), Atlantic salmon (*Salmo salar*), Dover sole (*Solea solea*), European sea bass (*Dicentrarchus labrax*), Pacific salmonids of the genus *Oncorhynchus* or turbot (*Scopthalmus maximus*), plus various tropical species from the Great Barrier Reef. Recent meta-analyses examining how the physiology of ectotherms might shape responses to global change all note the lack of physiological information on a vast majority of fish species and geographical areas (Sunday *et al.*, 2012; Seebacher *et al.*, 2015; Lefevre, 2016; Killen *et al.*, 2016b).

It is essential that physiologists pursue cooperative research projects progressively to fill knowledge gaps (Cattano et al., 2016; Ferreira-Martins, 2016; di Santo *et al.*, 2016), to investigate the ecophysiology of many more species, which may be harder to obtain and/or keep in captivity. Focus could be on key elements in food chains and species with specific or rare (Mouillot *et al.*, 2013) ecological functions. Studies are needed to evaluate ranges for tolerance of major environmental factors, such as temperature, dissolved gases (hypoxia and hypercarbia), acidification and salinity, within which different marine fish species function effectively, and the thresholds beyond which performance is impaired and survival or reproduction is at risk.

Researchers must keep in mind the complexities inherent to the physiology of marine fishes. For example, the environmental physiology of populations can vary markedly across a species' geographical range (Conover *et al.*, 2006; Gardiner *et al.*, 2010), and specific life stages can be critical bottlenecks for population or species persistence in the presence of ongoing global change (Petitgas *et al.*, 2013; Ferreira-Martins, 2016). Physiological research on minute marine fish embryos and larvae is technically very challenging (Peck and Moyano, 2016), but these life stages may be the most sensitive to environmental stressors (Killen *et al.*, 2007). This Theme Section has two research articles on the ecophysiology of early life stages, with a study of tolerance of little skate (*Leucoraja erinacea*) embryos to hypoxia (di Santo *et al.*, 2016) and a study of tolerance of larvae of the ocellated wrasse (*Symphodus ocellatus*) to natural acidification at Mediterranean volcanic seeps (Cattano *et al.*, 2016).

A database is currently being prepared for the public domain, as an output of Action FA1004. At present, it comprises effects of temperature on aerobic metabolic scope, digestive energetics (specific dynamic action) and growth, as well as an analysis of hypoxia tolerance, in marine and euryhaline fish species. In this Theme Section, Rogers *et al.* (2016) analyse the database of hypoxia tolerance, measured by respirometry as the critical oxygen partial pressure for regulation of aerobic metabolism ($P_{crit}$). This revealed that $P_{crit}$ was, as expected, highly variable among species but was also influenced by temperature, $CO_2$, acidification, toxic metals and feeding, as well as by the method used to measure it, especially if $CO_2$ accumulated in the respirometry system. The database will provide an open repository for a progressive accumulation of physiological trait data, which can be used towards conservation objectives; for example, directly in terms of tolerance thresholds that can be 'biomarkers' of environmental stress or to parameterize ecological models, as described below.

## Applications for physiological biomarkers

Biomarkers of environmental pressures hold promise for conservation research in marine fishes (Cooke and O'Connor, 2010); for example, to evaluate ecological quality of habitats within the context of the EU Water Framework Directive or to establish the 'health' of populations in particular habitats of interest. A prime example of an endocrine biomarker is plasma vitellogenin, which, in freshwater fishes, has been established as a key indicator of exposure to endocrine-disrupting chemicals (Sumpter and Jobling, 1995; Tyler *et al.*, 1996).

If endocrine, cellular and molecular biomarkers are to be useful for conservation research, it is important to understand their limitations, which can include a lack of mechanistic basis for their interpretation, complicated response patterns in wild animals and unclear links to Darwinian fitness (Forbes *et al.*, 2006; McKenzie *et al.*, 2007; Dantzer *et al.*, 2014). The most promising approaches are multifactorial and use a combination of indicators at different levels of biological organization. These allow relationships to be established among measures of functional integrity, such as condition factor or otolith growth rates, and the endocrine, cellular and molecular biomarkers. This, in turn, can then be related to differences in the biotic and abiotic quality of habitats (Cooke and Suski, 2008; Adams and Ham, 2011; Jeffrey *et al.*, 2015; King *et al.*, 2016; Madliger *et al.*, 2016). In developing such suites of biomarkers for conservation research, the focus should be on reliable and user-friendly measures that combine field and experimental approaches and provide ecological relevance (Adams and Ham, 2011; Jeffrey *et al.*, 2015; Madliger *et al.*, 2016).

Revealing generalized 'stress' in natural populations can be very informative. The stress hormones, glucocorticoids, measured in feathers, hair, moulted skin or scat, are widely used in conservation physiology research in wild tetrapods (Dantzer *et al.*, 2014). There is evidence that scales can be used in this manner in fishes (Aerts *et al.*, 2015), which opens up this practice for conservation research. Cortisol can be measured in fish eggs as a biomarker of maternal stress levels; in tropical reef fishes, increased egg cortisol was linked to poor reproductive success and reduced offspring size (Gagliano and McCormick, 2009). A blood sample can also provide a wealth of biomarker information in fishes. A major problem with wild marine fishes is accounting for the acute stress of capture, but various biomarkers are presumably not sensitive to this, such as some oxidative stress indicators, stress proteins and the expression of stress-related genes in nucleated teleost red blood cells (Beaulieu and Costantini, 2014; Chadwick *et al.*, 2015; Jeffrey *et al.*, 2015; Madliger *et al.*, 2016).

Simple measures of condition factor and energy reserves are informative physiological biomarkers of population health (Claireaux *et al.*, 2004; Jeffrey *et al.*, 2015) that, in this Theme Section, are applied as biomarkers for effects of parasitism in a small pelagic species, the European anchovy (*Engraulis encrasicolus*), in the northwest Mediterranean (Ferrer-Maza *et al.*, 2016). Some studies have measured physiological indicators of whole-animal performance, such as swimming ability, or hypoxia and thermal tolerance, including measurements on fishes in mesocosms or caged at specific sites (Claireaux *et al.*, 2004; McKenzie *et al.*, 2007; Roze *et al.*, 2013). A multifactorial approach has a number of obvious applications in evaluating impacts of major environmental stressors on marine fishes, and to predict the relative sensitivity of different species.

The critical thermal maximum is a physiological 'biomarker' of incipient lethal thermal tolerance (Lutterschmidt and Hutchison, 1997), which has been related to ecological phenomena caused by global change (Pörtner and Peck, 2010; Sunday *et al.*, 2011, 2012). In natural populations, habitat warming and extreme thermal events can generate sublethal molecular biomarker responses, notably heat shock proteins. Some freshwater Arctic charr (*Salvelinus alpinus*) populations inhabit water bodies that, as a result of global change, now exceed the fish's seasonal thermal optimum. These populations exhibit constitutive heat shock protein and glucose stress responses (Chadwick *et al.*, 2015). A history of exposure to extreme warming events can affect glucocorticoid responsiveness to acute stress in coral reef fishes (Mills *et al.*, 2015), and the thermal regime during development and incubation may have marked influences on offspring (Zambonino-Infante *et al.*, 2013; Moyano *et al.*, 2016). There is an opportunity, therefore, to develop databases on lethal thresholds, evaluated as the critical thermal maximum, but also to investigate how evidence of sublethal thermal stress in populations of interest might relate to functional indicators, such as indicators of bioenergetic or nutritional status. Beyond direct management applications, this sort of information would also inform projections of species sensitivity to predicted patterns of global warming.

Loss of equilibrium during progressive hypoxia has been used as an indicator of incipient lethal hypoxic threshold (Anttila *et al.*, 2013; Roze *et al.*, 2013; Claireaux and Chabot, 2016), and the $P_{crit}$ (or $O_{2crit}$) is a physiological 'biomarker' of sublethal hypoxia tolerance (Claireaux and Chabot, 2016; Rogers *et al.*, 2016). Above the $O_{2crit}$, fishes can show reduced aerobic scope, which can be linked to impairments to physiological performance and reduced appetite. Although these physiological effects of hypoxia may be understood mechanistically, evaluating the impact of environmental hypoxia upon fish ecology and evolution remains difficult in practice. Overcoming this challenge is becoming increasingly important in the face of growing marine hypoxic zones. The constitution of a database gathering key information about species' oxygen requirements and susceptibility to reduced oxygen availability (Rogers *et al.*, 2016) is an important first step. Once again, a multifactorial approach holds promise. Knowledge of hypoxia thresholds, based upon laboratory experiments, could be compared with biochemical, physiological, bioenergetic, nutritional and behavioural indicators in populations of interest, in order to gain insight into the ecological consequences of prevailing hypoxic stress.

Noise pollution can be a major environmental stress for marine fishes, which has been studied for its effects on their behaviour (Slabbekoorn *et al.*, 2010; Simpson *et al.*, 2016) and their physiology (Nichols *et al.*, 2015; Sierra-Flores *et al.*, 2015; Celi *et al.*, 2016; Simpson *et al.*, 2016). In experimental conditions, noise can cause physiological stress responses and upregulation of stress proteins (Nichols *et al.*, 2015; Celi *et al.*, 2016). Research in aquaculture has shown that Atlantic cod exposed to daily, low-level noise pollution during the spawning window accumulated cortisol in their eggs and had lower egg production and fertilization rates (Sierra-Flores *et al.*, 2015). It seems evident, therefore, that further research should be performed to investigate responses to noise pollution in natural populations, applying a multifactorial approach (Adams and Ham, 2011; Jeffrey *et al.*, 2015) that could also include maternal effects, such as accumulation of cortisol in eggs and the potential downstream effects on larvae. Research should distinguish between the type and intensity of noise, on a continuum from not detectable to chronic but allowing habituation to acute and damaging.

In Europe, reforms to the Common Fisheries Policy have made discard of unwanted bycatch an important policy issue, and one where physiological biomarkers clearly have useful applications. This is a very active area of research, so this is restricted here to some generalizations about management needs that are useful to highlight for anyone who is starting out, as follows: (i) characterize the relative sensitivity of different species relative to gear types, environmental conditions and handling procedures; (ii) predict mortality (and sublethal fitness impacts) of discarded fishes (Davis, 2010); and (iii) identify strategies for reducing stress, injury and mortality and improving welfare (Metcalfe, 2009). Bearing in mind that all captured fishes will experience some level of stress and injury, these effects are related and, therefore, may be difficult to disentangle. It is difficult to generalize across species and capture method, and impacts may vary seasonally and ontogenetically. Even fishes that escape capture may suffer impacts of some kind.

## Is there a universal paradigm linking physiological function to ecological performance?

Understanding the physiological mechanisms that determine how marine fishes perform, in relationship to environmental conditions, should contribute to conservation activities by providing insights into current and future species abundance and distribution (Pörtner and Farrell, 2008; Pörtner, 2010; Jørgensen *et al.*, 2012; Teal *et al.*, 2015; Marras *et al.*, 2015a). One major hypothesis to define how environmental conditions affect performance, with implicit consequences for population dynamics and habitat selection, focuses on the ability of fishes to increase their rate of oxygen uptake to meet the metabolic demands of essential activities. First formulated by F. E. J. Fry, hence called the 'Fry paradigm' (Fry, 1971, 1947; Priede, 1985; Kerr, 1990), it is based upon scope for aerobic activity (Fig. 2). Aerobic scope is the integrated capacity of the cardiovascular and respiratory systems to provide oxygen for essential activities beyond vital basal metabolic processes, i.e. activities such as locomotion (e.g. evading predators, foraging, social interactions, migration), digestion and somatic and gonadal growth (Fry, 1971; Priede, 1985; Claireaux and Lefrançois, 2007; Pörtner and Farrell, 2008; Schulte, 2015; Baktoft *et al.*, 2016). Thus, the hypothesis provides a mechanistic link from the structural, biochemical and physiological components of metabolism to ecologically relevant performance measures. There is evidence that large-scale failures in upriver spawning migrations of adult Pacific salmon (*Oncorhynchus* species) may have occurred because abnormally high summer temperatures impaired swimming performance through reduced aerobic scope, which is one of the most prominent examples of conservation physiology research for fishes (Eliason *et al.*, 2011; Patterson *et al.*, 2016).

An important potential strength of the Fry paradigm is that it can integrate effects of major environmental stressors (Fig. 2), notably hypoxia, ocean acidification and pollutants, because they can all constrain aerobic scope (Claireaux and Lagardère, 1999; Lefrançois and Claireaux, 2003; Claireaux and Lefrançois, 2007; Ishimatsu *et al.*, 2008; Claireaux and Chabot, 2016). This focus on aerobic metabolism of fishes was the stimulus for a previous journal special issue by COST FA1004, which provided definitions and methods for measures of metabolic rate in fishes, and case studies illustrating the relevance of metabolic rate in management of fishing and environmental changes (Chabot *et al.*, 2016).

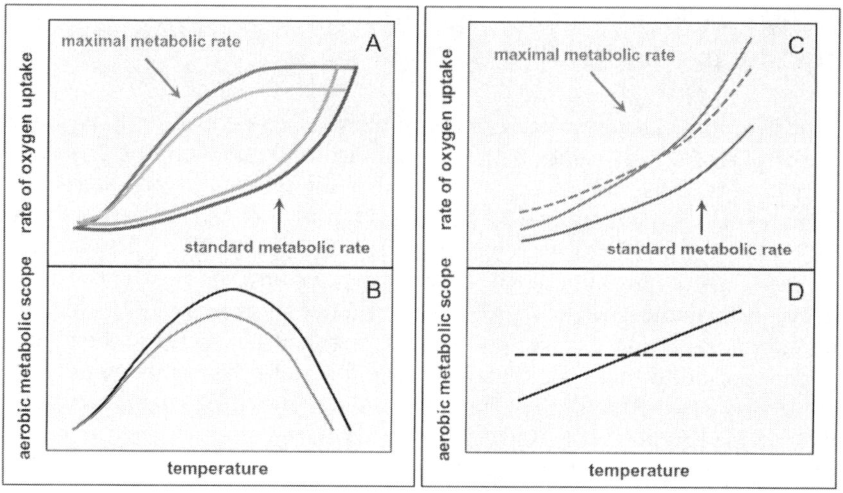


**Figure 2:** (**A**) The Fry or oxygen- and capacity- limited thermal tolerance (OCLTT) paradigm, a conceptual model of how environmental factors influence aerobic metabolic scope (AS) of fishes (Fry, 1971; Pörtner, 2010). Acclimatization temperature, on the abscissa, is a factor that controls the rates of all metabolic processes in ectothermic marine fishes. The blue and red lines in (A) model how standard (minimal) metabolic rate (SMR) and maximal metabolic rate (MMR) vary as a function of temperature. The difference between SMR and MMR is AS, the capacity of the cardiorespiratory system to provide oxygen for all activities above maintenance, i.e. locomotion, digestion and tissue growth. The SMR, the cost of vital basal processes, increases exponentially with temperature, owing to direct acceleratory effects of heat. At low temperatures, MMR is low because of depressive effects of cold on biochemistry and physiology, including mitochondrial oxygen supply and ATP production. As a result, AS is small. As temperature rises, AS increases as biochemical and physiological rate capacity increase. At a certain temperature, however, MMR reaches an asymptote, attributable to intrinsic limitations to cardiorespiratory capacity. At temperatures above this, SMR rises inexorably until basic maintenance requires the entire cardiorespiratory capacity, so AS falls to zero. (**B**) The resultant relationship of AS with acclimatization temperature, with a clear bell-shaped performance curve and a temperature range where AS is maximal. The form of this curve is expected to be species (or even population) specific and to reflect evolutionary history. The other lines in (A) show effects of loading factors, such as stress (light blue), and limiting factors, such as hypoxia (orange), with the resultant reduction in AS shown in (B) in grey. Figure redrawn from McKenzie (2011). (**C**) The relationship between SMR, MMR and temperature that is shown by many species, with (**D**) showing the resultant

relationship of AS with temperature (Lefevre *et al.*, 2016). All fishes show a similar effect of temperature on SMR, but some show a parallel increase in MMR, such that scope is the same across all temperatures (dashed MMR and AS lines in C and D, respectively). Other species show a greater increase in MMR than in SMR, such that scope rises progressively with temperature (continuous MMR and AS lines in C and D, respectively).

The Fry paradigm was the basis for the more mechanistic oxygen- and capacity- limited thermal tolerance (OCLTT) hypothesis, which posits that the decline in aerobic scope outside an optimal temperature range is caused by an impaired capacity of mitochondria to use oxygen at low temperatures or of the cardiorespiratory system to supply oxygen at high temperatures (Pörtner and Knust, 2007; Pörtner and Farrell, 2008; Pörtner, 2010; Schulte, 2015). An influential study by Pörtner and Knust (2007)indicated a correlation across laboratory measurements of aerobic scope at various temperatures with thermal tolerance, growth rate and local abundance of a wild population of the eel pout (*Zoarces viviparous*). The Fry and OCLTT paradigms have attracted considerable research attention for fishes (Clark *et al.*, 2013; Gräns *et al.*, 2014; Norin *et al.*, 2014; Wang *et al.*, 2014; Brijs *et al.*, 2015) and other ectotherms (Overgaard *et al.*, 2012; Ern *et al.*, 2014; Fobian *et al.*, 2014; Verberk *et al.*, 2016), especially to investigate the mechanistic tenets of the OCLTT hypothesis. In this Theme Section, Baktoft *et al.* (2016) found no phenotypic covariance between aerobic scope of European perch (*Perca fluviatilis*), as measured in the laboratory, and their spontaneous swimming activity in the wild, suggesting that other factors may override any links between scope and fish performance in routine 'benign' conditions (Killen *et al.*, 2013). Careful laboratory studies on various fishes have failed to find evidence that aerobic scope declines when they are near their upper lethal temperature or that oxygen delivery is the factor defining tolerance of acute warming (Norin *et al.*, 2014; Wang *et al.*, 2014; Brijs *et al.*, 2015). A study by Claësson *et al.* (2016) on European eel (*Anguilla anguilla*), in this Theme Section, reports that aerobic scope increases with acute warming, underpinned by increases in cardiac output, until temperatures that are almost lethal. Thus, the universality of the Fry and OCLTT paradigms has been questioned, and this remains an active debate (Clark *et al.*, 2013; Schulte, 2015; Farrell, 2016).

Also in this Theme Section, Lefevre (2016) presents a comprehensive review and analysis of the effects of temperature and mild hypercarbia (reflecting projected increases in water $CO_2$), alone and in combination, on aerobic scope in fishes and other marine ectotherms. This revealed more variation in the response of aerobic scope to elevated temperature and $CO_2$ than would be predicted by the Fry and OCLTT paradigms. Although some species exhibited an aerobic performance curve that rose and then declined as a function of increasing temperature, a considerable number of species did not. Some exhibited no change or a decrease in aerobic scope, whereas many exhibited a constant increase, without any mortality, as they were warmed towards their lethal threshold (Fig. 2; Lefevre, 2016). The effects of elevated $CO_2$ also varied among species, often being without effect or sometimes increasing aerobic scope. In cases where hypercarbia suppressed aerobic scope, high temperature sometimes had a synergistic effect, but a simple additive effect was the most common (Lefevre, 2016).

Overall, although it is intuitive that physiological energetics will be of ecological significance for aquatic ectotherms (Fry, 1947, 1971; Ware, 1982; Priede, 1985; Kerr, 1990; Jørgensen *et al.*, 2012), it would also be unwise blindly to assume that the Fry and OCLTT paradigms hold for all marine fish species (Lefevre, 2016). The effects of temperature on aerobic scope may depend upon a species' ecology and history of exposure to diurnal or seasonal temperature variations (Norin *et al.* 2014; Lefevre, 2016). A lasting impact of the paradigms, whatever might be learned about their universality, is that they have focused attention on how thermal performance curves can provide a mechanistic link between physiology and ecology for fishes (Schulte, 2015). In particular, they can be integrated into ecological models to provide insights for management and conservation.

## Integrating physiology into ecological models

Models are now important tools for projecting the impact of global change on abundance and distribution of marine fishes. The ability to transfer knowledge of historical observations and make robust projections of future distributions is essential to provide sound advice for management decisions (Jørgensen *et al.*, 2012; Cooke *et al.*, 2014; Peck *et al.*, 2016). This is an area where physiology is perceived to hold great promise for conservation research, through integration into mechanistically based models of habitat suitability, which should provide increased confidence in projections (Hollowed *et al.*, 2011; Jørgensen *et al.*, 2012, 2016; Teal *et al.*, 2012, 2015; Cooke *et al.*, 2013, 2014; Deutsch *et al.*, 2015; Marras *et al.*, 2015a; Peck *et al.*, 2016).

Species distribution models are a common ecologically based approach, which use associations between aspects of habitat and known occurrences of species in order to define sets of conditions in which species are likely to occur (Ben Rais Lasram *et al.*, 2010; Albouy *et al.*, 2012, 2013). The correlative approach has contributed significantly to projections of the potential effects of climate change on marine fish distributions. Its practical advantages are simplicity and flexibility in data requirements, and the range of biotic/abiotic interactions that can be incorporated (Kearney and

Porter, 2009). Such correlative approaches are not, however, underpinned by mechanistic causalities, which is a prerequisite for confident projections of species range shifts (Jørgensen *et al.*, 2012; Teal *et al.*, 2015).

Physiology-based models should be able to deal with these issues of extrapolation because the organismal response is measured in the laboratory in controlled environmental conditions. Furthermore, physiology-based models overcome the circularity of predicting species response to climate change using range filling of potential distributional areas (Teal *et al.*, 2015; Peck *et al.*, 2016). Models that incorporate physiology typically focus on energetics because of the intuitive link to ecological performance (Jørgensen *et al.*, 2012). These models vary in the assumptions and structure of the physiology that is included, from the empirically driven Wisconsin school of bioenergetics modelling (Hewett and Johnson, 1987; Hanson *et al.*, 1997) to dynamic energy budget (DEB) models that strive for a universal description of organismal energetics derived from first principles (Kooijman, 1993, 2010). That being said, the integration of physiology into models is an area where further research and input are vitally needed. There is a need for reliable knowledge about how fishes function, in order to ensure that 'universal' traits of energetics are valid and are correctly represented in model parameterizations. This is essential to improve confidence in predictions about effects of climate change (Brander, 2015; Peck *et al.*, 2016). The debate surrounding the Fry and OCLTT paradigms' performance curves has already been mentioned, and the physiological principles underlying some other influential model projections (Pauly, 1981; Cheung *et al.*, 2011, 2012) have also been questioned (Brander, 2015).

Aerobic scope can be a useful physiological parameter for models that link individual energetics to processes at higher biological levels (Fig. 3), and which can incorporate interactions among stressors (Jørgensen *et al.*, 2012). One approach has used aerobic scope to define habitat suitability, based on laboratory measurements of it as a function of acclimatization temperature in target species, coupled with oceanographic modelling. The model outputs include 'metabolic maps' (Del Raye and Weng, 2015; Deutsch *et al.*, 2015; Marras *et al.*, 2015a,b; see also Martin *et al.*, 2015) based on the hypothesis that scope is an indicator of relative fitness potential (Fig. 4). Models based on aerobic scope can be useful in studies of invasive species, by projecting the relative performance of a native species and its competitor counterpart, thus estimating the 'winners' and 'losers' under climate-driven change for various locations and at different times (Marras *et al.*, 2015a). Other applications could include studies on key predators or prey species, in order to evaluate possible effects of global change on trophic relationships and food webs.

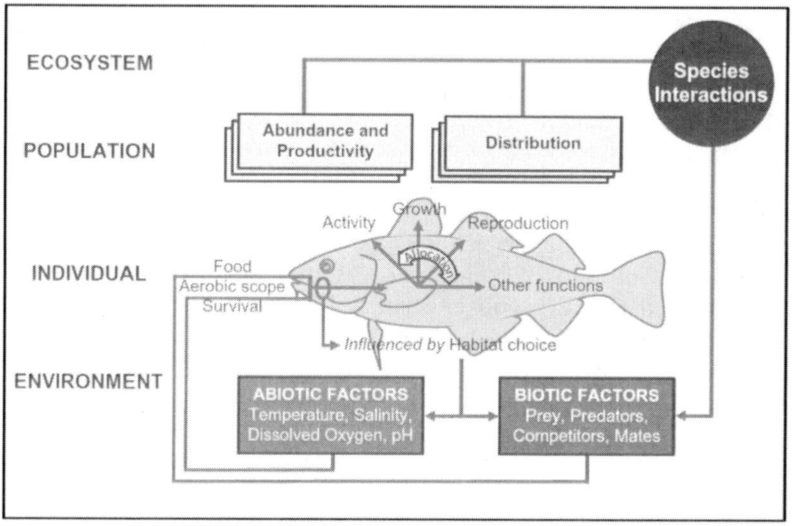


**Figure 3:** A schematic diagram of how aerobic scope can be integrated into mechanistically based models. It can be used to form a performance curve with temperature, to describe habitat suitability (metabolic maps, e.g. Marras *et al.*, 2015a) and it can be a constraint for oxygen allocation to competing activities in life-history models (Holt and Jørgensen, 2014, 2015). See Jørgensen *et al.* (2012) for more details.

A

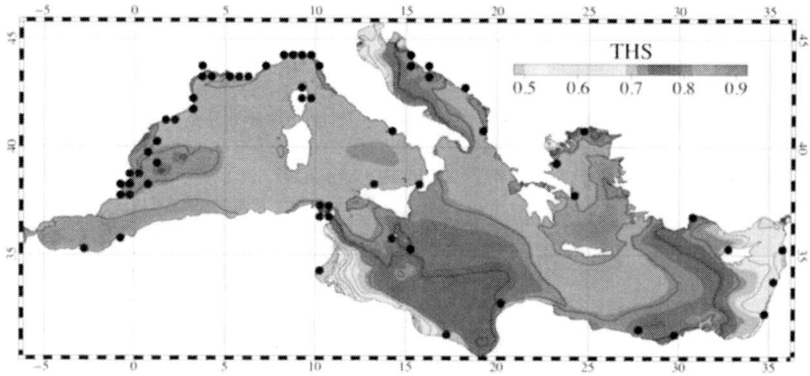

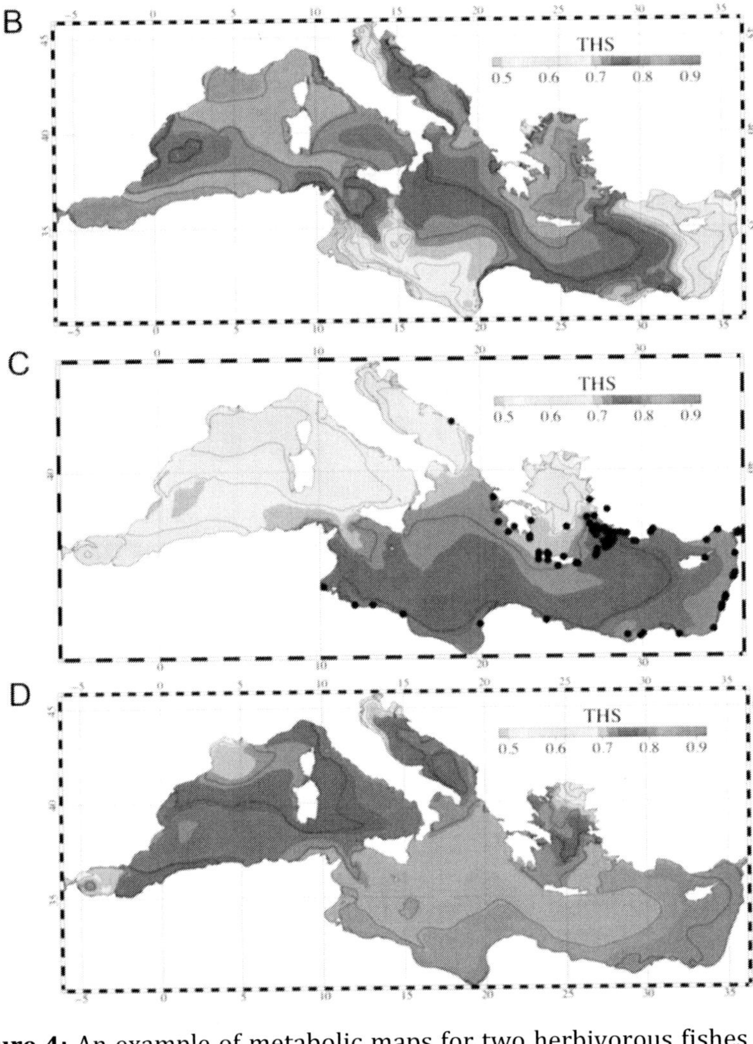


**Figure 4:** An example of metabolic maps for two herbivorous fishes in the Mediterranean, developed by combining an aerobic scope performance curve with oceanographic data (for details, see Marras *et al.*, 2015a). Thermal habitat suitability (THS) was computed for the whole Mediterranean Sea from the basin-scale model results. (**A**) Thermal habitat suitability of a native species, salema (*Salpa salpa*), based on present-day simulation results. (**B**) Thermal habitat suitability of the salema projected for a future scenario. (**C**) Thermal habitat suitability of an invasive lessepsien species, the marbled spinefoot (*Siganus rivulatus*), based on present-day simulation. (**D**) Thermal habitat suitability of the marbled spinefoot projected for the future scenario. Black dots represent the sites where the fish species have been observed.

There are also life-history models that incorporate aerobic scope as a constraint in life-history evolution, in order to explore its links to fitness (Holt and Jørgensen, 2014, 2015; Jørgensen *et al.*, 2016). They integrate the physiology of oxygen uptake and use with foraging and digestion and with life-history traits, such as growth, survival and reproduction (Holt and Jørgensen, 2014, 2015). When these characteristics of an individual were optimized together in a model for Atlantic cod, simulations suggested that fitness would rapidly decline at high temperatures as a result of energy-budgeting conflicts (Holt and Jørgensen, 2014, 2015; Jørgensen *et al.*, 2016), driven in part by increased food requirements (Johansen *et al.*, 2015). These models are interesting because their projections appear to be relatively robust to the shape of the aerobic scope performance curve near the lethal limit, because fitness peaked at cooler temperatures (Holt and Jørgensen, 2014, 2015; Jørgensen *et al.*, 2016). Thus, irrespective of any doubts about the universality of the Fry or OCLTT paradigms, these life-history models suggest that oxygen budgets may well define a main constraint for future projections of marine fishes under environmental change (Holt and Jørgensen, 2014, 2015; Johansen *et al.*, 2014, 2016).

Dynamic energy budget theory has also been used as a mechanistic basis to model habitat suitability (Teal *et al.*, 2012, 2016; Raab *et al.*, 2013). The theory is grounded in the idea that metabolism is organized in the same way within all organisms, including fishes (Fig. 5). It derives from a number of assumptions that can describe empirical patterns, such as the van Bertlanffy growth curve or Kleiber's rule (for a list, see Sousa *et al.*, 2008), which are consistent throughout the animal kingdom. The advantage is that the standard DEB model can be applied to all organisms and therefore all fish species, with each described by a set of species-specific parameters. Although parameterization requires empirical data, if data are lacking the model can still provide useful insights with data from related species for which more is known. The potential applications of the DEB-based models are similar to the aerobic scope models, but their particular value is that they provide outputs of growth and fecundity in relationship to environmental conditions, such as temperature or food availability. Dynamic energy budget modelling, in combination with ecosystem models that provide spatial and temporal data on environmental conditions, has been used to develop maps of optimal habitats for growth of marine flatfishes and to project these under climate-driven warming (Teal *et al.*, 2012). The DEB theory can also be used to investigate effects on energetic pathways of other stressors, such as hypoxia, acidification or pollutants, if data are available.

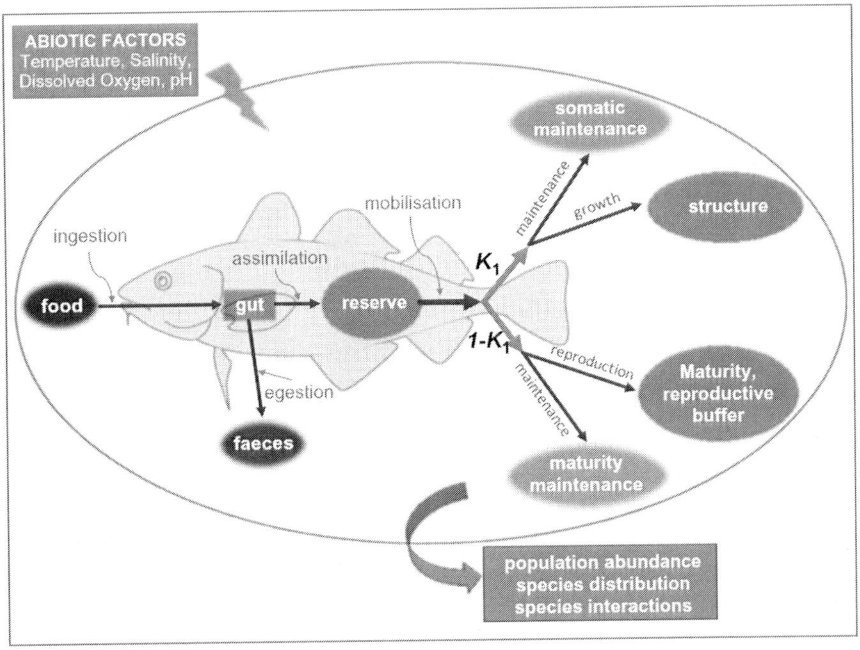


**Figure 5:** Schematic representation of the standard dynamic energy budget (DEB) model showing the paths of energy flow through a fish (or any organism). Sources or sinks of energy are shown as green, brown and orange ovals; the blue ovals are the three state variables describing the organism. Processes affecting energy flows are indicated by black arrows. A defining feature of DEB models is the existence of reserves, from which allocation rules (proportion $K_1$) define the partition of energy among processes such as maintenance (somatic or gonadal), somatic growth and reproduction. Dynamic energy budget models can be parameterized to account for effects of abiotic variables, and their universal principles allow for interspecific comparisons of parameter estimates. See Teal *et al.* (2012) for details of an application to evaluate and project marine fish habitat suitability.

Physiological models based on aerobic scope or DEB need to be integrated with physical ocean models and validated against population- and community-level data, so that they can achieve their promise. This would allow them to contribute to, for example, the Intergovenmental Panel on Climate Change (IPCC) predictions for effects of warming on global marine fisheries (Cheung *et al.*, 2009, 2013; Sumaila *et al.*, 2011). The ability to investigate how other processes, such as hypoxia, ocean acidification and trophic disruption, will interact with warming is now recognized as a research priority (Gunderson *et al.*, 2016) and is a major strength of the models that incorporate aerobic scope or DEB. This has been highlighted as an area of great uncertainty in other physiology-based models (Brander, 2015). Once again, however, application of physiology in models requires more information on many more marine fish species. Parameterizing any mechanistically based model with valid physiological data could be a major undertaking, requiring significant long-term studies, use of facilities and personnel. International collaboration and funding are therefore required to coordinate development of laboratory and field measurements of physiology and physiologically based models.

Embedding physiological knowledge of species within models representing the spatial dynamics of marine food webs can provide concrete advice for fish conservation. This is especially true in light of the current emphasis on ecosystem-based fisheries and ecosystem management in Europe and elsewhere. Ecophysiologists and modellers can collaborate to create new tools, beyond well-established models, such as the Ecopath with Ecosim (Christensen and Walters, 2004) or species distribution models based on bioclimate envelopes (Peck *et al.*, 2016), to enhance understanding of fishes and their responses to global change and to provide knowledge and tools to support adaptive management (Williams, 2011; Elliott *et al.*, 2015; Queirós *et al.*, 2016).

## The significance of individual variation

A major challenge for conservation of biodiversity is to understand the capacity of species to acclimatize and, ultimately, to adapt genetically to ongoing global change (Seebacher and Franklin, 2012; Crozier and Hutchings, 2014; Seebacher *et al.*, 2015). A core issue is to understand the different facets of intraspecific diversity, i.e. the differences among individuals within a population or species that are the substrate for evolution by natural selection. Of particular interest, from a conservation perspective, is the individual variation in physiological sensitivity to environmental conditions that exists within a given species, including how this may vary among populations across their geographical range, as a result of local adaptation. This variation needs to be understood in itself, as an indication of the potential resilience of a given population or species to environmental change and habitat modification. Ultimately, the goal is to understand how such variation links to life-history variation, to adaptation and evolution of the population or species, and so to underlying heritable genetic variation. Such associations are far from being understood in marine fishes, even for the most intensively studied species, such as Atlantic cod or Atlantic salmon.

The causes and consequences of individual variation in physiology are currently major areas of research, and there are many fundamental and 'mechanistic' physiological questions with conservation implications, such as the significance for an ecosystem approach to fisheries (Killen *et al.*, 2015; Ward *et al.*, 2016). Intraspecific diversity can have a genetic basis but it can also vary with life stages and sex and be affected by transgenerational maternal effects and early life experience (Gore and Burggren, 2012; Ho and Burggren, 2012; Miller *et al.*, 2012; Zambonino-Infante *et al.*, 2013). One area of individual physiological variation that is of major interest is 'metabolic phenotypes', meaning animals with different metabolic rates and aerobic scopes, and the ecological and evolutionary significance of this (Metcalfe *et al.*, 2016b). An associated core issue, which transcends physiology, is to understand how major physiological, behavioural and life-history traits might co-vary, whether they might associate into syndromes, and how these might be maintained by ecological trade-offs (Réale *et al.*, 2010; Careau and Garland, 2012; Killen *et al.*, 2013). These various questions are far too broad and complex to be reviewed adequately here, but they are very poorly understood in marine fishes as a whole. A perspectives paper in this Theme Section considers these issues and, therefore, the reasons why individual variation should be taken into account in the ecosystem approach to fisheries (Ward *et al.*, 2016).

Although physiological traits are often attributed ecological and evolutionary significance, there is a need to investigate trait repeatability in wild fish populations and whether the temporal stability of traits may be affected by changing environmental conditions. Temporal stability of physiological traits, plus a genetic component to the observed intraspecific variation, is a prerequisite for a trait to be a target for natural selection. This would influence the ability of species to evolve the trait in response to environmental conditions. Changing environments may erode or enhance trait repeatability, possibly changing which traits are under direct and correlated selection. At present, investigation of potential effects of climate change in marine fishes have primarily examined how warming or ocean acidification can influence population means for variables such as locomotory capacity, metabolism or behaviour (Seebacher *et al.*, 2015; Lefevre, 2016; Nagelkerken and Munday, 2016). The current lack of information about how such environmental disturbances affect trait repeatability is a crucial gap that hinders the ability to predict how populations can cope through evolutionary responses. Ongoing advances in respirometry and biotelemetry/biologging, in particular, should increase understanding of trait repeatability in marine fishes and its response to changing environments. The repeatability of traits, and the extent to which this is context dependent, is the topic of a review by Killen *et al.* (2016a) in this Theme Section, with consideration of the implications for management and conservation of fish populations.

It is worthwhile to consider whether ecological models can incorporate individual variation, how these might be parameterized, and whether new models might be needed. Unstructured population models, such as the Lotka–Volterra competition or predator–prey model, assume that all individuals are equal. Age-structured models, such as the well-known Leslie matrix model (Caswell, 2000), take ontogenetic stage into account but ignore other potential sources of variation, assuming a constant and similar environment for all individuals and that animals of the same age remain exactly the same across time. Physiologically structured population models consider the potential for a variable environment to introduce variation, such that animals of a similar age may differ significantly, allowing for richer ecological interactions (de Roos, 1997; de Roos and Persson, 2013). The basic physiology and behaviour of each individual in these models is characterized by a parameter vector, which thus represents the genotype, and the vector of state variables represents the phenotype. The DEB model of the individual can serve as a building block for these physiologically structured models, with an example provided by van der Meer (2016) in this Theme Section. All individuals usually have the same parameter vector and thus the same genetic constitution in physiologically structured population models (Kooi and van der Meer, 2010). Studies to investigate the evolutionary stability of populations require genetic variability. Adaptive dynamics models (Dieckmann and Law, 1996), for example, aim to find a population where mutants (with a slightly different vector to residents) can no longer invade (Metz *et al.*, 1992). The approach normally assumes clonal reproduction, but sexual reproduction can be incorporated, such that diversity among sexes can be considered. In other models (e.g. Giske *et al.*, 2014), the genotype is explicitly modelled, which allows for emergent genetic variation and coexistence of different genotypes and phenotypes (e.g. behavioural strategies).

Many current population and community models ignore all potential sources of individual variation. Given that there is a huge difference between an Atlantic cod larva and full-grown adult, and a huge difference in growth rate between a well-fed or starved cod, ontogenetic stage and environmental history should, at least, be incorporated into population models. Beyond that, certain phenomena may not be understood and may not be predicted well if such variation is not considered (Ward *et al.*, 2016). Furthermore, the variation may itself be of interest and provide insights into the underlying biological processes that have produced and maintained it. Given that parameterizing models to account for the various different potential facets of individual variation is a major long-term undertaking, in terms of facilities and personnel, pragmatic alternatives are worthy of careful investigation. One such alternative is pattern-oriented modelling (Grimm *et al.*, 2005), which focuses on empirically quantifying processes and key trade-offs that cause or constrain variation, then using a model to predict individual variation and compare it with observed variation.

## Understanding what fishes do in nature

It remains a central problem to relate the physiology of fishes, measured in the laboratory, to the habitats and conditions they experience (and will select) in their natural environment. There are immense technical difficulties in following fishes in the vast three-dimensional marine realm, let alone in measuring physiological variables or estimating their physiological state and whether they occupy habitats that optimize some element of their physiology (Freitas *et al.*, 2016).

Active and passive acoustic tracking is already widely used in marine fishes, in order to follow them and estimate variables such as swimming speed and distance, plus two- and three-dimensional positioning. The acoustic signal carries over only relatively limited spatial scales, but the ongoing development of networks of acoustic receivers along coastlines, such as the Ocean Tracking Network (http://oceantrackingnetwork.org/ research/ canadian-projects/), will provide extremely valuable information about, for example, habitat use or migration patterns of marine fishes, that has major applications for conservation research and policy. Rapidly evolving techniques of measurement of physiological variables from free-living animals, including fishes, have been suggested to provide 'answers to questions that we did not know we should ask'.

The first biotelemetric measurement was probably performed by Marey (1896). Biotelemetry and biologging (Fig. 6) are now starting to provide information on the physiology of animals in the field. Together with tracking data, they are providing a better picture of the life cycles of some economically important species, plus information about the structure of their populations (Metcalfe, 2006; Rutz and Hays, 2009; Block *et al.*, 2011; Metcalfe *et al.*, 2012; Whitlock *et al.*, 2015). These tools can record an animal's physiology while simultaneously recording environmental conditions around it, in order to investigate assumptions based on laboratory experiments. For example, some acoustic telemetry tags can measure oxygen content in the water surrounding a fish and transmit this in real time (Svendsen *et al.*, 2006). There are emerging techniques to collect physiological information on free-swimming fishes, which can then be used to estimate energetics as a function of prevailing environmental conditions (Gräns *et al.*, 2010; Wright *et al.*, 2014; Metcalfe *et al.*, 2016a). Third-generation biotelemetry systems are being developed for simultaneous measurement of multiple physiological variables; for example, blood flow, blood pressure, electrocardiograms, electromyograms, three-dimensional acceleration and temperature. These can have a bidirectional radio frequency link that allows the implant to send data and accept commands to perform tasks. The signal from the implant can be viewed online, with a transmission range of ~10 m in air. This is, however, reduced in water, especially sea water, where alternative strategies are required, such as acoustic signalling or biologging.

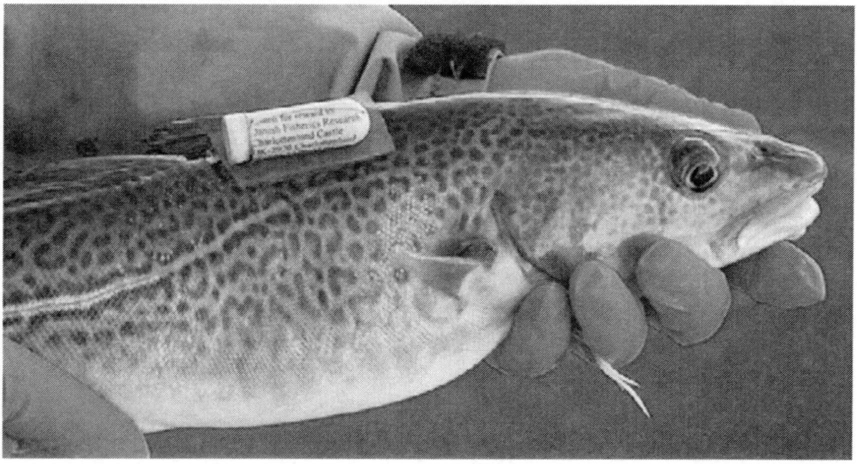

**Figure 6:** An Atlantic cod (*Gadus morhua*) carrying a data storage tag that records pressure, temperature and salinity. Photograph: Stefan Neuenfeldt, DTU Aqua.

Biologging, where the physiological data are stored in the tag/implant and then recovered, can be used on fishes released into open water (Fig. 6). Recovery of biologging tags remains a constraint, in particular for species that are not fished commercially or are under a fishing moratorium. A low recovery rate can make this method very costly, not only for the initial investment in tags but for the effort to implant them. Biologgers can collect and store both physiological variables (e.g. electrocardiogram, acceleration) and environmental parameters (e.g. pressure, temperature) that can be used to reconstruct migration pathways (Metcalfe and Arnold, 1997; Hunter *et al.*, 2004), link behaviour to environmental conditions (Righton *et al.*, 2001, 2010; Sims *et al.*, 2003), characterize population structure (Metcalfe, 2006) and estimate energetic costs of different behaviours or interactions with humans (reviewed by Cooke *et al.* 2016). Pop-up tags that store data are also now widely used in studies on marine fishes, with the major advantage that data can be recovered via satellite. There are limitations to the size of fish that can carry the tags, and the tags can be expensive. Thus, most research has been on large and economically valuable pelagic or demersal fishes where, however, pop-ups have provided valuable knowledge for management and conservation (e.g. Block *et al.*,

2011; Whitlock *et al.*, 2015). In most cases, these tags are not 'biologgers'; they store data only on environmental parameters, such as temperature and pressure.

An exciting development is the application of three-axis accelerometer tags (both in telemetry and biotelemetry/logger platforms) to monitor energy expenditure (Metcalfe *et al.*, 2016a), activity and state. Movement is one of the four main bodily functions that incur energetic costs in animals. The energy expenditure is governed by muscle contractions and is typified by variable acceleration of the body (Gleiss *et al.*, 2010), so records of the tri-axial acceleration of fishes should provide a useful proxy for activity-specific energy costs. Recent studies have correlated dynamic tri-axial body acceleration with rates of oxygen uptake in various aquatic species, including hammerhead sharks (*Sphyrna lewini*; Gleiss *et al.* 2010) and European sea bass (*Dicentrarchrus labrax*; Wright *et al.*, 2014). Bi-axial and tri-axial acceleration, root mean square acceleration and acoustically transmitted acceleration data have also provided some exciting insights into fish behaviour and physiology (Clark *et al.*, 2010; Wilson *et al.*, 2013; Marras *et al.*, 2015b). High-frequency accelerometry can be used to distinguish among various behaviours, such as feeding strikes and anti-predator escapes (Broell *et al.*, 2013).

There are potential applications for biotelemetry, especially biologging, that can be highlighted for marine fish conservation research (Metcalfe *et al.*, 2012). Tracking can improve our understanding of seasonal movements and space use and would be invaluable for evaluating the design and effectiveness of marine protected areas and to identify potential spawning aggregations. The addition of measurements of pressure, temperature and acceleration to the tracking devices can enable investigation of ontogenetic changes in the behaviour of pelagic species and evaluation of options for selective fishing strategies. Biotelemetric/logging data can also be combined with other assays, in particular of blood chemistry, to determine post-release survival of bycaptured (or sport-fished) animals. Environmental data collected by tracking or biotelemetric/logger tags can be used to define behavioural thresholds for critical habitat parameters, such as temperature (Neat and Righton, 2007; Righton *et al.*, 2010), oxygen concentration (Prince *et al.*, 2010) and salinity. This remains, therefore, a very exciting area of research and technological development.

## Making physiology relevant to decision-making

The value of mechanistic physiological information is currently not widely appreciated by resource managers and policymakers, not least because physiologists have not made a consistent effort to promote their science in this regard (Cooke and O'Connor, 2010; Horodysky *et al.*,

2015, 2016; Patterson *et al.*, 2016). There is a real opportunity to develop fish environmental physiology as a discipline, by contributing to conservation research (Cooke *et al.*, 2013; Madliger *et al.*, 2016). Physiology can reveal mechanisms, which can be used to explain ecological patterns, which may then support evidence-based predictions and management decisions (Cooke *et al.*, 2013; Madliger *et al.*, 2016). Physiological tools and knowledge have already contributed to conservation goals for marine fishes; for example, to the management of migrating Pacific salmon, to improving survival from bycatch in specific fisheries, or to reducing the impact of tourism on some natural fish populations (Madliger *et al.*, 2016; Patterson *et al.*, 2016). Physiology should also be able to inform policy decisions about the following: limiting mortalities from discards from many further fisheries; the design of marine protected zones; adaptation to global change; predicting potential for invasive species; and many other things.

In this Theme Section, Horodysky *et al.* (2016) and Patterson *et al.* (2016) provide thoughtful analyses of how physiological research and the research process relate to the needs of resource managers and their decision-making process. Physiology must contribute to a broader toolbox or conceptual framework within which policy operates. Although mechanistic insight can be very useful for managers and physiology can provide a component of this, it cannot be the only source of information; it must be considered alongside genetics, behavioural ecology, trophic webs, physical oceanography, and so on (Horodysky *et al.*, 2015, 2016). Patterson *et al.* (2016)synthesize the reasons why physiological research on sockeye salmon (*Oncorhynchus nerka*) migration contributed successfully to management decisions in British Columbia (Canada). A main driver was an existing political motivation, based on observations of reduced salmon runs that seemed linked to rising river temperatures, which then funded targeted research; that is, there was a direct connection between a management problem and funding of physiologically based solutions. The collaboration was then successful because of accountability, legal clarity, effective institutional environments, good personal relationships and peer acceptance. Interactions between researchers and stake-holders were crucial, so that the people most affected by decisions were familiar with the research and so that personal relationships improved overall trust. Patterson *et al.* (2016) urge researchers to be aware of the need to provide confident predictions regarding future outcomes, which are tailored to specific management objectives; in particular, to be able to quantify uncertainty to the level desired by managers or other knowledge practitioners.

Fish physiologists generally lack contact with key policymakers and do not have direct information channels to attune and balance their research with policy decisions. Thus, a pervasive challenge is to integrate with other disciplines and scale up from physiology to decision-making. Fishery biologists, by the nature of their work, do have direct contacts and therefore represent an important link to policy for physiologists. Figure 7 is a flow diagram of how physiological research can inform management and policy decisions, and the feedbacks to research that can be used for adaptive management strategies (Williams, 2011). Monitoring of biomarkers, at immediate time scales, can provide advice for specific local conservation decisions; for example, to assess ecological status of coastal zones or for early warning of impacts of global change. This can feed back to elicit more focused biomarker research and monitoring at the local scale, but also feed forward to national and international monitoring; for example, in the context of the EU Water Framework Directive. As mentioned, however, another route to influence policy is through increased interactions with fishery biologists. Information about biomarkers of survival after discard can be provided to fishery biologists, in support of advice they might provide at a national level (Fig. 7). Information from field and laboratory physiology studies can populate databases, in order to then parameterize models to project ecological consequences at the level of populations, species and assemblages. These can influence wider-scale policy decisions, over longer time scales (Cooke *et al.*, 2014). This flow of information may reveal knowledge gaps and longer-term policy priorities that, in turn, can feed back to drive more research; for example, in large international collaborative projects.

It is worthwhile to consider how detailed knowledge of the physiology, ecology and life history of marine fishes might be distilled for easy application in wide-scale decision-making on sensitivity of communities and assemblages to environmental change. Biological traits analysis has often been applied to assess the impact of environmental change on terrestrial and freshwater communities but, so far, has relatively few applications to marine fishes (Elleouet *et al.*, 2014). Biological traits analysis holds promise because physiological, ecological and life-history trait data exist for many marine fish species, and an analysis of available information is the first step towards constructing a trait-based index of climate change sensitivity, in order to identify which aspects would be needed to develop the index as a tool. This would then be assessed against existing community-level data from observational studies, at sites that have been subject to recent and documented environmental change. Ideally, in the future, such functional sensitivity indices could be used to construct simple models of species extinctions and predict the likely impact of climate change on biodiversity and community function.

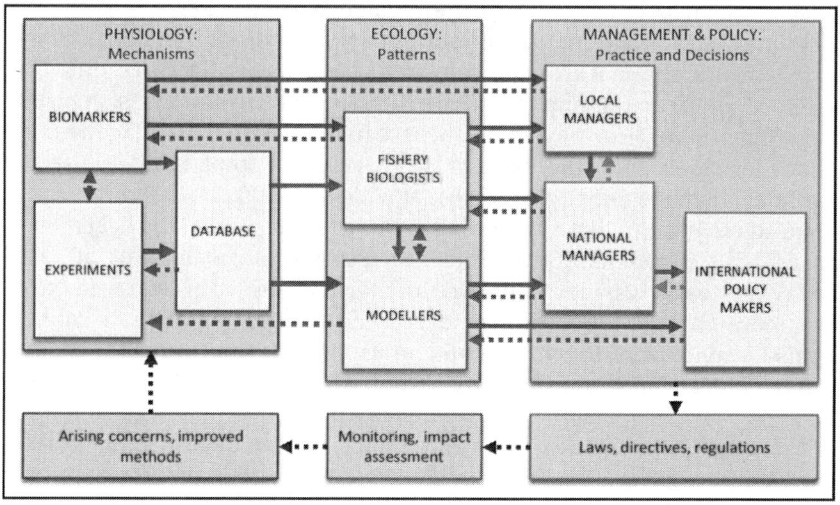


**Figure 7:** Flow diagram of how physiological information can inform management and/or policy decisions (continuous lines) for marine fishes, and how analysis of the information can be fed back to develop targeted research activities (dotted lines). Red dotted arrows show flow of policy decisions (in the European Union). Biomarker information can be used directly for local management (in particular, early warning and evaluation of ecological status). Physiological information can also influence national and international management/policy indirectly, by interactions with ecologists; for example, biomarkers of bycatch survival to inform fishery biologists, or physiological databases for use in modelling of population dynamics or effects of global change. The number of dotted lines feeding back to physiology reveal the many contributions that physiological research could make to adaptive management programmes, including large-scale and long-term research in response to, for example, EU or Intergovenmental Panel on Climate Change recommendations.

Initial attempts at bridging gaps among marine fish physiologists, ecologists, modellers and policymakers have been made, which bear reporting here. Roundtable discussions at a conference, funded by Action FA1004, were aimed at understanding some of the barriers to knowledge exchange between physiological and advisory processes, how to refine policy-management issues so they can be reflected better in conservation research, and whether fish physiologists have sufficiently considered the impact of their research on stakeholder and policy advice. A diversity of views was expressed, and the discussions are best summarized as the following general themes.

i. The need for commonality of language. Dialogue is needed to achieve common understanding among physiologists, modellers and policy advisors. The simplest of terms can have a different meaning among scientists from different fields, and among stakeholders. This needs to be overcome without diminishing the autonomy of the various disciplines. A glossary of common terms, linked to the database of physiological information, would be useful.

ii. Temporal scales are often different for research and policy. Physiology may not be able to provide rapid advice to support a pressing policy decision. Fishery discards are a prime example, where policy changes resulted from societal pressure and not scientific understanding. Robust scientific underpinning would have required detailed and complex studies, achieved too slowly for policymakers. Such policy can, of course, then fund *post hoc*research to investigate physiological impacts of discarding and the likelihood of surviving it for the relevant species. A troubling example of temporal asynchrony is the lack of immediate concrete policy responses to evidence of profound effects on marine ecosystems of gradual ongoing climate change (Elliott *et al.*, 2015).

iii. Don't give me the details, just the summary. Physiologists are interested in mechanistic detail, the responses of individuals and populations to changes in environmental conditions, typically of one particular model species. Such details often, however, run counter to effective stakeholder engagement and/or the advice needed for policy, which require synthesis of information into tangible effects. For example, the implications for fishing and fishermen, for future scenarios on fishing areas and species (in the context of climate change) or for the number and size of marine protected areas. Physiologists must learn to present their information in a holistic and understandable manner, including presentation to others in the scientific community, such as ecologists and modellers, who could translate, interpret and summarize the consequences for policy advice.

iv. Physiologists need to champion their cause. Physiologists must recognize that their research has impact, particularly through interaction with other related disciplines. It is not sufficient for physiologists to provide the data to parameterize models, then to dissociate themselves from the modelling outcomes. They must contribute to interpretation of results, in order to influence policy decisions. Physiologists need to understand the realms of policy work and policy decisions better, and the linkages from physiology to ecology then policy, in order to influence outcomes through co-production of knowledge and transdisciplinary research.

## CONCLUSIONS

There is much potential for physiological research to contribute to conservation of marine fish biodiversity and fisheries, which strengthens fish environmental physiology as a discipline. There is a clear need to increase the overall knowledge base about marine fish environmental physiology, especially tolerance thresholds for major environmental stressors and how such stressors affect performance within their tolerated range. Physiologists should explore avenues for international collaborative research, in order to avoid duplication of effort and cover as broad a range of species as possible. A particular application of such data would be to improve the reliability of models in order to gain a better understanding of what defines current fish distribution and abundance and, therefore, to increase confidence in projections of the effects of ongoing global change. Increased interaction with researchers using other tools, notably fishery biologists, ecologists and modellers, will provide a very fruitful avenue to increase the scope and impact of marine fish conservation physiology research and to make such research relevant to policy decisions.

## ACKNOWLEDGEMENTS

The authors are grateful to the Society for Experimental Biology and Oxford University Press for publishing the Theme Section and for waiving publication fees. They are also grateful to the COST Office Scientific Officer of Action FA1004, Dr Ioanna Stavridou, to the Action Financial Officer, Dr Susana Moreira, and to the Action Grant Holder, CIIMAR Porto. The COST Action FA1004 Final Conference and associated workshop on linking marine fish conservation physiology to policy was supported in part by the Région Languedoc Roussillon and by the Centre for Marine Biodiversity Exploitation and Conservation (UMR9190 Marbec) in Montpellier, France.

## FUNDING

This publication arose from conferences and workshops funded by EU COST Action FA1004 Conservation Physiology of Marine Fishes.

# REFERENCES


1. Adams SM, Ham KD (2011) Application of biochemical and physiological indicators for assessing recovery of fish populations in a disturbed stream. Environ Manag 47: 1047–1063.
2. Aerts J, Metz JR, Ampe B, Decostere A, Flik G, De Saeger S (2015) Scales tell a story on the stress history of a fish. PLoS One 10: e0123411.
3. Albouy C, Guilhaumon F, Araújo MB, Mouillot D, Leprieur F (2012) Combining projected changes in species richness and composition reveals climate change impacts on coastal Mediterranean fish assemblages. Glob Chang Biol 18: 2995–3003.
4. Albouy C, Guilhaumon F, Leprieur F, Ben Rais Lasram F, Somot S, Aznar R, Velez L, Le Loc'h F, Mouillot D (2013) Projected climate change and the changing biogeography of coastal Mediterranean fishes. J Biogeogr 40: 534–547.
5. Anttila K, Dhillon RS, Boulding EG, Farrell AP, Glebe BD, Elliott JAK, Wolters WR, Schulte PM (2013) Variation in temperature tolerance among families of Atlantic salmon (Salmo salar) is associated with hypoxia tolerance, ventricle size and myoglobin level. J Exp Biol 216: 1183–1190.
6. Arnold SJ (1983) Morphology, performance and fitness. Am Zool 23: 347–361.
7. Baktoft H, Jacobsen L, Skov C, Koed A, Jepsen N, Berg S, Boel M, Aarestrup K, Svendsen JC (2016) Phenotypic variation in metabolism and morphology correlating with animal swimming activity in the wild: relevance for the OCLTT (oxygen- and capacity-limitation of thermal tolerance), allocation and performance models. Conserv Physiol 4: cov055; https://doi.org/10.1093/conphys/cov055.
8. Beaulieu M, Costantini D (2014) Biomarkers of oxidative status : missing tools in conservation physiology. Conserv Physiol 2: cou014; https://doi.org/ 10.1093 /conphys/cou014.
9. Ben Rais Lasram F, Guilhaumon F, Albouy C, Somot S, Thuiller W, Mouillot D (2010) The Mediterranean Sea as a 'cul-de-sac' for endemic fishes facing climate change. Glob Chang Biol 16: 3233–3245.
10. Block BA, Jonsen ID, Jorgensen SJ, Winship AJ, Shaffer SA, Bograd SJ, Hazen EL, Foley DG, Breed GA, Harrison A-L et al. (2011) Tracking apex marine predator movements in a dynamic ocean. Nature 475: 86–90.
11. Brander K (2015) Improving the reliability of fishery predictions under climate change. Curr Clim Chang Rep 1: 40–48.
12. Brijs J, Jutfelt F, Clark TD, Gräns A, Ekström A, Sandblom E (2015) Experimental manipulations of tissue oxygen supply do not affect warming tolerance of European perch. J Exp Biol 218: 2448–2454.
13. Broell F, Noda T, Wright S, Domenici P, Steffensen JF, Auclair J-P, Taggart CT (2013) Accelerometer tags: detecting and identifying activities in fish and the effect of sampling frequency. J Exp Biol 216: 1255–1264.
14. Buckley LB, Hurlbert AH, Jetz W (2012) Broad-scale ecological implications of ectothermy and endothermy in changing environments. Glob Ecol Biogeogr 21: 873–885.
15. Careau V, Garland T (2012) Performance, personality, and energetics: correlation, causation, and mechanism. Physiol Biochem Zool 85: 543–571.
16. Caswell H (2000) Prospective and retrospective perturbation analyses: their roles in conservation biology. Ecology 81: 619–627.
17. Cattano C, Giomi F, Milazzo M (2016) Effects of ocean acidification on embryonic respiration and development of a temperate wrasse living along a natural CO2 gradient. Conserv Physiol 4: cov073; https://doi.org/10.1093/conphys/cov073.
18. Celi M, Filiciotto F, Maricchiolo G, Genovese L, Quinci EM, Maccarrone V, Mazzola S, Vazzana M, Buscaino G (2016) Vessel noise pollution as a human threat to fish: assessment of the stress response in gilthead sea bream (Sparus aurata, Linnaeus 1758). Fish Physiol Biochem 42: 631–641.

19. Chabot D, McKenzie DJ, Craig JF (2016) Metabolic rate in fishes: definitions, methods and significance for conservation physiology. J Fish Biol 88: 1–9.
20. Chadwick JG, Nislow KH, McCormick SD (2015) Thermal onset of cellular and endocrine stress responses correspond to ecological limits in brook trout, an iconic cold-water fish. Conserv Physiol 3: cov017; https://doi.org/ 10.1093/ conphys/cov017.
21. Chave J (2013) The problem of pattern and scale in ecology: what have we learned in 20 years. Ecol Lett 16: 4–16.
22. 3 Cheung WWL, Lam VWY, Sarmiento JL, Kearney K, Watson R, Pauly D (2009) Projecting global marine biodiversity impacts under climate change scenarios. Fish Fish 10: 235–251.
23. Cheung WWL, Dunne J, Sarmiento JL, Pauly D (2011) Integrating ecophysiology and plankton dynamics into projected maximum fisheries catch potential under climate change in the Northeast Atlantic. ICES J Mar Sci 68: 1008–1018.
24. Cheung WWL, Sarmiento JL, Dunne J, Frölicher TL, Lam VWY, Deng Palomares ML, Watson R, Pauly D (2012) Shrinking of fishes exacerbates impacts of global ocean changes on marine ecosystems. Nat Clim Chang 2: 1–5.
25. Cheung WWL, Pauly D, Sarmiento JL (2013) How to make progresss in projecting climate change impacts. ICES J Mar Sci 70: 1069–1074.
26. Christensen V, Walters CJ (2004)Ecopath with Ecosim: methods, capabilities and limitations. Ecol Model 172: 109–139.
27. Claësson D, Wang T, Malte H (2016 ) Maximal oxygen consumption increases with temperature in the European eel (Anguilla anguilla) through increased heart rate and arteriovenous extraction. Conserv Physiol 4. https://doi.org/ 10.1093/conphys/cow027
28. Claireaux G, Chabot D (2016) Responses by fishes to environmental hypoxia: integration through Fry's concept of aerobic metabolic scope. J Fish Biol 88: 232–251.
29. Claireaux G, Lagardère JP (1999) Influence of temperature, oxygen and salinity on the metabolism of the European sea bass. J Sea Res 42: 157–168.
30. Claireaux G, Lefrançois C (2007) Linking environmental variability and fish performance: integration through the concept of scope for activity. Philos Trans R Soc Lond B Biol Sci 362: 2031–2041.
31. Claireaux G, Désaunay Y, Akcha F, Aupérin B, Bocquené G, Budzinski H, Cravedi J-P, Davoodi F, Galois R, Gilliers C et al. (2004) Influence of oil exposure on the physiology and ecology of the common sole Solea solea: experimental and field approaches. Aquat Living Resour 17: 335–351.
32. Clark TD, Sandblom E, Hinch SG, Patterson DA, Frappell PB, Farrell AP (2010) Simultaneous biologging of heart rate and acceleration, and their relationships with energy expenditure in free-swimming sockeye salmon (Oncorhynchus nerka). J Comp Physiol B Biochem Syst Environ Physiol 180: 673–684.
33. Clark TD, Sandblom E, Jutfelt F (2013) Aerobic scope measurements of fishes in an era of climate change: respirometry, relevance and recommendations. J Exp Biol 216: 2771–2782.
34. Conover DO, Clarke LM, Munch SB, Wagner GN (2006) Spatial and temporal scales of adaptive divergence in marine fishes and the implications for conservation. J Fish Biol 69: 21–47.
35. Cooke SJ, O'Connor CM (2010) Making conservation physiology relevant to policy makers and conservation practitioners. Conserv Lett 3: 159–166.
36. Cooke SJ, Suski CD (2008) Ecological restoration and physiology : an overdue integration. Bioscience 58: 957–968.
37. Cooke SJ, Sack L, Franklin CE, Farrell AP, Beardall J, Wikelski M, Chown SL (2013) What is conservation physiology? Perspectives on an increasingly integrated and essential science. Conserv Physiol 1: cot001; https://doi.org/10.1093/ conphys/cot001.

38. Cooke SJ, Killen SS, Metcalfe JD, McKenzie DJ, Mouillot D, Jørgensen C, Peck MA (2014) Conservation physiology across scales : insights from the marine realm. Conserv Physiol 2: cou024; https://doi.org/10.1093/conphys/cou024.
39. Cooke SJ, Brownscombe JW, Raby GD, Broell F, Hinch SG, Clark TD, Semmens JM (2016). Remote bioenergetics measurements in wild fish: opportunities and challenges. Comp Biochem Physiol A Mol Integr Physiol . In press. https://doi.org/10.1016/j.cbpa.2016.03.022.
40. Crozier LG, Hutchings JA (2014) Plastic and evolutionary responses to climate change in fish. Evol Appl 7: 68–87.
41. Dantzer B, Fletcher QE, Boonstra R, Sheriff MJ (2014) Measures of physiological stress: a transparent or opaque window into the status, management and conservation of species. Conserv Physiol 2: cou023; https://doi.org/ 10.1093/ conphys/cou023.
42. Davis MW (2010) Fish stress and mortality can be predicted using reflex impairment. Fish Fish 11: 1–11.
43. De Roos AM (1997) A gentle introduction to physiologically structured population models. In: Structured-population Models in Marine, Terrestrial, and Freshwater Systems (Tuljapurkar, S. & Caswell, H., eds), pp. 119–204. New York, USA: Chapman & Hall.
44. De Roos AM, Persson L (2013) Population and community ecology of ontogenetic development . Princeton University Press. 538pp
45. Del Raye G, Weng KC (2015) An aerobic scope-based habitat suitability index for predicting the effects of multi-dimensional climate change stressors on marine teleosts. Deep Res Part II Top Stud Oceanogr 113: 280–290.
46. Deutsch C, Ferrel A, Seibel B, Pörtner HO, Huey RB (2015) Ecophysiology. Climate change tightens a metabolic constraint on marine habitats. Science 348: 1132–1135.
47. Dieckmann U, Law R (1996) The dynamical theory of coevolution: a derivation from stochastic ecological processes. J Math Biol 34: 579–612.
48. di Santo V, Tran AH, Svendsen JC (2016) Progressive hypoxia decouples activity and aerobic performance of skate embryos. Conserv Physiol 4: cov067; https://doi.org/10.1093/conphys/cov067.
49. Duarte C (2014) Global change and the future ocean: a grand challenge for marine sciences. Front Mar Sci 1: 1–16.
50. Eliason EJ, Clark TD, Hague MJ, Hanson LM, Gallagher ZS, Jeffries KM, Gale MK, Patterson DA, Hinch SG, Farrell AP (2011) Differences in thermal tolerance among sockeye salmon populations. Science 332: 109–112.
51. Elleouet J, Albouy C, Ben Rais Lasram F, Mouillot D, Leprieur F (2014) A trait-based approach for assessing and mapping niche overlap between native and exotic species: the Mediterranean coastal fish fauna as a case study. Divers Distrib 20: 1333–1344.
52. Elliott M, Borja A, McQuatters-Gollop A, Mazik K, Birchenough S, Andersen JH, Painting S, Peck M (2015) Force majeure: will climate change affect our ability to attain good environmental status for marine biodiversity. Mar Pollut Bull 95: 7–27.
53. Ern R, Huong DTT, Phuong NT, Wang T, Bayley M (2014) Oxygen delivery does not limit thermal tolerance in a tropical eurythermal crustacean. J Exp Biol 217: 809–814.
54. FAO (2014) The State of World Fisheries and Aquaculture . Food and Agriculture Oraganization of the United Nations.
55. Farrell AP (2016) Pragmatic perspective on aerobic scope: peaking, plummeting, pejus and apportioning. J Fish Biol 88: 322–343.
56. Feder ME, Bennett AF, Huey RB (2000) Evolutionary physiology. Annu Rev Ecol Syst 31: 315–341.
57. Ferreira-Martins D, Coimbra J, Antunes C, Wilson JM (2016) Effects of salinity on upstream-migrating , spawning sea lamprey, Petromyzon marinus. Conserv Physiol 4: cov064; https://doi.org/10.1093/conphys/cov064.

58. Ferrer-Maza D, Lloret J, Faliex E, Sasal P (2016) Links between parasitism, energy reserves and fecundity of European anchovy, Engraulis encrasicolus, in the northwestern Mediterranean Sea. Conserv Physiol 4: cov069; https://doi.org/10.1093/conphys/cov069.
59. Fobian D, Overgaard J, Wang T (2014) Oxygen transport is not compromised at high temperature in pythons. J Exp Biol 217: 3958–3961.
60. Forbes VE, Palmqvist A, Bach L (2006) The use and misuse of biomarkers in ecotoxicology. Environ Toxicol Chem 25: 272–280.
61. Freitas C, Olsen EM, Knutsen H, Albretsen J, Moland E (2016) Temperature-associated habitat selection in a cold-water marine fish. J Anim Ecol 85: 628–637.
62. Fry FEJ (1947) The effects of the environment on animal activity. Univ Toronto Stud, Biol Ser 55: 1–62.
63. Fry FEJ (1971) The effect of environmental factors on the physiology of fish. In: WS Hoar, DJ Randall, eds, Fish Physiology , Vol 6. Academic Press, New York, pp 1–98.
64. Gagliano M, McCormick MI (2009) Hormonally mediated maternal effects shape offspring survival potential in stressful environments. Oecologia 160: 657–665.
65. Gardiner NM, Munday PL, Nilsson GE (2010) Counter-gradient variation in respiratory performance of coral reef fishes at elevated temperatures. PLoS One 5: e13299.
66. Garland T, Carter PA (1994) Evolutionary physiology. Annu Rev Physiol 56: 579–621.
67. Giske J, Eliassen S, Fiksen Ø, Jakobsen PJ, Aksnes DL, Mangel M, Jørgensen C (2014) The emotion system promotes diversity and evolvability . Proc Biol Sci 281: 20141096.
68. Gleiss AC, Dale JJ, Holland KN, Wilson RP (2010) Accelerating estimates of activity-specific metabolic rate in fishes: testing the applicability of acceleration data-loggers. J Exp Mar Biol Ecol 385: 85–91.
69. Gore M, Burggren WW (2012) Cardiac and metabolic physiology of early larval zebrafish (Danio rerio) reflects parental swimming stamina. Front Physiol 3: 35.
70. Gräns A, Olsson C, Pitsillides K, Nelson HE, Cech JJ Jr, Axelsson M (2010) Effects of feeding on thermoregulatory behaviours and gut blood flow in white sturgeon (Acipenser transmontanus) using biotelemetry in combination with standard techniques. J Exp Biol 213: 3198–3206.
71. Gräns A, Jutfelt F, Sandblom E, Jönsson E, Wiklander K, Seth H, Olsson C, Dupont S, Ortega-Martinez O, Einarsdottir I et al. (2014) Aerobic scope fails to explain the detrimental effects on growth resulting from warming and elevated CO2 in Atlantic halibut. J Exp Biol 217: 711–717.
72. Grimm V, Revilla E, Berger U, Jeltsch F, Mooij WM, Railsback SF, Thulke H-H, Weiner J, Wiegand T, DeAngelis DL (2005) Pattern-Oriented Modeling of Agent-Based Complex Systems: Lessons from Ecology. Science 310: 987–991.
73. Gunderson AR, Armstrong EJ, Stillman JH (2016) Multiple stressors in a changing world: the need for an improved perspective on physiological responses to the dynamic marine environment. Ann Rev Mar Sci 8: 357–378.
74. Hanson PC, Johnson TB, Schindler DE, Kitchell JF (1997) Fish bioenergetics 3.0. University of Wisconsin, Sea Grant Institute. Sea Grant Tech Report, WISCUT-97-001, Madison, Wisconsin.
75. Heffernan JB, Soranno PA, Angilletta MJ, Buckley LB, Gruner DS, Keitt TH, Kellner JR, Kominoski JS, Rocha AV, Xiao J et al. (2014) Macrosystems ecology: understanding ecological patterns and processes at continental scales. Front Ecol Environ 12: 5–14.
76. Hewett SW, Johnson BL (1987) A Generalized Bioenergetics Model of Fish Growth for Microcomputers . University of Wisconsin, Sea Grant Institute, Madison, Wisconsin.
77. Ho DH, Burggren WW (2012) Parental hypoxic exposure confers offspring hypoxia resistance in zebrafish (Danio rerio). J Exp Biol 215: 4208–4216.

78. Hollowed AB, Barange M, Ito SI, Kim S, Loeng H, Peck MA (2011) Effects of climate change on fish and fisheries: forecasting impacts, assessing ecosystem responses, and evaluating management strategies. ICES J Mar Sci 68: 984–985.
79. Holt RE, Jørgensen C (2014) Climate warming causes life-history evolution in a model for Atlantic cod (Gadus morhua). Conserv Physiol 2: cou050; https://doi.org/10.1093/conphys/cou050.
80. Holt RE, Jørgensen C (2015) Climate change in fish : effects of respiratory constraints on optimal life history and behaviour. Biol Lett 11: 20141032.
81. Horodysky AZ, Cooke SJ, Brill RW (2015) Physiology in the service of fisheries science: why thinking mechanistically matters. Rev Fish Biol Fish 25: 425–447.
82. Horodysky AZ, Cooke SJ, Graves JE, Brill RW (2016) Fisheries conservation on the high seas: linking conservation physiology and fisheries ecology for the management of large pelagic fishes. Conserv Physiol 4: cov059; https://doi.org/10.1093/conphys/cov059.
83. Huey RB, Kearney MR, Krockenberger A, Holtum JAM, Jess M, Williams SE (2012) Predicting organismal vulnerability to climate warming: roles of behaviour, physiology and adaptation. Philos Trans R Soc Lond B Biol Sci 367: 1665–1679.
84. Hunter E, Metcalfe JD, O'Brien CM, Arnold GP, Reynolds JD (2004) Vertical activity patterns of free-swimming adult plaice in the southern North Sea. Mar Ecol Prog Ser 279: 261–273.
85. Hutchinson GE (1957) Concluding remarks. Cold Spring Harb Lab Press 22: 415–427.
86. IPCC (2014) Part A: Global and Sectoral Aspects. (Contribution of Working Group II to the Fifth Assessment Report of the Intergovernmental Panel on Climate Change). Clim Chang 2014 Impacts, Adapt Vulnerability 1132.
87. Ishimatsu A, Hayashi M, Kikkawa T (2008) Fishes in high-CO2, acidified oceans. Mar Ecol Prog Ser 373: 295–302.
88. Jeffrey JD, Hasler CT, Chapman JM, Cooke SJ, Suski CD (2015) Linking landscape-scale disturbances to stress and condition of fish: implications for restoration and conservation. Integr Comp Biol 55: 618–630.
89. Johansen JL, Messmer V, Coker DJ, Hoey AS, Pratchett MS (2014) Increasing ocean temperatures reduce activity patterns of a large commercially important coral reef fish. Glob Chang Biol 20: 1067–1074.
90. Johansen JL, Pratchett MS, Messmer V, Coker DJ, Tobin AJ, Hoey AS (2015) Large predatory coral trout species unlikely to meet increasing energetic demands in a warming ocean. Sci Rep 5: 13830.
91. Jørgensen C, Peck MA, Antognarelli F, Azzurro E, Burrows MT, Cheung WWL, Cucco A, Holt RE, Huebert KB, Marras S et al. (2012) Conservation physiology of marine fishes: advancing the predictive capacity of models. Biol Lett 8: 900–903.
92. Jørgensen C, Enberg K, Mangel M (2016) Modelling and interpreting fish bioenergetics: a role for behaviour, life-history traits and survival trade-offs. J Fish Biol 88: 389–402.
93. Kearney M, Porter W (2009) Mechanistic niche modelling: combining physiological and spatial data to predict species' ranges. Ecol Lett 12: 334–350.
94. Kerr SR (1990) The Fry paradigm: its significance for contemporary ecology. Trans Am Fish Soc 119: 779–785.
95. Killen SS, Costa I, Brown JA, Gamperl AK (2007) Little left in the tank: metabolic scaling in marine teleosts and its implications for aerobic scope. Proc Biol Sci 274: 431–438.
96. Killen SS, Marras S, Metcalfe NB, McKenzie DJ, Domenici P (2013) Environmental stressors alter relationships between physiology and behaviour. Trends Ecol Evol 28: 651–658.
97. Killen SS, Nati JJH, Suski CD (2015) Vulnerability of individual fish to capture by trawling is influenced by capacity for anaerobic metabolism. Proc Biol Sci 282: 20150603.

98. Killen SS, Adriaenssens B, Marras S, Claireaux G, Cooke SJ (2016a) Context dependency of trait repeatability and its relevance for management and conservation of fish populations. Conserv Physiol 4: cow007; https://doi.org/10.1093/conphys/cow007.

99. Killen SS, Glazier DS, Rezende EL, Clark TD, Atkinson D, Willener AST, Halsey LG (2016b) Ecological influences and morphological correlates of resting and maximal metabolic rates across teleost fish species. Am Nat 187: 592–606.

100. King GD, Chapman JM, Cooke SJ, Suski CD (2016) Stress in the neighborhood: tissue glucocorticoids relative to stream quality for five species of fish. Sci Total Environ 547: 87–94.

101. Kooi BW, van der Meer J (2010) Bifurcation theory, adaptive dynamics and dynamic energy budget-structured populations of iteroparous species. Philos Trans R Soc Lond B Biol Sci 365: 3579–3590.

102. Kooijman SALM (1993) Dynamic Energy Budgets in Biological Systems: Theory and Applications in Ecotoxicology . New York, USA: Cambridge University Press.

103. Kooijman SALM (2010) Dynamic Energy Budget Theory for Metabolic Organization , 3rd edn. Cambridge, UK: Cambridge University Press.

104. Lefevre S (2016) Are global warming and ocean acidification conspiring against marine ectotherms? A meta-analysis of the respiratory effects of elevated temperature, high CO2 and their interaction. Conserv Physiol 4: cow009; https://doi.org/10.1093/conphys/cow009.

105. Lefrançois C, Claireaux G (2003) Influence of ambient oxygenation and temperature on metabolic scope and scope for heart rate in the common sole Solea solea. Mar Ecol Prog Ser 259: 273–284.

106. Le Quesne WJF, Pinnegar JK (2012) The potential impacts of ocean acidification: scaling from physiology to fisheries. Fish Fish 13: 333–344.

107. Lutterschmidt WI, Hutchison VH (1997) The critical thermal maximum: history and critique. Can J Zool 75: 1561–1574.

108. McKenzie DJ (2011) Swimming and Other Activities | Energetics of Fish Swimming. Encyclopedia of Fish Physiology . Elsevier Inc, London, United Kingdom.

109. McKenzie DJ, Garofalo E, Winter MJ, Ceradini S, Verweij F, Day N, Hayes R, van der Oost R, Butler PJ, Chipman JK et al. (2007) Complex physiological traits as biomarkers of the sub-lethal toxicological effects of pollutant exposure in fishes. Philos Trans R Soc Lond B Biol Sci 362: 2043–2059.

110. McNeil BI, Sasse TP (2016) Future ocean hypercapnia driven by anthropogenic amplification of the natural CO2 cycle. Nature 529: 383–386.

111. Madliger CL, Cooke SJ, Crespi EJ, Funk JL, Hultine KR, Hunt KE, Rohr JR, Sinclair BJ, Suski CD, Willis CKR et al. (2016) Success stories and emerging themes in conservation physiology. Conserv Physiol 4: cov057; https://doi.org/10.1093/conphys/cov057.

112. Marey M (1896) Phenomenon of flight in the animal kingdom. In: Smithsonian Annual Report. US Government Print Office, Washington, DC, pp 226–285.

113. Marras S, Cucco A, Antognarelli F, Azzurro E, Milazzo M, Bariche M, Butenschön M, Kay S, Di Bitetto M, Quattrocchi G et al. (2015a) Predicting future thermal habitat suitability of competing native and invasive fish species: from metabolic scope to oceanographic modelling. Conserv Physiol 3: cou059; https://doi.org/10.1093/conphys/cou059.

114. Marras S, Noda T, Steffensen JF, Svendsen MBS, Krause J, Wilson ADM, Kurvers RHJM, Herbert-Read J, Boswell KM, Domenici P (2015b) Not so fast: swimming behavior of sailfish during predator–prey interactions using high-speed video and accelerometry. Integr Comp Biol 55: 719–727.

115. Martin BT, Nisbet RM, Pike A, Michel CJ, Danner EM (2015) Sport science for salmon and other species: ecological consequences of metabolic power constraints. Ecol Lett 18: 535–544.

116. Metcalfe JD (2006) Fish population structuring in the North Sea: understanding processes and mechanisms from studies of the movements of adults. J Fish Biol 69: 48–65.
117. Metcalfe JD (2009) Welfare in wild-capture marine fisheries. J Fish Biol 75: 2855–2861.
118. Metcalfe JD, Arnold G (1997) Tracking fish with electronic tags. Nature 387: 665–666.
119. Metcalfe JD, Le Quesne WJF, Cheung WWL, Righton DA (2012) Conservation physiology for applied management of marine fish: an overview with perspectives on the role and value of telemetry. Philos Trans R Soc Lond B Biol Sci 367: 1746–1756.
120. Metcalfe JD, Wright S, Tudorache C, Wilson RP (2016a) Recent advances in telemetry for estimating the energy metabolism of wild fishes. J Fish Biol 88: 284–297.
121. Metcalfe NB, Van Leeuwen TE, Killen SS (2016b) Does individual variation in metabolic phenotype predict fish behaviour and performance. J Fish Biol 88: 298–321.
122. Metz JA, Nisbet RM, Geritz SA (1992) How should we define 'fitness' for general ecological scenarios. Trends Ecol Evol 7: 198–202.
123. Miller GM, Watson S-A, Donelson JM, McCormick MI, Munday PL (2012) Parental environment mediates impacts of increased carbon dioxide on a coral reef fish. Nat Clim Chang 2: 858–861.
124. Mills SC, Beldade R, Chabanet P, Bigot L, O'Donnell JL, Bernardi G (2015) Ghosts of thermal past: reef fish exposed to historic high temperatures have heightened stress response to further stressors. Coral Reefs 34: 1255–1260.
125. Mouillot D, Bellwood DR, Baraloto C, Chave J, Galzin R, Harmelin-Vivien M, Kulbicki M, Lavergne S, Lavorel S, Mouquet N et al. (2013) Rare species support vulnerable functions in high-diversity ecosystems. PLoS Biol 11: e1001569.
126. Moyano M, Illing B, Peschutter P, Huebert KB, Peck MA (2016) Thermal impacts on the growth, development and ontogeny of critical swimming speed in Atlantic herring larvae. Comp Biochem Physiol A Mol Integr Physiol 197: 23–34.
127. Nagelkerken I, Munday PL (2016) Animal behaviour shapes the ecological effects of ocean acidification and warming: moving from individual to community-level responses. Glob Chang Biol 22: 974–989.
128. Neat F, Righton D (2007) Warm water occupancy by North Sea cod. Proc Biol Sci 274: 789–798.
129. Nichols TA, Anderson TW, Širović A (2015) Intermittent noise induces physiological stress in a coastal marine fish. PLoS One 10: 1–13.
130. Norin T, Malte H, Clark TD (2014) Aerobic scope does not predict the performance of a tropical eurythermal fish at elevated temperatures. J Exp Biol 217:244–251.
131. Overgaard J, Andersen JL, Findsen A, Pedersen PBM, Hansen K, Ozolina K, Wang T (2012) Aerobic scope and cardiovascular oxygen transport is not compromised at high temperatures in the toad Rhinella marina. J Exp Biol 215: 3519–3526.
132. Patterson DA, Cooke SJ, Hinch SG, Robinson KA, Young N, Farrell AP, Miller KM (2016). A perspective on physiological studies supporting the provision of scientific advice for the management of Fraser River sockeye salmon (Oncorhynchus nerka). Conserv Physiol 4: cow026; https://doi.org/10.1093/conphys/cow026.
133. Pauly D (1981) The relationships between gill surface area and growth performance in fish : a generalization of von Bertalanffy's theory of growth. Berichte der Dtsch Wissenchaftlichen Kommission fur Meeresforsch 28: 251–282.
134. Pauly D, Zeller D (2016) Catch reconstructions reveal that global marine fisheries catches are higher than reported and declining. Nat Commun 7: 10244.
135. Pauly D, Christensen V, Guenétte S, Pitcher TJ, Sumaila UR, Walters CJ, Watson R, Zeller D (2002) Towards sustainability in world fisheries. Nature 418: 689–695.
136. Persson L, De Roos AM (2006) Food-dependent individual growth and population dynamics in fishes. J Fish Biol 69: 1–20.

137. Peck MA, Moyano M (2016) Measuring respiration rates in marine fish larvae: challenges and advances. J Fish Biol 88: 173–205.
138. Peck MA, Arvanitidis C, Butenschön M, Canu DM, Chatzinikolaou E, Cucco A, Domenici P, Fernandes JA, Gasche L, Huebert KB et al. (2016) Projecting changes in the distribution and productivity of living marine resources: a critical review of the suite of modelling approaches used in the large European project VECTORS. Estuar Coast Shelf Sci . https://doi.org/10.1016/j.ecss.2016.05.019.
139. Perry AL, Low PJ, Ellis JR, Reynolds JD (2005) Climate change and distribution shifts in marine fishes. Science 308: 1912–1915.
140. Petitgas P, Rijnsdorp AD, Dickey-Collas M, Engelhard GH, Peck MA, Pinnegar JK, Drinkwater K, Huret M, Nash RDM (2013) Impacts of climate change on the complex life cycles of fish. Fish Oceanogr 22: 121–139.
141. Pörtner H-O (2010) Oxygen- and capacity-limitation of thermal tolerance: a matrix for integrating climate-related stressor effects in marine ecosystems. J Exp Biol 213: 881–893.
142. Pörtner H-O, Farrell AP (2008) Physiology and climate change. Science 322: 690–692.
143. Pörtner H-O, Knust R (2007) Climate change affects marine fishes through the oxygen limitation of thermal tolerance. Science 315:95–97.
144. Pörtner H-O, Peck MA (2010) Climate change effects on fishes and fisheries: towards a cause-and-effect understanding. J Fish Biol 77: 1745–1779.
145. Priede IG (1985) Metabolic scope in fishes. In: P Tytler, P Calow, eds, Fish Energetics, New Perspectives . Croom-Helm, London, pp 33–64.
146. Prince ED, Luo J, Goodyear CP, Hoolihan JP, Snodgrass D, Orbesen ES, Serafy JE, Ortiz M, Schirripa MJ (2010) Ocean scale hypoxia-based habitat compression of Atlantic istiophorid billfishes. Fish Oceanogr 19:448–462.
147. Prosser CL (1950) Comparative Animal Physiology . Saunders, Philadelphia.
148. Queirós A, Huebert K, Keyl F, Fernandes J, Stolte W, Maar M, Kay S, Jones M, Teal L, Hamon K et al. (2016). Solutions for ecosystem-level protection of ocean systems against climate change. Glob Chang Biol . In press. https://doi.org/10.1111/gcb.13423.
149. Raab K, Llope M, Nagelkerke LAJ, Rijnsdorp AD, Teal LR, Licandro P, Ruardij P, Dickey-Collas M (2013) Influence of temperature and food availability on juvenile European anchovy Engraulis encrasicolus at its northern boundary. Mar Ecol Prog Ser 488: 233–245.
150. Réale D, Garant D, Humphries MM, Bergeron P, Careau V, Montiglio P-O (2010) Personality and the emergence of the pace-of-life syndrome concept at the population level. Philos Trans R Soc Lond B Biol Sci 365: 4051–4063.
151. Righton D, Metcalfe JD, Connolly P (2001) Fisheries. Different behaviour of North and Irish Sea cod. Nature 411: 156.
152. Righton D, Andersen K, Neat F, Thorsteinsson V, Steingrund P, Svedäng H, Michalsen K, Hinrichsen H, Bendall V, Neuenfeldt S et al. (2010) Thermal niche of Atlantic cod Gadus morhua: limits, tolerance and optima. Mar Ecol Prog Ser 420: 1–13.
153. Rijnsdorp AD, Peck MA, Engelhard GH, Mo C, Pinnegar JK (2009) Resolving the effect of climate change on fish populations. ICES J Mar Sci 66: 1570–1583.
154. Rogers NJ, Urbina MA, Reardon EE, McKenzie DJ, Wilson RW (2016) A new analysis of hypoxia tolerance in fishes using a database of critical oxygen level (Pcrit). Conserv Physiol 4: cow012; https://doi.org/10.1093/conphys/cow012.
155. Roze T, Christen F, Amerand A, Claireaux G (2013) Trade-off between thermal sensitivity, hypoxia tolerance and growth in fish. J Therm Biol 38: 98–106.
156. Rutz C, Hays GC (2009) New frontiers in biologging science. Biol Lett 5: 289–292.
157. Schmidt-Neilsen K (1982) Animal Physiology. Adaptation and Environment . Cambridge University Press, New York.
158. Schulte PM (2015) The effects of temperature on aerobic metabolism: towards a mechanistic understanding of the responses of ectotherms to a changing environment. J Exp Biol 218: 1856–1866.

159. Seebacher F, Franklin CE (2012) Determining environmental causes of biological effects: the need for a mechanistic physiological dimension in conservation biology. Philos Trans R Soc Lond B Biol Sci 367: 1607–1614.
160. Seebacher F, White CR, Franklin CE (2015) Physiological plasticity increases resilience of ectothermic animals to climate change. Nat Clim Chang 5: 61–66.
161. Sierra-Flores R, Atack T, Migaud H, Davie A (2015) Stress response to anthropogenic noise in Atlantic cod Gadus morhua L. Aquac Eng 67: 67–76.
162. Simpson SD, Radford AN, Nedelec SL, Ferrari MCO, Chivers DP, McCormick MI, Meekan MG (2016) Anthropogenic noise increases fish mortality by predation. Nat Commun 7: 10544.
163. Sims DW, Southall EJ, Richardson AJ, Reid PC, Metcalfe JD (2003) Seasonal movements and behaviour of basking sharks from archival tagging: no evidence of winter hibernation. Mar Ecol Prog Ser 248: 187–196.
164. Slabbekoorn H, Bouton N, Van Opzeeland I, Coers A, Cate C, Popper AN (2010) A noisy spring: the impact of globally rising underwater sound levels on fish. Trends Ecol Evol 25: 419–427.
165. Sousa T, Domingos T, Kooijman SALM (2008) From empirical patterns to theory: a formal metabolic theory of life. Philos Trans R Soc Lond B Biol Sci 363: 2453–2464.
166. Sumaila UR, Cheung WWL, Lam VWY, Pauly D, Herrick S (2011) Climate change impacts on the biophysics and economics of world fisheries. Nat Clim Chang 1: 449–456.
167. Sumpter JP, Jobling S (1995) Vitellogenesis as a biomarker for estrogenic contamination of the aquatic environment. Environ Health Perspect 103: 173–178.
168. Sunday JM, Bates AE, Dulvy NK (2011) Global analysis of thermal tolerance and latitude in ectotherms. Proc Biol Sci 278: 1823–1830.
169. Sunday JM, Bates AE, Dulvy NK (2012) Thermal tolerance and the global redistribution of animals. Nat Clim Chang 2: 686–690.
170. Svendsen JC, Aarestrup K, Steffensen JF, Herskin J (2006) A Novel Acoustic Dissolved Oxygen Transmitter for Fish Telemetry. Mar Technol Soc J 40: 103–108
171. Teal LR, van Hal R, van Kooten T, Ruardij P, Rijnsdorp AD (2012) Bio-energetics underpins the spatial response of North Sea plaice (Pleuronectes platessa L.) and sole (Solea solea L.) to climate change. Glob Chang Biol 18: 3291–3305.
172. Teal LR, Marras S, Peck MA, Domenici P (2016) Physiology-based modelling approaches to characterize fish habitat suitability: their usefulness and limitations. Estuar Coast Shelf Sci . https://doi.org/10.1016/j.ecss.2015.11.014.
173. Tyler CR, Van Der Eerden B, Jobling S, Panter G, Sumpter JP (1996) Measurement of vitellogenin, a biomarker for exposure to oestrogenic chemicals, in a wide variety of cyprinid fish. J Comp Physiol B Biochem Syst Environ Physiol 166: 418–426.
174. Van Der Meer J (2016) A paradox in individual-based models of populations. Conserv Physiol 4: 1–12.
175. Verberk WCEP, Overgaard J, Ern R, Bayley M, Wang T, Boardman L, Terblanche JS (2016) Does oxygen limit thermal tolerance in arthropods? A critical review of current evidence. Comp Biochem Physiol A Mol Integr Physiol 192: 64–78.
176. Wang T, Overgaard J (2007) Ecology. The heartbreak of adapting to global warming. Science 315: 49–50.
177. Wang T, Lefevre S, Iversen NK, Findorf I, Buchanan R, McKenzie DJ (2014) Anaemia only causes a small reduction in the upper critical temperature of sea bass: is oxygen delivery the limiting factor for tolerance of acute warming in fishes. J Exp Biol 217: 4275–4278.
178. Ward TD, Algera DA, Gallagher AJ, Hawkins E, Horodysky A, Jørgensen C, Killen SS, McKenzie DJ, Metcalfe JD, Peck MA et al. (2016) Understanding the individual to implement the ecosystem approach to fisheries management. Conserv Physiol 4: cow005; https://doi.org/10.1093/conphys/cow005.
179. Ware DM (1982) Power and evolutionary fitness of teleosts. Can J Fish Aquat Sci 39: 3–13.

180. Whitlock RE, Hazen EL, Walli A, Farwell C, Bograd SJ, Foley DG, Castleton M, Block BA (2015) Direct quantification of energy intake in an apex marine predator suggests physiology is a key driver of migrations. Sci Adv 1: e1400270.
181. Williams BK (2011) Adaptive management of natural resources—framework and issues. J Environ Manag 92: 1346–1353.
182. Wilson RW (2014) Chapter 3.6 Fish. In: D Laffoley, J Baxter, F Thevenon, J Oliver, eds, The Significance and Management of Natural Carbon Stores in the Open Ocean. Full report . IUCN, Gland, Switzerland, pp 81–94. ISBN: 978-208317-1692-3.
183. Wilson RW, Millero FJ, Taylor JR, Walsh PJ, Christensen V, Jennings S, Grosell M (2009) Contribution of fish to the marine inorganic carbon cycle. Science 323: 359–362.
184. Wilson SM, Hinch SG, Eliason EJ, Farrell AP, Cooke SJ (2013) Calibrating acoustic acceleration transmitters for estimating energy use by wild adult Pacific salmon. Comp Biochem Physiol A Mol Integr Physiol 164: 491–498.
185. Wright S, Metcalfe JD, Hetherignton S, Wilson RP (2014) Estimating activity-specific energy expenditure in a teleost fish, using accelerometer loggers. Mar Ecol Prog Ser 496: 19–32.
186. Zambonino-Infante JL, Claireaux G, Ernande B, Jolivet A, Quazuguel P, Sévère A, Huelvan C, Mazurais D (2013) Hypoxia tolerance of common sole juveniles depends on dietary regime and temperature at the larval stage: evidence for environmental conditioning. Proc Biol Sci 280: 20123022.

# CHAPTER 12

# Effects of Habitat Enrichment and Food Availability on The Foraging Behaviour of Juvenile Atlantic Cod (*Gadus Morhua L*)

**Olav Moberg & Victoria A. Braithwaite Knut Helge Jensen & Anne Gro Vea Salvanes**

*Department of Biology, University of Bergen, PO Box 7800 5020 Bergen, Norway*

*School of Forrest Resources & Department of Biology, Penn State University, Pennsylvania, PA 16827, USA*

## ABSTRACT

The environment can play an important role in shaping how an animal behaves, and how well the animal performs in a particular environment can be influenced by early experiences. The tradition of releasing captive-reared juveniles into the wild in an effort to strengthen wild fish populations has often had little success owing to high post-release mortality. Fish reared under standard hatchery conditions are provided with fewer stimuli and they receive excess quantities of pellet food that are easy to handle and consume. Captive reared fish therefore appear to be under-stimulated and overfed. Several studies have demonstrated that simple structural enrichment in the rearing facilities promotes flexible behaviour compared to fish reared in plain, standard hatchery tanks. Less attention has been given to the effects of the diet. Here we use a cross-factored design to test the relative role of food ration and spatial enrichment on foraging behaviour. Our results show that fish from enriched environments, regardless of previous food-ration size, were more reluctant to start feeding on the first day in a novel arena. On day two and three, however, fish with prior experience of a low food ration showed greater foraging activity and efficiency than fish fed on full rations. On the second and third day, prior experience with enrichment was less important. We discuss how early feeding experience in combination with structural enrichment may contribute in producing fish that are better suited for release into the wild.

**Keywords:** Atlantic cod *Gadus morhua* Enrichment Food ration Foraging behaviour Restocking

# INTRODUCTION

Following a decline in several important fish-stocks, particularly marine species, extensive restocking programmes have been undertaken, but often these have had limited success because of the high post-release mortality (Olla et al. 1994, 1998; Brown and Day 2002; Chan et al. 2003). For example, analysis of long-term data-sets from released larval cod along the Norwegian coast demonstrates little, or limited effect on populations, despite initial effects positively influencing the 0+ class (Salvanes et al. 1994; Svasand et al. 2000; Chan et al. 2003). There are, however, examples where larger juvenile cod have had higher post-release survival compared to smaller fish. This success may have been due to size-dependant predation or the fact the larger fish were reared in big, enclosed ponds which them gave a more natural rearing environment than conventional hatchery facilities (Kristiansen et al. 2000).

While in some cases failure of these programmes may be because the ecosystem is unable to support the large numbers of fish released, it is generally assumed that hatchery-reared fish are less fit for survival compared to fish that develop in the wild environment (Brown and Day 2002; Salvanes and Braithwaite 2006). As brood fish used for restocking purposes are commonly taken from local stock (so that local genetic adaptations are conserved), it is reasonable to assume that the major differences between wild and artificially reared fish are caused by the hatchery rearing environment. Fish reared for restocking are usually produced using commercial fish breeding techniques, where the focus is on rapid production of large numbers of individuals. The main difference between hatchery facilities and the wild is the homogenous nature of parameters that normally vary in a natural environment, such as temperature, salinity, light-conditions, fish density, feeding and the level of structural complexity.

It has been reported for many species, over several taxa, that early experience plays an important role in shaping adult behaviour (Sackett et al. 1999; Rosenzweig 2003; Knudsen 2004; Poirier et al. 2004; Braithwaite and Salvanes 2005). The kinds of early experience that can have an effect include social interactions, exposure to predators, experience of foraging on live prey and, more generally, interacting with structural complexity such as enrichment in the captive environment. A large body of literature has shown that introducing enrichment into the rearing environment induces behavioural changes. For example, in mammals, rats with structural enrichment display better problem solving capabilities (Rosenzweig and Bennett 1996), in birds the experience of a variety of airborne odours in early life predisposes pigeons to develop olfactory maps (Wiltschko et al. 1989). In fish shoaling responses are more varied in individuals reared with structural enrichment compared to fish reared in plain environments (Salvanes et al. 2007). These examples illustrate how the addition of enrichment can affect the development of the behavioural phenotype, and this may, in part, explain why hatchery fish behave differently to their wild counterparts.

To increase post-release survival one can try to mimic a natural environment in the hatchery. It is believed that a variable rearing environment may help with this by priming fish to behave in a more flexible and adaptive manner (Berejikian et al. 2001; Brown and Day 2002; Salvanes and Braithwaite 2006; Lee and Berejikian 2008). Most of the research investigating the effects of enrichment in fish rearing environments has focused on the role of structural enrichment. However, other stochastic environmental factors could be manipulated. For example, food availability in the wild is not evenly distributed, but rather it is patchy in time and space, unlike the hatchery where it is easy to catch pellet food that is continuously available.

In terms of post-release mortality, foraging and predation are considered to be major factors (Olla et al. 1994). Foraging is a risky activity, and animals need to be aware of the associated dangers—knowing when to be cautious is adaptive (Kelley and Magurran 2003). As such, the development of behaviour associated with the caution/boldness temperament axis is likely to have an impact on survival. A previous experiment in which cod were reared with variable enrichment produced fish that showed a greater propensity to explore a novel area, that were faster at recovering from stressor and that had a faster transition from pellets to live prey (Braithwaite and Salvanes 2005). In this study the timing of food availability was varied on a temporal and spatial basis, and this appeared to create fish that were bolder i.e. faster at emerging into a novel area (Braithwaite and Salvanes 2005). As yet, however, no studies have investigated the effect of how much food is available during rearing.

Here we use a cross factored design to determine the effects of a reduced food regime on hatchery fish reared with or without environmental enrichment. It is already known that the addition of physical enrichment into a hatchery environment generates fish with more adaptive behaviour. Our aim with this study was to determine whether the level of food availability interacts with physical enrichment to influence behaviour. For instance, fish on a lower ration may be forced to take more risks and develop bolder behaviour, but fish that experience variability in their physical environment as well as a lower food ration, may learn to be cautious under some circumstances and yet be bold under others.

# MATERIALS AND METHODS

## Fish

A total of 72 fish were tested with an additional 25 fish used for social stimuli. The social stimuli fish were used to encourage the test fish to feed while reducing the stress of social deprivation when single test fish were in observation tanks (Nordeide and Svasand 1990). As the fish were used for behavioural studies in aquaria, sub-samples were size-matched to standardize the fish-size relative to experimental aquaria and prey, but sampled randomly from all replicate rearing tanks. We used length as our size criteria, and the size of the fish selected were intermediate—i.e. not the longest or the shortest for any of the groups.

Fish were obtained as fertilized eggs from a commercial cod-farmer using local, wild-caught brood-stock (Bømlo, Western Norway). The eggs were spawned at the same day, and hatched on 4th of April 2006. At day one post hatching larvae were transported to the High Technology Centre in Bergen. Larvae were start-fed wild-caught, natural zooplankton and housed in plain, 1 × 1 × 0.6 m tanks from arrival until they were 90 days old. The cod were then introduced to commercially available dry food pellets (EWOS, MARIN) used throughout the rearing and during the experiment (see Table 1). The fish were kept on 12–12 h light regime, and flow-through seawater at 10 ± 0.5°C. The behavioural screening took place in November 2006.

**Table 1** Calculations of ration size by fish size for the full ration (100%) food group. Food-factor calculations are based on natural growth in wild cod, and known food efficiency and temperature in the rearing tanks

| Fish size (g) | Daily growth (%) | Food-factor (%) |
|---|---|---|
| 0.70 | 7.00 | 5.60 |
| 3.00 | 4.50 | 3.60 |
| 10.00 | 2.50 | 2.40 |
| 15+ | 2.50 | 2.00 |

All work was conducted with the approval of the departmental research ethics board at the Department of Biology, University of Bergen, under the Norwegian Veterinary Authorities site licence no 18.

## Rearing environments

In July 2006, five enriched tanks (1 × 1 × 0.6 m) were set up as described in Braithwaite and Salvanes (2005), and a further five plain tanks were set up as controls. The introduction of pebbles, rocks and artificial kelp created the enrichment, while control environments consisted of standard, plain conventional rearing tanks. In the enriched tanks, rocks and pebbles covered approximately 65% of the bottom area. Light was provided by fluorescent tubes (40 W), placed 1.5 m above the water level. The tanks were flushed for debris every second day, and cleaned every week. Two of the enriched tanks and two of the plain were given reduced food-rations. This left three tanks with full rations in the enriched group and a further 3 in the plain group. The additional enriched and plain tanks in the full ration groups were reared for a different experiment, but some of those fish also took part in this experiment (i.e. fish in the full ration group were sampled from all three tanks for both plain and enriched treatments). Standard food rations were calculated from growth rates of wild cod (Hawkins et al. 1985), and adjusted for growth, using known food efficiency of the commercially available cod food (EWOS, MARIN) (See Table 1). Fish in the full ration food group (100%), were hand-fed with a full food-ration, delivered once a day, in the morning. The reduced-food groups (50%) were also fed a full ration, but every second day. To control for handling differences on the day without feeding, the tanks on reduced food-rations had a cup of sea-water added to their tank on days without feeding. These arrangements generated four treatment groups; (*1*) Enriched, 50% food ration, (*2*) Enriched, 100% food ration, (*3*) Plain, 50% food ration, (*4*) Plain, 100% food ration. The fish were reared under these conditions for 16 weeks before they were tested for behavioural differences.

## Test-aquaria

Screening took place in eight test-aquaria (95 l–70 × 38 × 36 cm); each sectioned into two equal partitions (Fig. 1), and separated by transparent glass. One side housed the test fish; the other side contained two social stimuli fish (Fig. 1). The stimuli fish were randomly sampled and distributed among the test-aquaria, and to avoid unnecessary stress from handling, they stayed in their designated experimental tanks during all replicate trials. The remaining nine stimuli fish were kept in a separate tank as replacements, to counter potential problems with mortality. All tanks were externally covered by black plastic on three sides, with the front covered by a removable, black plastic curtain that remained closed except for periods of filming or when we needed to observe the number of uneaten food-items. Each compartment had continuous flow-through of sea-water, from the same source, and with the same physical qualities as the water circulated in the rearing tanks.

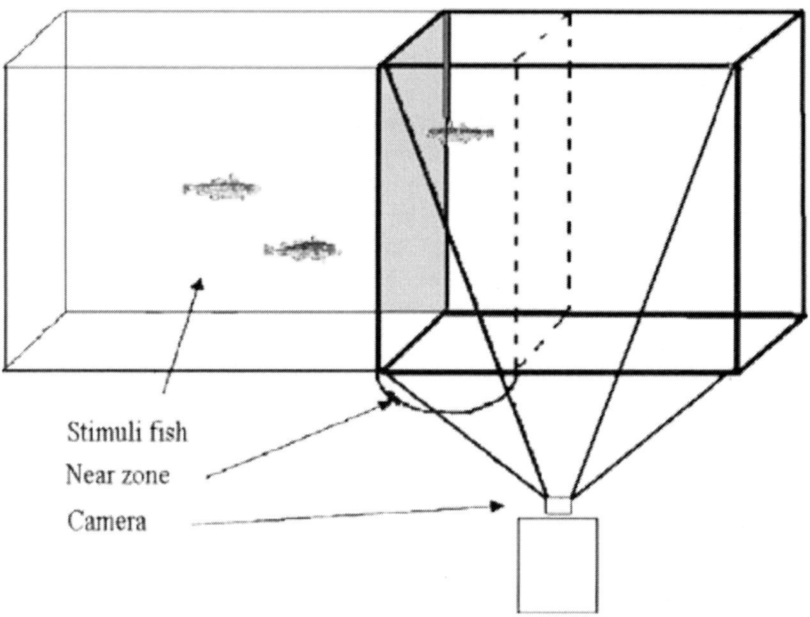


**Figure. 1** Observation tanks with camera and defined near zone

## Experimental protocol

A trial consisted of 3 days of observations, with eight parallel trials; two fish from each of the four treatment groups. For each trial, a single test fish were placed into one compartment, with a pair of social stimuli fish in the opposite compartment to reduce the effects of social deprivation when transferred to the experimental environment (Nordeide and Svasand 1990). Position of test fish in the left or right compartment of the test-aquaria was counterbalanced across trials. Thus, through the experiment, all groups were evenly distributed between left or right test compartments. The next morning (day 1) observations began. To stimulate the test fish into feeding on the novel prey, the stimuli fish were fed five live mysids *(Praunus flexuosus)* twice and at 180 and 90 min before the start of an observation. Prey were delivered through a hole in the tank lid, centrally positioned over each compartment. Test observations began by introducing five mysids into both the social stimuli fish and the test fish. For the first 15 min, the front curtain was raised and the tank was filmed (Canon MV700*i* VCR) to allow us to record foraging behaviour, movement and social preferences. After 15 min had elapsed, the front of the test-aquaria was covered with the black curtain, and the fish were left undisturbed, except when we counted the number of uneaten mysids at 45, 75 and 135 min after the beginning of each observation. All observations and counts were done by carefully lifting a bottom corner of the plastic cover in the front of the test-compartment corner. After the last count at 135 min, any uneaten mysids were removed from the test-aquaria.

On days 2 and 3, we followed the same procedure as described for day 1, but on these days we did not pre-feed the social stimuli fish, but fed them five mysids at the same time as the test fish. Thus, on days 2 and 3 the test fish and social stimuli fish were all fed five mysids at the start of the observations.

To quantify social interactions and activity, the compartment housing the test fish was divided with an imaginary line, creating a 'near' zone next to social stimuli fish (chosen as the section closest to the stimuli fish) when analysing the video of activity (Fig. 1). Activity and social preferences were scored manually from video recordings, using software from EthoLog (V 2.25) (Ottoni 2000). The activity was measured as the number of times a fish entered or left the 'near' zone (Fig. 1). The social preference was measured as time (s) spent in the near zone.

## Statistical analysis

The probability of capturing a pursued prey was quantified from successful catches of individual prey at the set intervals. Social preferences were scored as time spent in the defined 'near' zone next to the social stimuli fish. Activity was scored as the number of times the fish moved in and out of the near zone to be close to the social stimuli fish.

All data were analysed using R 2.11.1 (Crawley 2007; R Development Core Team 2010). We used generalized linear mixed effects models (glmmPQL) (Zuur et al. 2009; Venables and Ripley 2002), with binomial error distribution to statistically assess the importance of rearing environment (enriched and plain), ration size (100% ration and 50% ration) and trial day (1, 2 and 3) on the probability of capturing prey. A glmmPQL with Poisson error distribution was used to test for the effects of rearing environment, ration size and day on the number of times individuals moved into or out of the 'near' zone, and on number of attacks against mysids (Zuur et al. 2009; Venables and Ripley 2002).

To test if time spent in the 'near' zone varied, we used a linear mixed effect model (lme), assuming a normal error distribution (Zuur et al. 2009). All of these analyses included rearing environment, ration size and day as main effects and individual ID as a random effect, and assuming a first-order autocorrelation structure in this repeated measures analysis. The variable *day* was treated as an ordered categorical variable. We included all main effects and interaction terms in the initial model, which was then simplified by sequentially removing all non-significant terms to achieve the minimal adequate model. The analysis of *number of attacks against mysids* the first 15 min of each day showed only an effect of day and food level and the model had no significant interactions.

To test for length, weight and condition factor depending on food level and rearing background (enriched vs. plain), we used a two-way ANOVA.

# RESULTS

Fish that experienced a 50% food ration increased their probability of capturing prey more rapidly over the 3 experimental days than fish that experienced a 100% food ration (glmmPQL; $t = 2.445$, $p = 0.015$, Fig. 2). On day 1, there was no difference in prey capture success between 50% and 100% food ration fish (glmmPQL; $t = 0.269$, $p = 0.789$), but plain-reared fish had a higher probability of capturing pursued prey than enriched fish (glmmPQL; $t = 2.796$, $p = 0.007$, Fig. 2).

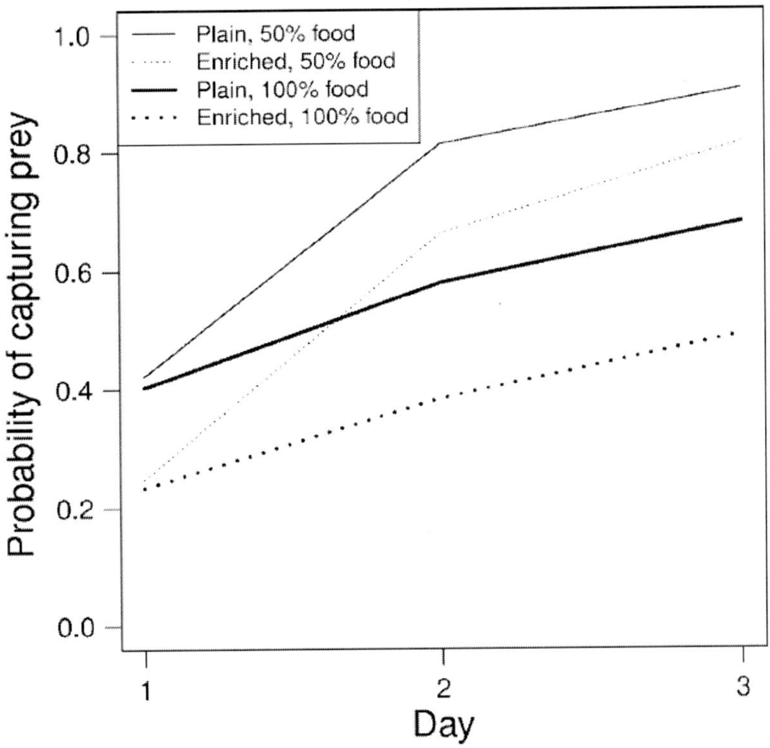


**Figure. 2** Estimated probability of capturing prey for 50% and 100% food ration groups for plain and enriched fish. The *lines* represent predicted values from the GlmmPQL model

The number of attacks increased over days (glmmPQL; $t = 3.924$, $p < 0.001$), and 50% food ration fish were more active in pursuing prey than 100% food ration fish (glmmPQL; $t = 2.974$, $p = 0.004$).

All treatment groups spent a similar amount of time in the 'near' zone next to the social stimuli fish (lme: rearing environment $F_{1,68} = 0.006$, $p = 0.941$; food ration, $F_{1,68} = 0.017$, $p = 0.896$; day, $F_{2,134} = 1.032$, $p = 0.359$). Fish activity (the number of times a fish entered or left the 'near' zone next to the social stimuli fish) increased over days (glmmPQL; $t = 3.684$, $p < 0.001$; Fig. 3). Plain fish were consistently more active than enriched fish (glmmPQL; $t = 2.291$, $p = 0.025$; Fig. 3).

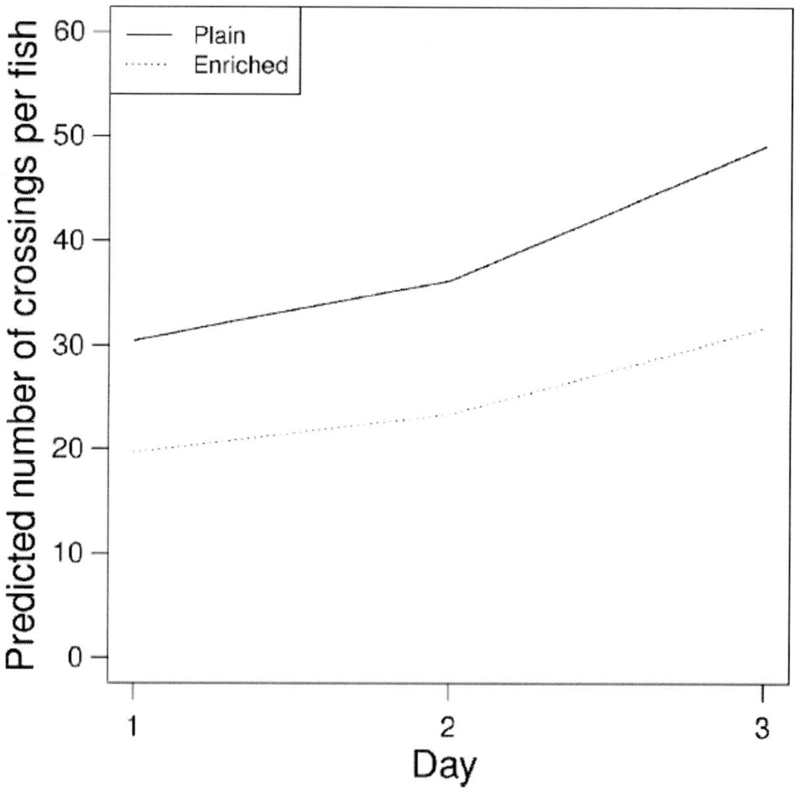


**Figure. 3** Predicted activity level estimated as the number of times the fish crossed the border between the far and the near zone (next to the social stimuli fish) as a function of experimental day and rearing environment (enriched and plain) the first 15 min within each day. The *lines* represent predicted values from the glmmPQL

There were no significant differences in weight, length and C-factor between the plain and enriched groups. Fish from 100% food groups tended to be longer (ANOVA; $F = 2.94$; $p = 0.091$, Table 2a), and they weighed significantly more (ANOVA: $F = 4.58$; $p = 0.036$ Table 2b) and had a higher C-factor (ANOVA; $F = 8.23$; $p = 0.005$, Table 2c).

**Table 2** Length (cm), weight (g) and condition factor for experimental fish after the experiment. Mean values are given with standard deviation (SD)

| | | Mean (SD) | Df | F-value | p-value |
|---|---|---|---|---|---|
| A) Length (cm) | 100% food | 11.81 (1.69) | 1.00 | 2.94 | 0.09 |
| | 50% food | 11.17 (1.42) | | | |
| B) Weight (g) | 100% food | 14.62 (7.26) | 1.00 | 4.58 | 0.04 |
| | 50% food | 11.42 (4.53) | | | |
| C) Condition factor | 100% food | 0.825 (0.06) | 1.00 | 8.23 | 0.01 |
| | 50% food | 0.786 (0.06) | | | |

## DISCUSSION

When cod were introduced to novel, live prey in the test environment, fish with experience of enrichment were more reluctant to start foraging compared to fish from plain tanks. This was the case regardless of the feeding ration they had experienced previously. On days 2 and 3 of the experiment, however, feeding ration had a bigger effect with the 50% fed fish, showing higher clearance rates of live mysids than 100% ration fed fish, regardless of habitat background. Together these results suggest that in a novel situation, the experience with enrichment has an immediate effect on foraging behaviour, seen here as an initial reluctance to forage on a novel type of prey. However, having been fed on a 50% limited ration also influences foraging motivation and how efficiently prey is consumed. We observed that fish reared in plain tanks generally had higher levels of activity, moving more frequently in and out of the 'near' zone, close to the clear partition separating the test fish from the visual stimuli fish. Previous research has found that structural enrichment promotes flexible behaviour in hatchery-reared fish in terms of foraging, aggression, shoaling, stress-recovery and social learning (Salvanes and Braithwaite 2005; Salvanes et al. 2007; Strand et al. 2010); all of these factors that seem relevant to the boldness/caution complex.

In the present study, experience with structural enrichment in early life appears to influence the development of cautious behaviour by creating less bold, more cautious individuals as seen on day one. In a previous study we found enrichment to affect shoaling responses relative to the test-environment (Salvanes et al. 2007). Fish from enriched rearing environments were more flexible and switched between shoaling in the simple plain test environment to a more individual behaviour in a structurally complex environment, while fish reared in plain tanks shoaled regardless of the environment. This can be explained as a shift towards territorial behaviour in the fish from the enriched environment, while fish reared in plain, open-water environment show a more fixed shoaling mode irrespective of test environment. Salvanes and Braithwaite (2005) found enrichment to produce cod with asymmetries in aggressive behaviour, and also to have a higher use of shelter after attacks. Enriched fish were less likely to flee from attacks, and also directed their attacks more towards fish from the plain environment, despite the plain fish being significantly larger, suggesting a territorial behaviour as well as a capacity to assess the other fish in a conflict.

Shoaling is a group behaviour related to foraging as well as anti-predator defence and there is a certain trade-off between the relative safety in the group, and the increased foraging competition (Hoare et al. 2000).

Differences in shoaling behaviour may affect the foraging on several levels. For example, fish from the plain environment may have adapted to more direct competition with shoaling conspecifics in the open water environment. Fish from enriched environments, however, are more likely to enter a territorial mode with a more individual behaviour than the plain controls with a more fixed shoaling mode, and this may affect activity levels and how it searches for, and approaches prey. Non-territorial individuals from plain environments may therefore, through both higher activity and different search-modes, have higher encounter-rates of prey in the experimental area, while the fish with territorial experience to larger degree may act as a "sit and wait" predator, displaying less activity and appearing to be more cautious. Another factor in an enriched environment is that left-over food on the bottom opens up the possibility to forage among the enrichment rocks and pebbles, thereby creating an alternative strategy to surface feeding. The continuous competition with conspecifics in surface-feeding may favour aggressive, competitive feeding behaviour in the plain tanks, as opposed to fish from the enriched rearing environments that may avoid conflicts by adjusting to a bottom-feeding strategy. Thus, both the social differences and the structural environment must be expected to affect the way the fish search for, and approach their prey.

Consequently, we suggest that the observed differences in activity, boldness and foraging behaviour seen between fish with and without enrichment result from fish having to adapt to different levels of competition and foraging responses appropriate to their respective rearing environments.

The level of food ration also played a role, and appeared to interact with boldness, as the fish from the 50% food ration groups displayed more active and efficient foraging behaviour from day two. While earlier work indicates a role for structural enrichment, little is known about how food availability affects the development of foraging behaviour in hatchery fish. One study that partially addressed this, showed that juvenile cod exposed to a variable feeding regime (full ration), were quicker to transfer to a live prey diet (Braithwaite and Salvanes 2005). Our present study demonstrates a novel result; that cod reared with a *low ration diet* were also better at consuming novel live prey.

Several factors may explain the observed differences in foraging efficiency, and factors like feeding motivation, hunger state and the size-relationship between predator and prey should be considered. Size-difference between experimental fish and the prey and between test groups could not explain our results. We selected fish such that they were within an approximate length range that was appropriate for both the prey and aquarium size. As such, motivation related to the profitability of prey (based on prey-size), or differences in the relative size of the search area were minimized. Consequently, differences in foraging motivation may be affected by different food-administrations, i. e differences in time between feeding, rather than amount of food. We could not ignore, however, potential differences in metabolic rates or hunger-state that could be owing to reduced prior food-regimes. This could have contributed to a difference in motivation and foraging efficiency.

Motivation and activity are essential parts in learning processes, and increasing success-rates must be expected to correlate to number of foraging attempts. Thus, the increased foraging efficiency suggests that the 50% food ration group learned to catch and handle the novel prey at a faster rate than the 100% food ration group, but that this increase in learning rate may arise because of motivation and experience rather than better cognitive abilities.

Both structural enrichment and pre-training fish with live prey has been suggested to increase post-release survival in fish produced for release. Some studies have investigated feeding live-prey to hatchery-reared fish just prior to their release. Such research has proved inconclusive as fish with prior experience of live prey may feed well on the live prey they were trained with (Ellis et al. 2002), but they may not necessarily forage so well when introduced to other *novel*prey species (Massee et al. 2007). One experiment which investigated the effect of prior exposure to live prey in combination with enriched environments in juvenile Atlantic salmon *(Salmo salar L.)*, did report that fish can benefit from such a combination (Brown et al. 2003), thus presenting the rearing enrichment itself as a factor in the development of foraging behaviour. Thus, we suggest that if fish are reared for release, a low food ration combined with an enriched rearing environment may provide the fish with more relevant experience to prepare the fish for the wild.

While reduced food-rations appear to have a positive effect on foraging behaviour, they should not be reduced to a level that affects growth to a degree that may induce compensatory growth responses. Fish, in general have a strong ability to compensate for slow growth by bursts of rapid growth (Ali et al. 2003), often linked with negative long term effects on fitness (Metcalfe and Monaghan 2001). Consequently, releasing fish during a period of compensatory growth may counter the otherwise beneficial effects that variable food regime and environmental enrichment may provide.

Previously it has been suggested that providing a diet that reflects naturally occurring food would be one way to improve survival in hatchery-reared fish (Maynard et al. 1996). But the reality is that such an approach will be impossible for large-scale production of fish, because rearing sufficient quantities of live invertebrate prey is too costly and too labour intensive. The results we report here, however, suggest that an alternative approach may be to vary food rations in combination with structural enrichment in otherwise standard rearing facilities. Together, these two methods may provide a feasible way of rearing hatchery fish with increased post-release survival.

## ACKNOWLEDGEMENTS

We thank Frank Midtøy and David Strand for their help in setting up and running the experiment, and members of the Aquatic Behavioural Group for valuable comments and discussions. This work was supported by the Department of Biology, University of Bergen and The Norwegian Research Council (Grant: 177888/v40)

## REFERENCES

1. Ali M, Nicieza A, Wootton RJ (2003) Compensatory growth in fishes: a response to growth depression. Fish Fish 4:147–190
2. Berejikian BA, Tezak EP, Riley SC, LaRae AL (2001) Competitive ability and social behaviour of juvenile steelhead reared in enriched and conventional hatchery tanks and a stream environment. J Fish Biol 59:1600–1613
3. Braithwaite VA, Salvanes AGV (2005) Environmental variability in the early rearing environment generates behaviourally flexible cod: implications for rehabilitating wild populations. P Roy Soc B-Biol Sci 272: 1107–1113.
4. Brown C, Day RL (2002) The future of stock enhancements: Lessons for hatchery practice from conservation biology. Fish Fish 3:79–94

5. Brown C, Davidson T, Laland K (2003) Environmental enrichment and prior experience of live prey improve foraging behaviour in hatchery-reared Atlantic salmon. J Fish Biol 63:187–196
6. Chan KS, Stenseth NC, Kittilsen MO, Gjosaeter J, Lekve K, Smith T, Tveite S, Danielssen D (2003) Assessing the effectiveness of releasing cod larvae for stock improvement with monitoring data. Ecol Appl 13:3–22
7. Crawley MJ (2007) The R Book. R: a language and environment for statistical computing. Wiley, Vienna
8. Ellis T, Hughes RN, Howell BR (2002) Artificial dietary regime may impair subsequent foraging behaviour of hatchery-reared turbot released into the natural environment. J Fish Biol 61:252–264
9. Hawkins AD, Soofiani NM, Smith GW (1985) Growth and feeding of juvenile cod (*Gadus morhua L*). J Conseil 42: 11–32
10. Hoare DJ, Krause J, Peuhkuri N, Godin JGJ (2000) Body size and shoaling in fish. J Fish Biol 57:1351–1366
11. Kelley JL, Magurran AE (2003) Learned predator recognition and antipredator responses in fishes. Fish Fish 4:216–226
12. Knudsen EI (2004) Sensitive periods in the development of the brain and behavior. J Cogn Neurosci 16:1412–1425
13. Kristiansen TS, Ottera H, Svasand T (2000) Size-dependent mortality of juvenile reared Atlantic cod released in a small fjord. J Fish Biol 56:792–801
14. Lee JSF, Berejikian BA (2008) Effects of the rearing environment on average behaviour and behavioural variation in steelhead. J Fish Biol 72:1736–1749
15. Massee KC, Kim J, Berejikian BA, Hardy RW (2007) Prey selection and efficiency of naive and experienced juvenile sockeye salmon. J Fish Biol 70:1213–1223
16. Maynard DJ, McDowell GC, Tezak EP, Flagg TA (1996) Effect of diets supplemented with live food on the foraging behavior of cultured fall chinook salmon. Prog Fish Cult 58: 187–191
17. Metcalfe NB, Monaghan P (2001) Compensation for a bad start: grow now, pay later? Trends Ecol Evol 16:254–260
18. Nordeide JT, Svasand T (1990) The behavior of wild and reared juvenile cod, (*Gadus-morhua L*). Towards a potential predator. Aquaculture and Fisheries Management 21: 317–326
19. Olla BL, Davis MW, Ryer CH (1994) Behavioural deficits in hatchery-reared fished: potential effects on survival following release. Aquacult Fish Manage 25:19–34
20. Olla BL, Davis MW, Ryer CH (1998) Understanding how the hatchery environment represses or promotes the development of behavioral survival skills. B Mar Sci 62: 531–550
21. Ottoni EB (2000) EthoLog 2.2: a tool for the transcription and timing of behavior observation sessions. Behav Res Meth Instrum Comput 32:446–449
22. Poirier R, Chichery R, Dickel L (2004) Effects of rearing conditions on sand digging efficiency in juvenile cuttlefish. Behav Process 67: 273–279
23. R Development Core Team (2010) R: a language and environment for statistical computing. R Foundation for Statistical Computing, Vienna
24. Rosenzweig MR (2003) Effects of differential experience on the brain and behavior. Dev Neuropsychol 24:523–540
25. Rosenzweig MR, Bennett EL (1996) Psychobiology of plasticity: effects of training and experience on brain and behavior. Behav Brain Res 78:57–65
26. Sackett GP, Novak B, Kroeker R (1999) Early experience effects on adaptive behavior: theory revisited. Ment Retard Dev Disabil Res Rev 5:30–40
27. Salvanes AGV, Braithwaite VA (2005) Exposure to variable spatial information in the early rearing environment generates asymmetries in social interactions in cod (Gadus morhua). Behav Ecol Sociobiol 59:250–257
28. Salvanes AGV, Braithwaite VA (2006) The need to understand the behaviour of fish reared for mariculture or restocking. ICES J Mar Sci 63:346–354

29. Salvanes AGV, Giske J, Nordeide JT (1994) Life-history approach to habitat shifts for coastal cod, *Gadus morhua* L. Aquac Fish Manage 25:215–228
30. Salvanes AGV, Moberg O, Braithwaite VA (2007) Effects of early experience on group behaviour in fish. Anim Behav 74:805–811
31. Strand DA, Utne-Palm AC, Jakobsen PJ, Braithwaite VA, Jensen KH, Salvanes AGV (2010) Enrichment promotes learning in fish. Mar Ecol Prog Ser 412:273–282. doi:10.3354/meps08682
32. Svasand T, Kristiansen TS, Pedersen T, Salvanes AGV, Engelsen R, Naevdal R, Nodtvedt M (2000) The enhancement of cod stocks. Fish Fish 1:173–205
33. Venables W, Ripley B (2002) Modern Applied Statistics with S. Springer, New York. ISBN: 0-387-95457-0
34. Wiltschko R, Schops M, Kowalski U (1989) Pigeon homing-wind exposistions determines the importance of olfactory input. Naturwissenschaften 76:229–231
35. Zuur AF, Ieno EN, Walker NJ, Saveliev AA, Smith GM (2009) Mixed effects models and extensions in ecology with R. Springer, New York

# CHAPTER 13

# Longitudinal Movement of Fish in Response to a Single-Day Flow Pulse

**Lisa C. Thompson & Sarah A. Cocherell & Stephanie N. Chun & Joseph J. Cech Jr. & A. Peter Klimley**

*Wildlife, Fish, & Conservation Biology Department, University of California Davis, 1 Shields Avenue, Davis, CA 95616-8751, USA*

## ABSTRACT

We studied whether fish were displaced longitudinally downstream by a 1-day whitewater kayaking pulse flow release from Camino Dam on Silver Creek, a tributary of the South Fork American River, California. On 15 September 2004 flows were increased from a base flow of 0.48 $m^3 \cdot s^{-1}$ to a peak of 18.48 $m^3 \cdot s^{-1}$ by midday, and decreased back to base levels. Rainbow trout (*Oncorhynchus mykiss*) and brown trout (*Salmo trutta*) were observed in snorkel surveys before and after the pulse. Counts of young-of-the-year and juvenile trout were 26% and 9% lower after the pulse, respectively. Counts of adult trout were 12% higher. Six adult trout were radio-tagged and were observed in the reach before, during, and after the pulsed flow. Our results suggest that most trout were able to remain in the study reach during the pulse, but that smaller fish may be more likely to be displaced downstream.

**Keywords:** Pulsed flow Fish movement Snorkel Telemetry Rainbow trout *Oncorhynchus mykiss* Brown trout *Salmo trutta*

# INTRODUCTION

Pulsed flows are relatively rapid increases in flows followed by relatively rapid decreases in these flows, with flow fluctuations lasting minutes, hours, or days (Hunter 1992). Reasons for pulsed flows include: (1) generating electricity, (2) flushing streambeds, (3) providing additional water for downstream diversion for irrigation, (4) preventing reservoirs from flooding, and (5) facilitating human recreation such as rafting and kayaking. Pulsed releases for kayaking are frequently negotiated during Californian dam re-licensing by the Federal Energy Regulatory Commission (FERC), and 7-fold changes in flow are not uncommon (Klimley et al. 2007). Remote river reaches may experience only a few days of pulses per year, but the impact of these infrequent pulses on fish are uncertain. Young-of-the year fish may be particularly affected, since recreational pulses are often requested during the warm summer months when newly hatched fish are rearing, whereas high flows in Californian rivers would commonly occur due to winter rains or spring snowmelt.

Strong pulsed flows may force juvenile salmonids downstream (McCrimmon 1954; Erman and Leidy 1975; Ottaway and Clarke 1981; Ottaway and Forrest 1983; Heggenes and Traaen 1988; Crisp 1991; Crisp and Hurley 1991; Pearsons et al. 1992). Longitudinal displacement of juvenile coho salmon (*Oncorhynchus kisutch*) has been observed in streams during winter periods when floods are common (Shirvell 1994; Giannico and Healey 1998). However, the presence of complex habitat or coarse substrate may provide shelter to juvenile salmonids and minimize downstream displacement (Heggenes 1988; Bell et al. 2001). Older, larger fish may be less likely to be displaced downstream due to their increased swimming ability, relative to that of smaller fishes (Webb 1971). Rainbow trout (*O. mykiss*) (20–25 cm TL) were not displaced downstream by a spring flushing flow that increased discharge eight-fold (Simpkins et al. 2000).

There may be interactions between the impacts of flow releases and water temperature, or dissolved oxygen (Orth and Maughan 1982). Fish may seek deep pools in winter when water temperatures are low, and may thereby be protected from high flows in winter and late spring. During summer there may be a greater potential for high temperatures and low overnight dissolved oxygen concentrations that could interact with low flows. In a review of the impacts of peaking power on the survival of juvenile fish in the Columbia River basin, Bell et al. (1976) noted that rates of predation by piscivorous fish may be positively related to the temperature of water released from dams.

There may also be interactions between the seasonal timing and magnitude of pulse flow releases compared with normal flows for a given time of year, and the life history stage of fish present (Hunter 1992; Bradford et al. 1995; Saltveit et al. 2001). However, we found no previous studies that have addressed the effects of a single pulse on the distribution of young-of-the-year trout which would never have experienced a pulsed flow. Furthermore, we found no studies that have addressed the effects of a single pulse, with timing that was unseasonable for a Mediterranean climate, on the distribution of trout.

We studied the effects on fish of a 1 day pulse release from Camino Dam as condition of a FERC re-licensing, designed to test the potential of Silver Creek for Class V whitewater kayaking. The aim of our study was to determine whether fish, particularly young-of-the-year, would be displaced downstream by this single pulsed flow that was over 38 times higher than the base flow and occurred in September, an unnatural time of year for a Mediterranean system.

## Study site

Silver Creek is a tributary of the South Fork American River, central California. Water reaches the study section from Camino Reservoir, via Camino Dam (Fig. 1). Water flows into this reservoir from the upper part of Silver Creek, and, via the Jaybird Tunnel, from Union Valley Reservoir and Junction Reservoir. Water is stored in Camino Reservoir, and most of it is diverted through a series of tunnels to the Camino Powerhouse, from which it flows into the South Fork American River just upstream of Slab Creek Reservoir. Typically, a flow of about 0.5 $m^3 \cdot s^{-1}$ is released from Camino Dam to Silver Creek to sustain fish, with occasional peaks up to 1–2 $m^3 \cdot s^{-1}$. Discharge from Camino Dam to Silver Creek ranged between 0.25 and 2.11 $m^3 \cdot s^{-1}$from 1 October 2003 to 30 September 2004.

The section of Silver Creek that we studied was a bedrock reach (Bisson and Montgomery 1996). Mean wetted width at along the snorkel survey reach at baseflow was 8.4 m (± 4.5 SD), and the stream gradient was 2.7%. The fifteen pools in the 300-m long snorkel survey reach were at bedrock, but often contained some gravel, finer sediment, and fallen leaves. The habitat was relatively complex, with numerous boulders at the upper and lower boundaries of pools. The average maximum depth of the pools was 1.6 m (± 0.6 SD; range, 0.7–3.3 m; Table 1). Mean pool length was 16.5 m (± 10.1 SD; range, 5.5–37.2 m), and average maximum pool width was 10.3 m (± 5.5 SD; range, 3.0–26.8 m). Mean pool area was 175 $m^2$ (± 159 SD; range, 15–490 $m^2$). The total area of pools in the snorkel section was 2622 $m^2$. The total length of all pools was 247 m, comprising 82% of the habitat, with the remainder being cascades and runs.

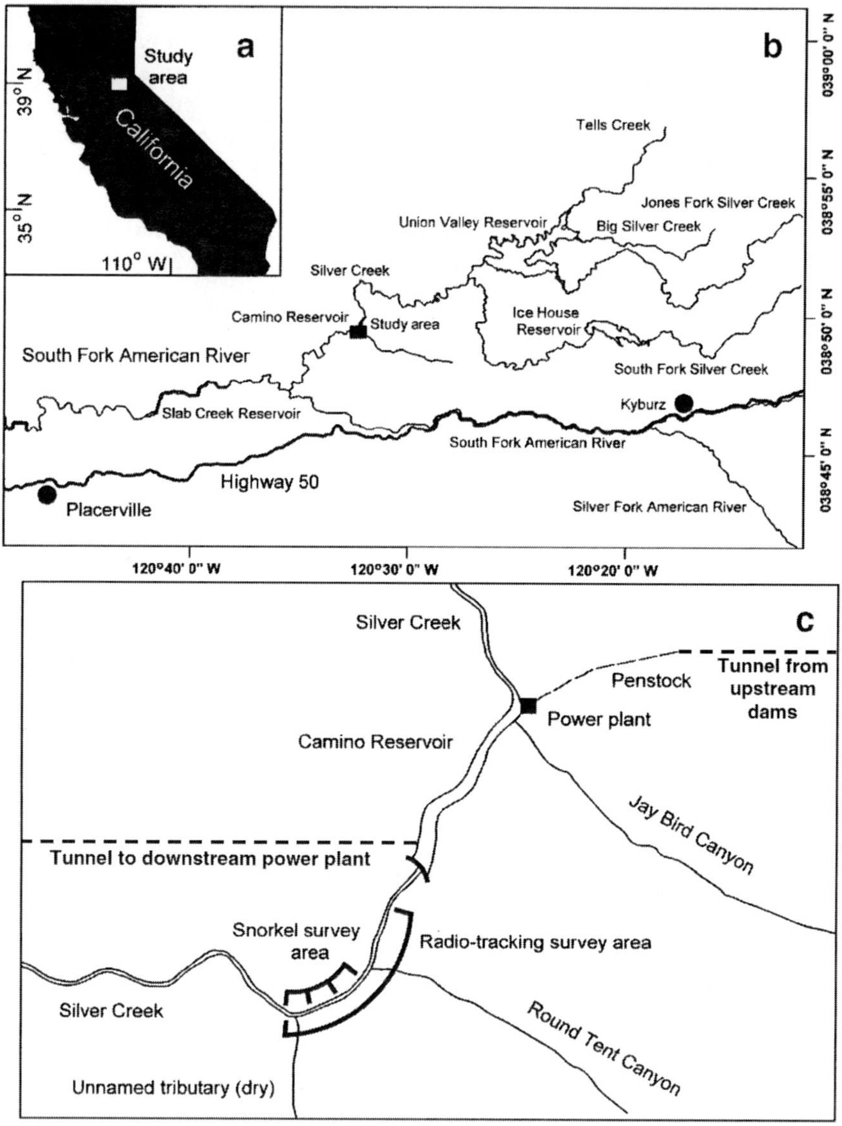


**Figure. 1** Location of study area on Silver Creek (**A**) within California, and **B** within the South Fork American River watershed. **C** Detail of study area showing sections of creek surveyed by radio-tracking and snorkeling

**Table 1** Numbers of fish observed in the 300-m snorkel study area prior to and after the pulse, and percent change in observations

| Age class | Pre-pulse | Post-pulse | Percent change |
|---|---|---|---|
| Young-of-the-year | 39 | 29 | −26 |
| Juvenile | 76 | 69 | −9 |
| Adult | 25 | 28 | 12 |
| Total | 140 | 126 | −10 |

## METHODS

We used a before-after study design and observed the distribution of trout before and after the pulse by snorkel surveys. This allowed us to observe all age classes, to determine whether smaller fish were more likely to be displaced downstream during a pulse. We also radio-tagged a small group of adult trout so that we could track the movement of individual fish before, during, and after the pulse. This allowed us to determine whether the fish observed in snorkel surveys after the pulse were likely to be the same individuals observed prior to the pulse (as opposed to fish washed into the study reach from further upstream). We marked some young-of-the-year and juvenile trout with visible implant elastomer, for the same purpose, but it was not possible to see the marks after the fish were released because of the distance between snorkelers and the fish (Klimley et al. 2005).

Flow values, measured at the outflow of Camino Dam, were provided by the Sacramento Municipal Utility District (SMUD). Daily average flow from Camino Dam into Silver Creek was obtained for the year from 1 October 2003 to 30 September 2004. Fifteen minute average flow was obtained for 15 September 2004. Flows in the study reach were probably slightly higher than the flow observed at the dam, since a small tributary joins Silver Creek just below the dam.

We measured water quality parameters in order to control for potential interactions with flow changes that might have affected fish distribution (Bell et al. 1976; Orth and Maughan 1982). Measurements of surface water temperature and dissolved oxygen concentration were made 500 m downstream of Camino Dam using a handheld meter (YSI, Model 550A) prior to, during, and after the 15 September 2004 pulsed flow. Triplicate water samples (250 mL each) were collected at the same location prior to, during, and after the pulsed flow. Samples were analyzed for total Kjeldahl nitrogen (TKN), ammonia-nitrogen ($NH_4$-N), nitrate-nitrogen ($NO_3$-N), soluble phosphorus, and total phosphorus. Water quality analyses were performed by the University of California Division of Agriculture and Natural Resources Analytical Laboratory, on the UC Davis campus.

A preliminary snorkel survey performed on 24 August 2004 of all the pools within 500 m downstream of the dam indicated that there were very few young-of-the-year or juvenile fish. No fish less than 14 cm total length were observed in this part of the creek, so the 300-m snorkel survey reach was located between 500 m and 800 m downstream of the dam. Within the snorkel survey area pool length and maximum pool width were measured with a tape measure. Maximum pool depth was measured using a stadia rod. Measurements were made on 16 September 2004, at base flow.

We recorded fish snorkel counts in the 15 study pools on 14 September 2004 using established procedures (Dolloff et al. 1996), moving upstream through the study reach. Fish were identified to species where possible, or listed as "unknown". Fish length was estimated by eye in order to classify fish into age/size categories. Young-of-the-year fish were defined as those with total length less than 10 cm. Juvenile fish were defined as those with total length of 10–18 cm. Adult fish were defined as those with total length greater than 18 cm. On 16 September 2004 the pools were again surveyed by snorkeling to determine post-pulse locations and numbers of fish.

In order to provide a test of whether trout observed before and after the pulse were the same fish, rather than fish washed into the reach from upstream, we individually marked a group of adult trout. We captured one rainbow trout and six brown trout (*Salmo trutta*) for radio tracking from a 600-m reach that began about 200 m downstream of Camino Dam (Fig. 1). We used a larger section of the stream than we snorkeled, in order to increase the capture of fish large enough to radio tag. Fish were captured from 9 to 14 September 2004, by angling using artificial lures and insects, and tagged in the field. Standard lengths ranged between 20.5–28.2 cm, and weights ranged between 118–360 g. Fish with a body mass between 105–225 g received a smaller tag (Lotek Wireless, NTC-4-2 L; 2.1 g, 8.3 mm diameter, 18.3 mm long; 93 d lifespan, 5 s pulse frequency) than fish with a body mass greater than 225 g (Lotek Wireless, NTC-6-2; 4.5 g, 9.1 mm diameter, 30.1 mm long; 126 d lifespan; 2.5 s pulse frequency). The tag weight did not exceed 2% of body weight (Jepsen et al. 2002).

Holding and recovery containers were treated with 10 $g \cdot L^{-1}$ rock salt to minimize fish stress by decreasing the osmotic gradient across the gills.

Temperature and dissolved oxygen were monitored with a handheld meter (YSI, Model 550A). Fish were anesthetized with carbon dioxide anesthetic in a bucket (Peake 1998), then through a tube into the mouth during surgery. Tags were inserted intraperitoneally, with the whip antenna extending through a separate hole in the body wall. Each fish was allowed to recover in a mesh live trap in the stream, then released to the same pool from which it was captured.

Radio-tagged fish were tracked by walking along the bank using a hand-held 3-element Yagi antenna and a manual-tracking receiver (Lotek Wireless Inc., SRX400). Fish locations were determined (1) at the time of release, (2) pre-pulse (the day before the pulse), (3) during the pulse-flow release, and (4) post-pulse (1 week after the pulse). Since our main interest was in whether these fish remained within the study reach, we recorded locations to the nearest pool. The location estimates have an error of approximately ± 13 m prior to and following the pulse release versus ± 25 m during the pulse flow period. At the base flow, there were 15 pools in the 300-m snorkel reach, and each pool was approximately 20 m long, thus the error was estimated to be ± 13 m. During the pulse the pools were larger and some merged together, so fish locations were recorded to within ± 25 m.

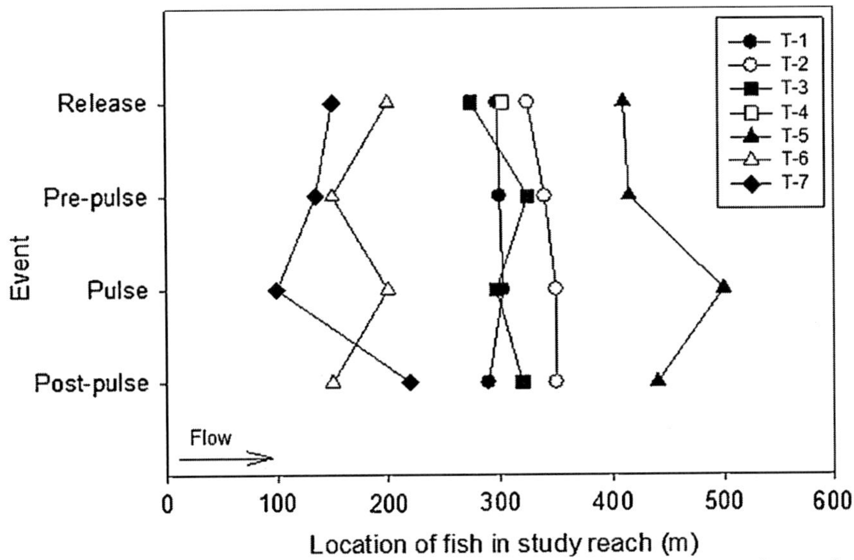


Figure. 2 Locations of radio-tagged fish at time of release, pre-pulse, pulse, and post-pulse. Symbols indicate location of fish along the study reach. Note: Camino Dam is 200 m upstream of the start of the study reach

One radio-tagged brown trout was not detected after being released. All other fish were observed during all subsequent tracking events. The distance moved by an individual fish between tracking events ranged from 0–120 m (Fig. 2). During the pulse flow event all six fish stayed within the 600-m study reach. Two fish moved upstream, three moved downstream, and one did not move.

## RESULTS

The flow out of Camino Dam was pulsed on 15 September 2004. The rate of discharge rose from 0.48 $m^3 \cdot s^{-1}$ at 0600 h to 18.49 $m^3 \cdot s^{-1}$ at 0945 h, stayed constant until 1445 h, and then dropped to 1.77 $m^3 \cdot s^{-1}$ by 1730 h, and 0.48 $m^3 \cdot s^{-1}$ by 2030 h (Fig. 3). The water level in the creek increased about 1 m at the peak flow, as measured by a gage at the midpoint of the study reach.

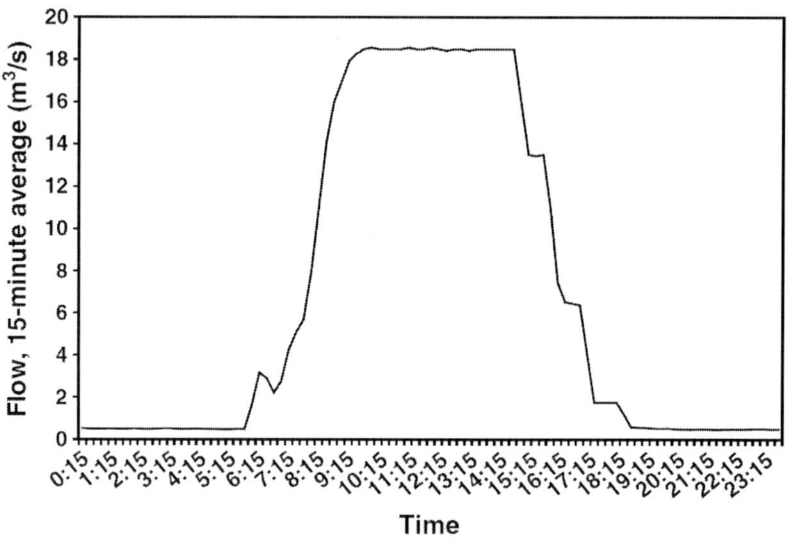


**Figure. 3** Discharge from Camino Dam to Silver Creek during the pulse flow event on 15 September 2004. Values are 15 min average flows

Water temperature before and after the pulse was 10.7°C. Values observed on the day of the pulse ranged between 10.2 and 10.9°C. Dissolved oxygen was approximately 11 $mg \cdot L^{-1}$ during the pulse, 9.64 $mg \cdot L^{-1}$ the following day, and 10.67 $mg \cdot L^{-1}$ on 21 September 2004, the final day of radio-tracking. Dissolved oxygen was not measured until 15 September at 1600 h due to a malfunction of the meter.

Total Kjeldahl nitrogen (TKN) values during the study ranged from below detection (<0.1 mg·L$^{-1}$) to 0.2 mg·L$^{-1}$. Ammonia-nitrogen ($NH_4$-N) and nitrate-nitrogen ($NO_3$-N) were below the detection limit (<0.05 mg·L$^{-1}$) in all samples. Soluble phosphorus (soluble P) was below the detection limit (<0.05 mg·L$^{-1}$) in all samples. Total phosphorus (total P) was below the detection limit (<0.1 mg·L$^{-1}$) in all samples.

Two species of fish were observed during snorkel surveys, rainbow trout, and non-native brown trout. Young-of-the-year, juveniles, and adults were observed for both species. A total of 140 fish were observed in the snorkel survey prior to the pulse flow release (0.47 fish/m). The number of young-of-the-year trout (of both species) observed within the 300 m section declined 26% between the pre-pulse and post-pulse snorkel surveys (Table 1). Juvenile numbers declined 9%, while adult numbers increased 12%. Total trout density in each pool did not appear to differ markedly before and after the pulse (Fig. 4). The numbers of young-of-the-year per pool tended to be lower after the pulse, but some groups of pools contained more young-of-the-year after the pulse (Fig. 5). Juvenile and adult fish numbers each increased in three pool groups, and declined in two pool groups, although the pattern was different for the two size classes. The total number of brown trout per pool ranged from 0-18 fish pre-pulse, and 0–7 fish post-pulse, with the highest number of brown trout occurring in Pool 13 both pre- and post-pulse. Rainbow trout were more abundant than brown trout in the study reach. While the total number of rainbow trout ranged from 1–20 fish pre-pulse, and 0–19 fish post-pulse, rainbow trout occurred in a higher proportion of pools than brown trout during both pre- and post-pulse.

## DISCUSSION

The majority of trout of all sizes observed in snorkel surveys did not experience catastrophic wash-out due to the pulse, or if they were displaced downstream, most had moved back upstream by the time of the post-pulse survey. Our results corroborate the observations of Salamunovich (2003) of fish response to pulsed recreational flow releases on the North Fork Feather River (California). Visual counts were made 24 h before and after the recreational flow releases. The study found little change in the observed species, number and sizes of fish before and after flow pulses.

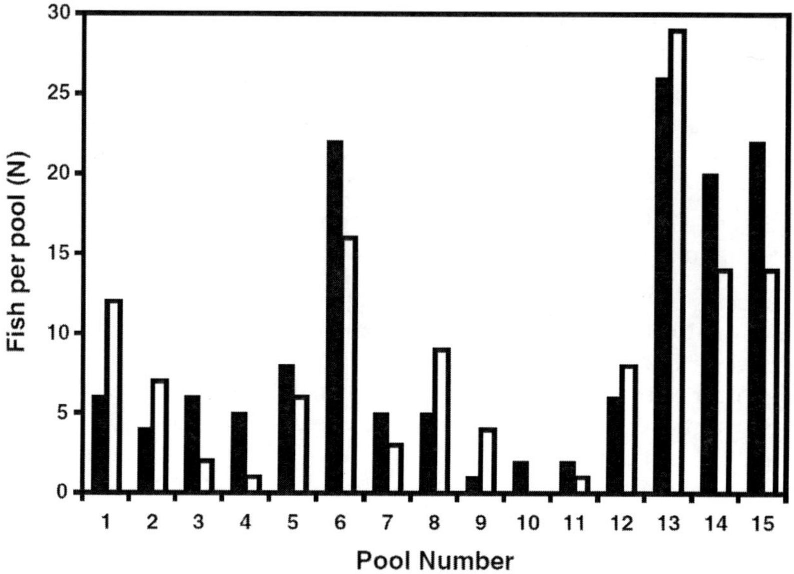


**Figure. 4** Total number of brown trout and rainbow trout observed in snorkel counts of Silver Creek pools before and after the kayaking pulse flow event on 15 September 2004. Counts from 14 September 2004 in black, and 16 September 2004 in white. The experimental reach contained a total of fifteen pools, with Pool 1 at the upstream end of the section, and Pool 15 at the downstream end of the section

However, our results also suggest that young-of-the-year and juvenile trout may have been more susceptible than adult trout to being displaced downstream, since counts of these size classes were lower in the post-pulse snorkel than the pre-pulse survey. Downstream displacement of small trout by high water velocities has been observed previously (Ottaway and Clarke 1981; Ottaway and Forrest 1983), and if smaller fish were indeed displaced downstream in our study this may be due to their swimming performance being lower than that of larger fish, as suggested by Webb (1971). Heggenes and Traaen (1988) also found that smaller fish were more likely to be displaced downstream by increased discharge. Our results for larger fish are consistent with those of Simpkins et al. (2000) where rainbow trout over 20 cm TL were not displaced downstream by a spring flushing flow.

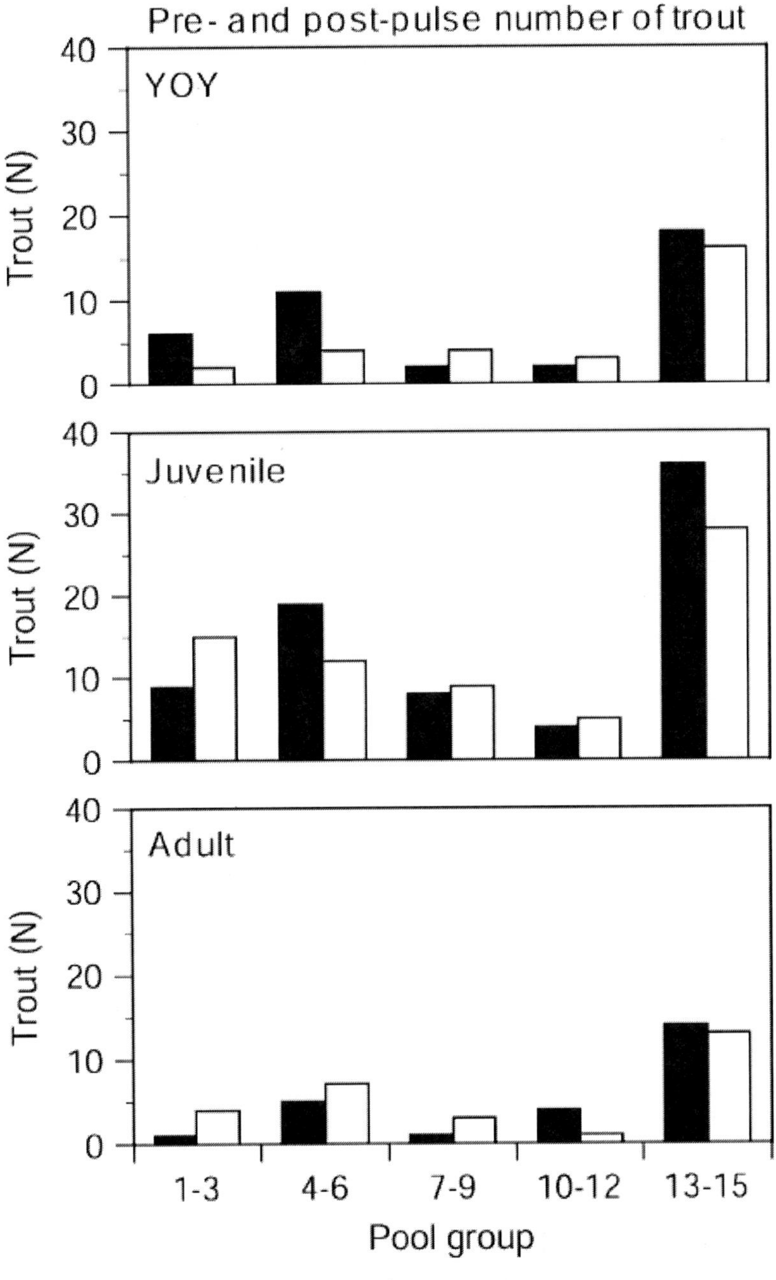
Pre- and post-pulse number of trout
YOY
Juvenile
Adult
Trout (N)
0
10
20
30
40
1-3
4-6
7-9
10-12
13-15
Pool group

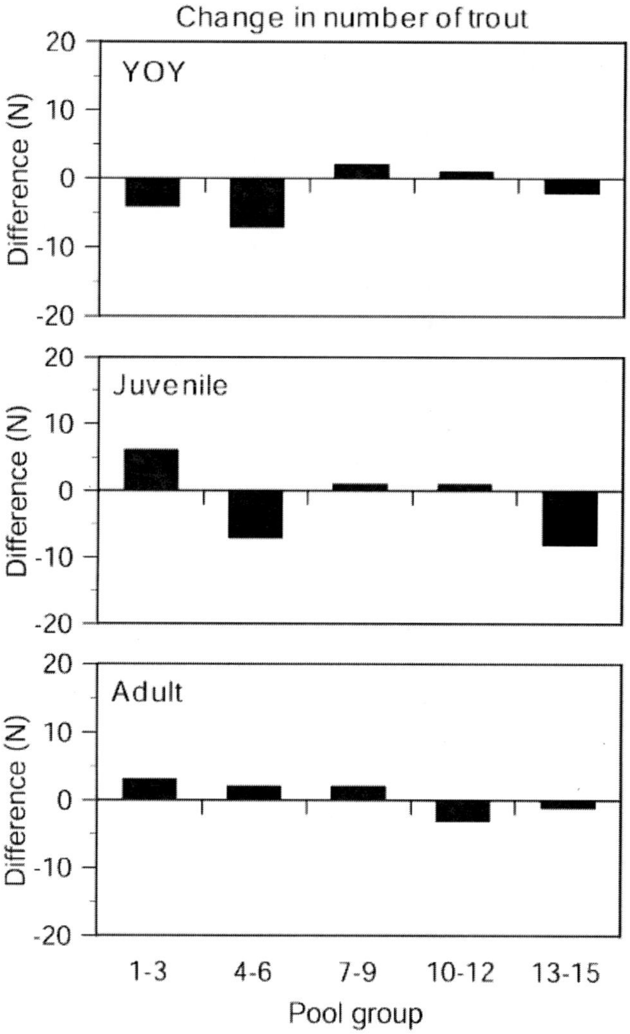


**Figure. 5** Column on the left shows the number of trout, grouped by age, observed in snorkel counts of Silver Creek pools before and after the kayaking pulse flow event on 15 September 2004. Counts from 14 September 2004 in black, and 16 September 2004 in white. Column on the right shows the change in numbers of trout, grouped by age, after the pulse, in comparison with pre-pulse observations. The experimental reach contained a total of fifteen pools, with Pool 1 at the upstream end of the section, and Pool 15 at the downstream end of the section. Counts for pools were summed in groups of three to avoid having zero values in the pre-pulse data

In our study water chemistry, temperature, and dissolved oxygen changed little during the pulse and were unlikely to have affected fish movement, leaving the flow change itself as the dominant factor. We caution that the complexity of habitat in Silver Creek, which has a large number of flow refugia (crevices, boulders, deep pools), may have mitigated the effects of the pulse, as has been observed in some prior studies (Bell et al. 2001; Nislow et al. 2002). However, there may have been indirect effects of the pulse in Silver Creek that we were unable to detect. The flow may have been sufficient to displace benthic macroinvertebrates, resulting in a decreased food supply for trout following the pulse (Lagarrigue et al. 2002). For safety reasons we were not able to observe the distribution of young-of-the-year trout during the pulse, but they may have been displaced out of preferred microhabitats, leading to decreased feeding and growth during the pulse (Korman and Campana 2009).

A source of uncertainty in our study is whether the fish we saw in the snorkel section during the post-pulse survey were same as those observed prior to the pulse, or fish that had washed into the snorkel section from further upstream. However, the presence of the 6 radio-tagged adult trout within the study area before, during, and after the pulse suggests that the fish observed in the snorkel surveys were the same individuals before and after the pulse. Also, on 24 August 2004 we conducted a snorkel survey of the stream between the dam and the snorkel reach. We observed only 20 trout, all between 14–24 cm long, in the 500-m long section (0.04 fish/m), suggesting that there were few fish that could have washed downstream to replace the fish that we counted prior to the pulse. We consider it unlikely that fish from upstream in Camino Reservoir or Silver Creek were entrained through Camino Dam during the pulse release and washed into the snorkel section. A site on Silver Creek, 1200 m upstream of Camino Reservoir, contained rainbow trout and brown trout in 2002, at densities similar to that of our snorkel reach (SMUD 2005b). However, during the pulse the water level in Camino Reservoir did not change appreciably, due to the combined operations of Jaybird Powerhouse and the outflow tunnel from the reservoir, so fish in upper Silver Creek would not have had any indication that flows were increasing out of the reservoir, and so would not have been attracted down into the reservoir. Little is known about the fish species and densities occurring within Camino Reservoir itself. During a SMUD hydropower re-licensing study conducted in 2002–2003 Camino Reservoir was removed from the study due to safety and access constraints (SMUD 2005a). The authors noted historic reports, some undated, which listed the fish species in Camino Reservoir as California roach, brook trout, brown trout, rainbow trout, riffle sculpin, Sacramento sucker, and “minnows”. Camino Reservoir has surface water temperatures and dissolved oxygen levels suitable for trout (SMUD 2004). However, SMUD staff who have fished for trout in both Camino Reservoir and in

Silver Creek downstream of Camino Dam reported that they have observed few trout in the reservoir.

A problem for our study was the lack of a robust experimental design, which limited our ability to make statistical inferences about our observations, as is common for studies of the effects of flow regulation on fish (Murchie et al. 2008). The observation of single pulses on multiple streams with comparable fish fauna, over the same time period, would have provided true replication, but such pulsed flows were not occurring on nearby streams. We could have observed multiple sub-reaches along Silver Creek, but since fish could have moved between the sub-reaches these "replicates" would not have been independent and would have resulted in a pseudoreplicated design (Hurlbert 1984). Furthermore, the sub-reaches would have been located along a gradient of distance from the dam, and of decreasing elevation, causing autocorrelation that would have further complicated analyses. Our design would also have benefited from one or more control reaches (Carpenter 1990), thus allowing a BACI design (Underwood 1992), but flows in the reach of Silver Creek immediately upstream of Camino Reservoir are influenced by other dams and reservoirs further upstream (see Fig. 1b). It would have been possible for us to conduct repeated daily snorkel surveys over several days prior to the pulsed flow, in order to have an estimate of the variability of our counts, and to see whether the count immediately after the pulse was statistically different. However, we were concerned that the repeated disturbance of the fish for several days prior to the pulse might have decreased their activity levels and feeding rates and affected their response to the pulse itself. As future FERC re-licensing studies occur in California streams opportunities may arise to repeat the observation of trout response to single-day pulses, and to pool the results of these studies for meta-analysis.

This study highlights several areas of uncertainty regarding the effects of single pulsed flows. First, the pulse flow release occurred in September, when young-of-the-year trout were well-developed and likely able to locate and move to low velocity areas to avoid the higher flows, allowing a majority to remain within the study reach. It would be informative to test pulses of similar magnitude at other times of the year, such as early summer, when fry may be newly emerged, and less able to locate and use flow refugia. Second, the habitat in Silver Creek is relatively complex and heterogeneous, which likely provided a wide variety of flow refugia for fish. It would be useful to test the impacts of a similar pulse in streams with poorer habitat to see whether fish are more likely to be displaced as habitat complexity declines. Third, the fish community in Silver Creek is very simple, and is likely compromised mainly of the two trout species that we observed. A similar magnitude pulse might have more dramatic impacts in streams with a greater number of species that prefer slower velocities than those preferred by trout. This could be tested in conjunction with tests of the effect of seasonality and habitat complexity.

# ACKNOWLEDGEMENTS

This study was funded by the Public Interest Energy Research Program of the California Energy Commission through the Pulsed Flow Program of the Center of Aquatic Biology and Aquaculture of the University of California, Davis. We also acknowledge support from the Division of Water Rights of the State Water Resources Control Board. We thank the Sacramento Municipal Utility District for providing information on creeks that were scheduled for pulsed flows, for assistance during the field work at Silver Creek, and for providing flow data for Camino Dam releases to Silver Creek. We received helpful comments on the manuscript from three anonymous reviewers.

# REFERENCES

1. Bell E, Duffy WG, Roelofs TD (2001) Fidelity and survival of juvenile coho salmon in response to a flood. Trans Am Fish Soc 130:450–458
2. Bell MC, Parkhurst ZE, Porter RG, Stevens M (1976) Effects of power peaking on survival of juvenile fish at lower Columbia and Snake River dams. U.S. Army Corps of Engineers, North Pacific Division (Contract No. DACW 57-75-C-0173), Portland, Oregon, USA. 35p
3. Bisson PA, Montgomery DR (1996) Valley segments, stream reaches, and channel units. In: Hauer FR, Lamberti GA (eds) Methods in stream ecology. Academic, San Diego, pp 23–52
4. Bradford MJ, Taylor GC, Allan JA (1995) An experimental study of the stranding of juvenile coho salmon and rainbow trout during rapid flow decreases under winter conditions. N Am J Fish Manage 15:473–479
5. Carpenter SR (1990) Large-scale perturbations: opportunities for innovation. Ecology 71:2038–2043
6. Crisp DT (1991) Stream channel experiments on downstream movement of recently emerged trout, *Salmo trutta* L. and salmon, *S. salar* L.—effects of developmental stage and day and night on dispersal. J Fish Biol 39:371–381
7. Crisp DT, Hurley MA (1991) Stream channel experiments on downstream movement of recently emerged trout, *Salmo trutta* L. and salmon, *Salmo salar* L.—1. Effect of four different water velocity treatments upon dispersal rate. J Fish Biol 39:347–361
8. Dolloff A, Kershner J, Thurow R (1996) Underwater observation. In: Murphy BR, Willis DW (eds) Fisheries techniques, 2nd edn. American Fisheries Society, Bethesda, Maryland, pp 533–554
9. Erman DC, Leidy GR (1975) Downstream movement of rainbow trout fry in a tributary of Sagehen Creek, under permanent and intermittent flow. Trans Am Fish Soc 104:467–473
10. Giannico GR, Healey MC (1998) Effects of flow and food on winter movements of juvenile coho salmon. Trans Am Fish Soc 127:645–651

11. Heggenes J (1988) Effects of short-term flow fluctuations on displacement of, and habitat use by, brown trout in a small stream. Trans Am Fish Soc 117:336–344
12. Heggenes J, Traaen T (1988) Downstream migration and critical water velocities in stream channels for fry of four salmonid species. J Fish Biol 32:717–727
13. Hunter MA (1992) Hydropower flow fluctuations and salmonids: a review of the biological effects, mechanical causes and options for mitigation. Washington Department of Fisheries Technical Report 119. Olympia, Washington
14. Hurlbert SH (1984) Pseudoreplication and the design of ecological field experiments. Ecol Monogr 54:187–211
15. Jepsen N, Koed A, Thorstad EB, Baras E (2002) Surgical implantation of telemetry transmitters in fish: how much have we learned? Hydrobiol 483:239–248
16. Klimley AP, Cech, Jr JJ, Thompson LC, Hamilton S, Chun SN (2005) Experimental and field studies to assess pulsed, water-flow impacts on the behavior and distribution of fishes in California rivers, annual report, 2004–05. Prepared for the PIER Program Area, California Energy Commission. 107 p http://www .energy.ca.gov /publications/displayOneReport.php?pubNum=CEC-500-2005-172
17. Klimley AP, Cech, Jr JJ, Thompson LC, Hamilton S (2007) Experimental and field studies to assess pulsed, water flow impacts on the behavior and distribution of fishes in the South Fork of the American River, II (Second Year). University of California, Davis, for the California Energy Commission, PIER Energy-Related Environmental Research. 85 p http://animalscience.ucdavis .edu/ PulsedFlow/Klimley%20500-09-067.pdf
18. Korman J, Campana SE (2009) Effects of hydropeaking on nearshore habitat use and growth of age-0 rainbow trout in a large regulated river. Trans Am Fish Soc 138:76–87
19. Lagarrigue T, Cereghino R, Lim P, Reyes-Marchant P, Chappaz R, Lavandier P, Belaud A (2002) Diel and seasonal variations in brown trout (*Salmo trutta*) feeding patterns and relationship with invertebrate drift under natural and hydropeaking conditions in a mountain stream. Aquat Living Resour 15(2):129–13
20. McCrimmon H (1954) Stream studies on planted Atlantic salmon. J Fish Res Bd Can 11:362–403
21. Murchie KJ, Hair KPE, Pullen CE, Redpath TD, Stephens HR, Cooke SJ (2008) Fish response to modified flow regimes in regulated rivers: research methods, effects and opportunities. River Res Applic 24:197–217
22. Nislow KH, Magilligan FJ, Folt CL, Kennedy BP (2002) Within-basin variation in the short-term effects of a major flood on stream fishes and invertebrates. J Freshwater Ecol 17(2):305–318
23. Ottaway EM, Clarke A (1981) A preliminary investigation into the vulnerability of young trout (*Salmo trutta* L.) and Atlantic salmon (*S. salar* L.) to downstream displacement by high water velocities. J Fish Biol 19:135–145
24. Ottaway EM, Forrest DR (1983) The influence of water velocity on the downstream movements of alevins and fry of brown trout, *Salmo trutta* L. J Fish Biol 23:221–436
25. Orth DJ, Maughan OE (1982) Evaluation of the incremental methodology for recommending instream flows for fishes. Trans Am Fish Soc 111:413–445
26. Peake S (1998) Sodium bicarbonate and clove oil as potential anesthetics for non-salmonid fishes. North Am J Fish Manage 18:919–924
27. Pearsons TN, Li HW, Lamberti GA (1992) Influence of habitat complexity on resistance to flooding and resilience of stream fish populations. Trans Am Fish Soc 121:427–436
28. SMUD (Sacramento Municipal Utility District) (2004) Sacramento Municipal Utility District's Upper American River Project (FERC Project No. 2101). Deepwater intake entrainment technical report, July 2004, 54 p
29. SMUD (Sacramento Municipal Utility District) (2005a) Sacramento Municipal Utility District's Upper American River Project (FERC Project No. 2101) and Pacific

Gas and Electric Company's Chili Bar Project (FERC Project No. 2155). Stream fisheries technical report, January 2005, version 2, 180 p
30. SMUD (Sacramento Municipal Utility District) (2005b) Sacramento Municipal Utility District's Upper American River Project (FERC Project No. 2101) and Pacific Gas and Electric Company's Chili Bar Project (FERC Project No. 2155). Reservoir fisheries technical report, April 2005, version 2, 83 p
31. Salamunovich T (2003) Rock Creek-Cresta (FERC No. 1962) recreation and pulse flow biological evaluation: stranding and displacement studies. Draft report of Thomas R. Payne and Associates, Arcata, CA to Pacific Gas and Electric Company, San Ramon, CA
32. Saltveit SJ, Halleraker JH, Arnekleiv JV, Harby A (2001) Field experiments on stranding in juvenile Atlantic salmon (*Salmo salar*) and brown trout (*Salmo trutta*) during rapid flow decreases caused by hydropeaking. Regul River 17:609–922
33. Shirvell CS (1994) Effect of changes in stream flow on the microhabitat use and movements of sympatric juvenile coho salmon (*Oncorhynchus kisutch*) and chinook salmon (*O. tshawytscha*) in a natural stream. Can J Fish Aquat Sci 51:1644–1652
34. Simpkins DG, Hubert WA, Wesche TA (2000) Effects of a spring flushing flow on the distribution of radio-tagged juvenile rainbow trout in a Wyoming tailwater. North Am J Fish Manage 20:546–551
35. Underwood AJ (1992) Beyond BACI: the detection of environmental impacts in the real, but variable, world. J Exp Mar Bio Ecol 161:145–178
36. Webb PW (1971) The swimming energetics of trout. I. Thrust and power at cruising speeds. J Exp Biol 55:489–500

# CHAPTER 14

# Environmental Influences on Juvenile Fish Abundances in a River-Dominated Coastal System

**L. Carassou,* B. Dzwonkowski, and F. J. Hernandez**

*Dauphin Island Sea Laboratory, 101 Bienville Boulevard, Dauphin Island, Alabama 36528, USA*

*Department of Marine Sciences, University of South Alabama, 307 University Boulevard, Life Science Building Room 25, Mobile, Alabama 36688, USA; and Dauphin Island Sea Laboratory, 101 Bienville Boulevard, Dauphin Island, Alabama 36528, USA*

*Alabama Department of Conservation and Natural Resources, Marine Resources Division, Post Office Box 189, 2 North Iberville Drive, Dauphin Island, Alabama 36528, USA*

## ABSTRACT

We investigated the influence of climatic and environmental factors on interannual variations in juvenile abundances of marine fishes in a river-dominated coastal system of the north-central Gulf of Mexico, where an elevated primary productivity sustains fisheries of high economic importance. Fish were collected monthly with an otter trawl at three stations nearMobile Bay from 1982 to 2007. Fish sizes were used to isolate juvenile stages within the data set, and monthly patterns in juvenile fish abundance and size were then used to identify seasonal peaks for each species. The average numbers of juvenile fish collected during these seasonal peaks in each year were used as indices of annual juvenile abundances and were related to corresponding seasonal averages of selected environmental factors via a combination of principal components analysis and co-inertia analysis. Factors contributing the most to explain interannual variations in juvenile fish abundances were river discharge and water temperature during early spring–early summer, wind speed and North Atlantic Oscillation index during late fall–winter, and atmospheric pressure and wind speed during summer–fall. For example, juvenile abundances of southern kingfish *Menticirrhus americanus* during summer–fall were positively associated with atmospheric pressure and negatively associated with wind speed during this period. Southern kingfish juvenile abundances during late fall–winter were also negatively associated with wind speed during the same period and were positively associated with river discharge during early spring–early summer. Juvenile abundances of the Atlantic croaker *Micropogonias undulatus* during early spring–early summer were negatively associated with river discharge and North Atlantic Oscillation during late fall–winter. Overall, the importance of river discharge for many of the species examined emphasizes the major role of watershed processes for marine fisheries production in coastal waters of the north-central Gulf of Mexico.

Long-term monitoring of many marine fish populations has revealed the importance of temporal variability at interannual and decadal scales (Hollowed et al. 2001; Lehodey et al. 2006). Interannual variations in adult fish abundances are mainly dependent on processes occurring during the early life stages (Cushing 1996; Fuiman and Werner 2002). In turn, survival rates of juvenile fish are a principal driver of variable year-class strength in the resulting adult population (Houde 1997; Miller and Kendall 2009). Identifying the factors that affect the interannual variability in juvenile fish abundances is thus critical for 411412 CARASSOU ET AL.a better understanding of variability in adult fish abundances and fisheries landings, and of fish population responses to a changing environment (Myers 1998; Brunel and Boucher 2007).

Among the factors affecting interannual patterns in juvenile fish abundances, climatic and local environmental variability plays an important role (Cushing 1996; Brunel and Boucher 2007). Juvenile abundances of a variety of fish species throughout the world have been related to indices of large-scale climate patterns, such as the Pacific Decadal Oscillation, the North AtlanticOscillation (NAO), or ElNiño–SouthernOscillation Index (SOI; Hollowed et al. 2001; Lehodey et al. 2006). These general climatic indices are synthetic representations of climate patterns at ocean basin scales, which affect local environmental conditions influencing juvenile fish abundances at the local habitat level. For example, minimum winter air temperature along the East Coast of the United States was shown to track larger-scale variations in NAO and was identified as a potential mechanism explaining juvenile abundances of the Atlantic croaker *Micropogonias undulatus* (Hare and Able 2007). Variability in sea surface temperatures (Ciannelli et al. 2005; Brunel and Boucher 2007), river discharge (Crecco et al. 1986; Martino and Houde 2010), and wind patterns (Daskalov 2003; Lloret et al. 2004) also participate in shaping variable estuarine–coastal hydrodynamic conditions that influence juvenile fish abundances.

The extent to which earlier studies can be generalized, however, remains uncertain because the intensity of climatic and environmental controls on juvenile fish abundances varies as a function of space and time (Myers 1998; Planque and Buffaz 2008). For example, correlations between environmental factors and juvenile fish abundances are generally more obvious and robust at the edges of the biogeographical ranges of fish species (Myers 1998) or during specific seasons or climatic phases (Ottersen et al. 2006; Planque and Buffaz 2008). Moreover, different components of environmental variability influence fish production at high versus low latitudes (Brander 2007). Biological factors, such as spawning stock biomass, have also been shown to affect the strength and significance of environmental controls on juvenile fish abundance patterns (Ottersen et al. 2006; Brander 2007). These spatial, temporal, and population-specfic variations emphasize a need for addressing the influence of environmental factors on juvenile fish abundances for multiple fish species in diverse ecosystems.This may provide crucial information on the consistency or variability of environment–juvenile abundance linkages for specific species and help in developing local tools for forecasting fish population responses to environmental changes.

Whereas many studies have addressed the effect of climatic and environmental factors on juvenile fish abundance dynamics along theU.S. East Coast (e.g., Lankford and Targett 2001; Hare and Able 2007) and West Coast (e.g., Kimmerer et al. 2001; Clark and Hare 2002), this question has rarely been examined in the northern Gulf ofMexico despite the economic importance of fisheries from this region (Browder 1993). The northern Gulf of Mexico is characterized by several coastal river systems that are known to enhance coastal primary productivity and support large finfish and penaeid shrimp fisheries (Browder 1993).Much of the research conducted in the region has focused on the Mississippi–Atchafalaya River system, which contributes 90% of the freshwater input to the Gulf of Mexico (Rabalais et al. 1996) and has been linked to fisheries production (Govoni 1997; Grimes 2001). However, relatively little research has focused on other Gulf river systems and their relationships to fisheries production. The Mobile Bay River system, in particular, which is formed at the confluence of the Tombigbee and Alabama rivers, drains an area of 115,000 km2 and represents the fourthlargest discharge in the USA and the second largest in the Gulf of Mexico (Schroeder 1978).

In the Mobile Bay area, published studies dealing with the ecology of fish early life stages had so far been limited to analyses of ichthyoplankton seasonality (Hernandez et al. 2010a, 2010b). Information about juvenile fish dynamics and responses to environmental factors is thus essential for a better understanding of interannual variability in fisheries production in this area. The objectives of the present study are thus to (1) describe interannual patterns in juvenile abundance displayed by common coastal marine fish species over a 26-year time series in coastal waters off Mobile Bay, Alabama, and (2) explore the relationships between these abundance patterns and a variety of climatic and local environmental factors.

# METHODS

*Data sources.*—Fish abundance data were provided by a fisheries-independent survey, the Fisheries Assessment and Monitoring Program (FAMP), conducted by the Alabama Department of Conservation and Natural Resources (ADCNR), Marine Resources Division (MRD). Sampling consisted of monthly otter trawl collections at a variety of sites along the Alabama coast from 1982 to 2007. The otter trawl had a 4.9-m opening and was made of 35-mm stretched mesh with a 4.5-mm cod end fittedwith a 4.7-mm liner. For the present study, data obtained at three coastal stations near Mobile Bay were compiled: Petit Bois Pass, Mobile Pass, and Perdido Pass (Figure 1). At each station and month (i.e., each sample), fish collected were identified and a maximum of 50 individuals were measured for each species (standard length, to the nearest 1 mm). Due to some modifications in the sampling design over the course of this long-term survey, 12 out of the 312 months of sampling were missing (no sample in October, November, or December 1998; January, June, July, August, or October 1999; January, March, or May 2000; or August 2005). In these instances, fish abundance values were replaced by the corresponding monthly averages over the 26-year period (i.e., 1982–2007).

Two general climate indices and seven local environmental factors (listed in Table 1) were obtained from National Oceanic and Atmospheric Administration (NOAA) National Weather Service (NWS) Climate Prediction Center (NOAA 2010a), NOAA National Data Buoy Center (NDBC; NOAA 2010b),

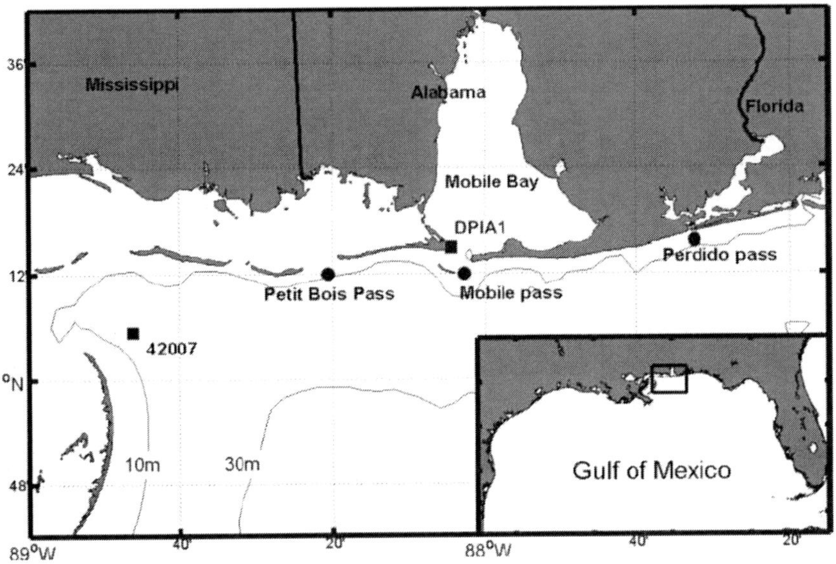


**FIGURE 1.** Locations of otter trawl stations (circles) of the Fisheries Assessment and Monitoring Program conducted by theAlabamaDepartment of Conservation and Natural Resources' Marine Resources Division, and locations of environmental stations (squares) of the National Oceanic and Atmospheric Administration's National Data Buoy Center (stations DPIA1 and 42007). Locations of the two U.S. Geological Survey gaging stations (Alabama and Tombigbee rivers; USGS 2010a, 2010b) are not shown because they are situated farther north on land. and U.S. Geological Survey (USGS) websites (USGS 2010a, 2010b). Data from NOAA–NWS were provided at monthly intervals. Data from NOAA–NDBC were collected at hourly intervals. Daily river discharge data were collected from two USGS gaging stations in the Alabama River (Clairborne Lock and Dam; USGS 2010a) and in the Tombigbee River (Coffeeville Lock and Dam; USGS 2010b). Their sum was used as a total freshwater discharge into Mobile Bay (Park et al. 2007).

*Fish data analysis.*—Due to the scarcity of information regarding relationships between juvenile fish abundances and environmental conditions in the study area, a multispecies approach was favored. We removed very rare species since their highly variable abundance and occurrence may confound multispecies patterns of interest (Wood and Austin 2009). Only the species contributing to at least 0.5% of the total fish abundance observed over the 26-year period were thus retained. Furthermore, fish age estimations are not available in the FAMP data set used in this study and published size-at-age relationships are not available for the retained species in the study region. We thus used size data to sort out juvenile stages in the data set.

**TABLE 1.** Climatic and environmental factors examined, with their respective units, sources, and codes. Measurement stations are depicted in Figure 1.

| Variable | Units | Source | Code |
|---|---|---|---|
| General climatic factors | | | |
| El Niño–Southern Oscillation Index | | NOAA 2010a | soi |
| North Atlantic Oscillation index | | NOAA 2010a | nao |
| Local environmental factors | | | |
| Air temperature | °C | NOAA 2010b (stations 42007 and DPIA1) | AT |
| Water temperature | °C | NOAA 2010b (stations 42007 and DPIA1) | WT |
| Wind speed | m/s | NOAA 2010b (stations 42007 and DPIA1) | WS |
| *u*-wind component (alongshore) | m/s | NOAA 2010b (stations 42007 and DPIA1) | uW |
| *v*-wind component (cross-shore) | m/s | NOAA 2010b (stations 42007 and DPIA1) | vW |
| Atmospheric pressure | bar | NOAA 2010b (stations 42007 and DPIA1) | AP |
| River discharge | $m^3/s$ | USGS 2010a (Clairborne Lock and Dam, Alabama River); USGS 2010b (Coffeeville Lock and Dam, Tombigbee River) | RD |

Variable Units Source Code General climatic factors El Ni˜no–Southern Oscillation Index NOAA 2010a soi North Atlantic Oscillation index NOAA 2010a nao Local environmental factors Air temperature °C NOAA 2010b (stations 42007 and DPIA1) AT Water temperature °C NOAA 2010b (stations 42007 and DPIA1) WT Wind speed m/s NOAA 2010b (stations 42007 and DPIA1) WS *u*-wind component (alongshore) m/s NOAA 2010b (stations 42007 and DPIA1) uW *v*-wind component (cross-shore) m/s NOAA 2010b (stations 42007 and DPIA1) vW Atm ospheric pressure bar NOAA 2010b (stations 42007 and DPIA1) AP River discharge m3/s USGS 2010a (Clairborne Lock and Dam, Alabama River); USGS 2010b (Coffeeville Lock and Dam, Tombigbee River) RD

We followed Miller and Kendall's (2009) definition of the juvenile stage: only fish larger than the size at metamorphosis, size at which squamation begins, size at which fin rays development is completed (depending on data availability in the literature), or a combination thereof, and smaller than the size at maturity, were considered as juveniles. Consequently, only species for which the latter parameters were available in the literature were finally retained (Table 2).

**TABLE 2**. Fish species commonly collected as juveniles in otter trawl samples at three stations in theMobile Bay area from 1982 to 2007, the respective juvenile size boundaries (standard length, mm), total number of juveniles (estimated *N*), 3-month peaks in juvenile abundance (2-month peaks for pinfish; see Figure 3), and corresponding seasonal groups and codes. Species are ordered alphabetically. Monthly patterns in juvenile abundance and mean size are depicted in Figure 3. See Methods for details on juvenile fish abundance estimations and on the determination of juvenile size boundaries and seasonal groups. Juvenile fish size distribution plots are provided in Figure 2.

| Species | Juvenile size boundaries (mm) | Estimated *N* | Peak months | Seasonal group | Code |
|---|---|---|---|---|---|
| Bay anchovy *Anchoa mitchilli* | 25–36 | 943 | May–Jul | Early spring–early summer | anmit(I) |
| Hardhead catfish *Ariopsis felis* (formerly *Arius felis*) | 60–125 | 2,172 | Nov–Jan | Late fall and winter | arfel(II) |
| Atlantic bumper *Chloroscombrus chrysurus* | 30–97 | 1,591 | Sep–Nov | Summer and fall | chchr(III) |
| Sand seatrout *Cynoscion arenarius* | 30–128 | 245 | Apr–Jun | Early spring–early summer | cyare(I) |
| | | | Nov–Jan | Late fall and winter | cyare(II) |
| Silver seatrout *Cynoscion nothus* | 30–159 | 278 | Sep–Nov | Summer and fall | cynot(III) |
| Fringed flounder *Etropus crossotus* | 20–84 | 565 | Dec–Feb | Late fall and winter | etcro(II) |
| | | | Jul–Sep | Summer and fall | etcro(III) |
| Pinfish *Lagodon rhomboides* | 30–105 | 161 | Dec, Jan | Late fall and winter | larho(II) |
| | | | Aug, Sep | Summer and fall | larho(III) |
| Spot *Leiostomus xanthurus* | 30–122 | 719 | Dec–Feb | Late fall and winter | lexan(II) |
| | | | Jun–Aug | Summer and fall | lexan(III) |
| Southern kingfish *Menticirrhus americanus* | 30–136 | 279 | Nov–Jan | Late fall and winter | meame(II) |
| | | | Jun–Aug | Summer and fall | meame(III) |
| Atlantic croaker *Micropogonias undulatus* | 30–139 | 1,936 | May–Jul | Early spring–early summer | miund(I) |
| Atlantic thread herring *Opisthonema oglinum* | 30–109 | 2,851 | Apr–Jun | Early spring–early summer | opogl(I) |
| Gulf butterfish *Peprilus burti* | 16–99 | 618 | Feb–Apr | Early spring–early summer | pebur(I) |
| Atlantic moonfish *Selene setapinnis* | 20–236 | 294 | Aug–Oct | Summer and fall | seset(III) |
| Blackcheek tonguefish *Symphurus plagiusa* | 20–90 | 772 | May–Jul | Early spring–early summer | sypla(I) |
| | | | Nov–Jan | Late fall and winter | sypla(II) |
| Hogchoker *Trinectes maculatus* | 20–99 | 247 | Jul–Sep | Summer and fall | trmac(III) |

Species-specific sizes at maturity ($S_{mat}$) were obtained from FishBase (Froese and Pauly 010) and Pattillo et al. (1997). When $S_{mat}$ estimates differed between the two references, the lower value was retained because using a lower $S_{mat}$ value reduces the likelihood that any mature individuals are included in the analysis (i.e., most conservative approach). Size at metamorphosis, size at which squamation begins, or size at which fin ray development is completed were obtained from Gallaway and Strawn (1974), Richards (2006), Fahay (2007), and Able and Fahay (2010) for hardhead catfish; Martin and Drewry (1978), Ditty and Truesdale (1983), and Rotunno and Cowen (1997) for Gulf butterfish; and Switzer (2003) for blackcheek tonguefish. When these latter estimates differed between different references for a given species, the larger value (i.e., most conservative) was retained. Juvenile size boundaries were then refined for each species by visualizing length frequency plots of all measured fish for each species (data not shown). Final juvenile size boundaries are shown in Table 2, and length frequency plots of measured juvenile fish for each species are shown in Figure 2. For each sample, the proportion of measured individuals comprised within the juvenile size boundaries was then calculated and applied to the total number of fish collected for each species, providing an estimate of the abundance of juveniles for each species in each sample (Table 2).



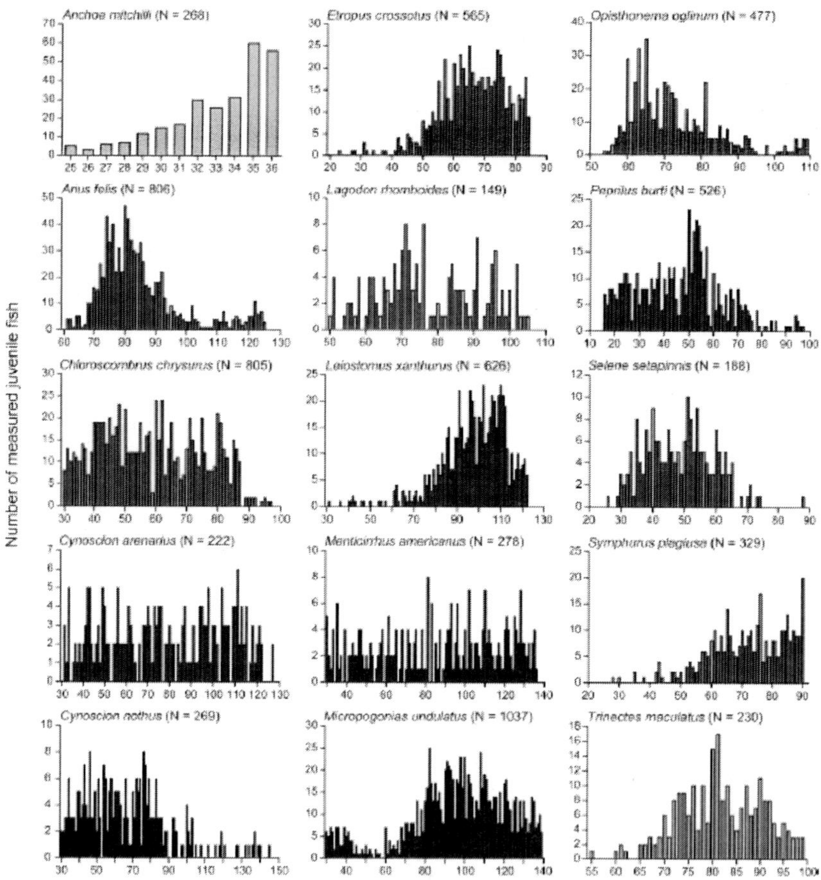


**FIGURE 2.** Size distribution of measured juvenile fish from 15 species collected between 1982 and 2007 with an otter trawl at three stations from the Mobile Bay area, Alabama. Common names of species are provided in Table 2.

Monthly patterns in juvenile fish abundances over the 26-year study period were examined in order to identify seasonal peaks for each species (Figure 3). Depending on species, one to two seasonal peaks were selected, a seasonal peak corresponding to the three consecutive months (two months in the case of pinfish) during which juvenile abundances were the highest (Figure 3). Based on these seasonal peaks, three groups of species were identified: (1) species for which juvenile



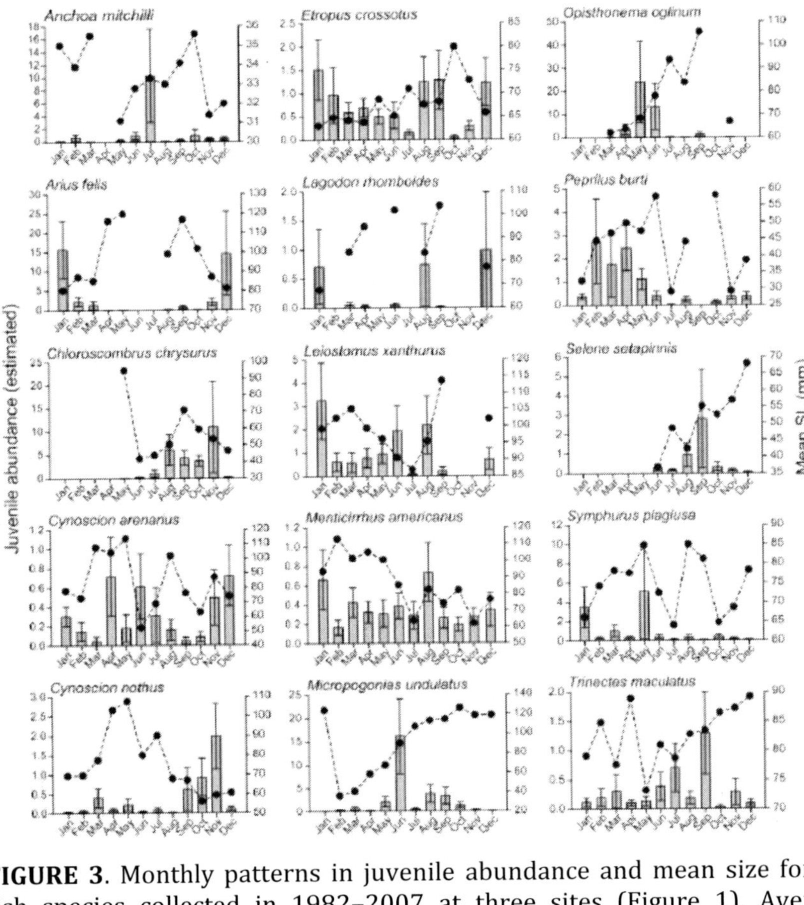


**FIGURE 3**. Monthly patterns in juvenile abundance and mean size for 15 fish species collected in 1982–2007 at three sites (Figure 1). Average (□}SE) juvenile abundances are shown with column charts and are associated with the left *y*-axes. Mean sizes ( tandard length [SL], mm) are represented by black shaded circles and are associated with the right *y*-axes. Species are presented in alphabetical order. Months selected for the calculation of annual juvenile abundance indices and corresponding seasonal groups for each species are shown in Table 2; common names of species are also provided in Table 2.

abundances peaked from early spring through early summer (i.e., group I, six species), (2) species for which juvenile abundances peaked in late fall and winter (group II, seven species), and (3) species for which juvenile abundances peaked during summer and fall (group III, eight species; Figure 3; Table 2). For each group, the average number of juvenile fish collected during the seasonal peak was used as the annual juvenile fish abundance index (JAI) for each species (JAIs were thus based on average numbers of juveniles collected in the two to three seasonal peak months × three stations = six to nine samples per year).

The JAIs were processed to obtain standardized annual anomalies by removing the mean and dividing with the SD over the 26-year period. Multispecies patterns in JAIs were thenanalyzed using centered principal components analysis (PCA), which is adapted to the treatment of variables expressed in similar units, and relies on the computation of covariances between variables (Legendre and Legendre 1998). The JAIs were log10($x$ + 1) transformed in order to clarify the projection of highly variable observations (years) and descriptors (species) on the factorial axes (principal components [PCs]), as recommended by Legendre and Legendre (1998) for Poisson distributed data. Three centered PCAs were conducted, one for each fish species group (I, II, III). The visualization of covariances between species (columns) and years (lines) on the two first PCs (PC1–PC2) provided a graphical synthetic representation of interannual patterns in juvenile abundances for each group of species. The absolute contributions (i.e., loadings) of each species on PC1 and PC2 were finally examined to isolate species that had a minor contribution (i.e., <5%) in driving interannual patterns in juvenile abundances for each group. This resulted in a total of three species from group III that were ignored in analyses of nvironment–juvenile abundance relationships.

*Environmental data analysis.*—Data for all limatic and environmental factors were processed to obtain monthly averages for each variable. hese monthly averages were obtained fromhigher resolution data for environmental factors minimum of 20 d of data for each monthly average) or directly provided for climatic factors Table 1). Short gaps in the NOAA–NDBC data (less than 13 h) were replaced with an estimated alue determined by linear interpolation between the two closest data points. Due to large gaps in emperature and wind data, two NDBC stations (42007 and DPIA1 in Figure 1) were merged into single time series. Gaps in the DPIA1 time series were filled using data from station 42007 that was adjusted using a linear fit to account for the minor magnitude differences for each parameter at the individual sites.

Monthly averages were then used to calculate seasonal averages for each factor. These seasonal averages were computed in accordance with the seasonal groups identified in fish data: (1) average of months included in

JAI calculations for fish species of group I, February–July (i.e., early spring–early summer); (2) average of months included in JAI calculations for fish species of group II, November–February (i.e., late fall–winter); and (3)verage of months included in JAI calculations for fish species of group III, June–November (i.e., summer–fall). Seasonal averages of environmental factors were then analyzed using normed PCA (Legendre and Legendre 1998), which is adapted to the treatment of variables expressed with different units and relies on the computation of correlations between variables (Legendre and Legendre 1998). Three normed PCAs were conducted, one for each seasonal group (I, II, III). The visualizationof correlations between environmental factors (columns) and years (lines) on the two first PCs (PC1–PC2) provided a graphical synthetic representation of interannual patterns in environmental conditions for each seasonal group. Moreover, correlations between variables on PC1–PC2 and absolute contributions (i.e., loadings) of variables were used to isolate a small number of independent factors that drove interannual patterns in environmental conditions at each season. Only variables with a contribution greater than 20% were retained, and when two variables were found highly correlated, only the one showing the highest contribution on PC1–PC2 was selected. This resulted in a total of 12 variables (four per seasonal group) that were retained for analyses of environment–juvenile fish abundance relationships.

*Analysis of relationships between environmental and fish data.*—The influence of environmental variables on interannual patterns in juvenile fish abundances was studied using a oinertia analysis (COIA). Co-inertia analysis is a two-table symmetric coupling method that provides great flexibility in identifying the common structure in a pair of data tables (Dol´edec and Chessel 1994; Dray et al. 2003). Co-inertia analysis is based on the statistic of co-inertia, which provides a measure of concordance between two data sets (Dray et al. 2003). The principle of COIA is to search for a vector in the environmental space and a vector in the faunistic space that maximizes the co-inertia between them (Thioulouse et al. 2004). These two vectors are used to define a new ordination plan on which environmental and faunistic variables are compared. Graphical results are then interpreted as in other multivariate methods: the distance of variables to the origin is indicative of their contribution on the ordination plan, and the angle between them measures their relationship (Legendre and Legendre 1998). In the present study, COIA was based on the matching between the coordinates of selected environmental factors on a new normed PCA and of selectedfish variables on a newcentered PCA (PCA–PCA–COIA; Dray et al. 2003). The normed PCA on environmental factors was based on a matrix composed of 26 lines (years) and 12 columns (four variables per seasonal group). The centered PCA on fish data was based on a matrix composed of 26 lines (years) and 18 columns (six species from group I, seven species from group II, and five species from group III). A Monte-Carlo test with 1,000 permutations of the observations was used to confirm the significance of the co-inertia results (fixed-*D* test; Dray et al. 2003). All multivariate analyses were performed with the ADE-4 software (Thioulouse et al. 2001).

# RESULTS

## Interannual Variations in Juvenile Fish Abundances

The two first PCs of the PCA conducted on fish group I (species for which juvenile abundances peaked in early spring through early summer; Figure 4a) explained 65.8% of interannual variability in juvenile abundances. Relatively high JAIs were observed for Gulf butterfish and Atlantic

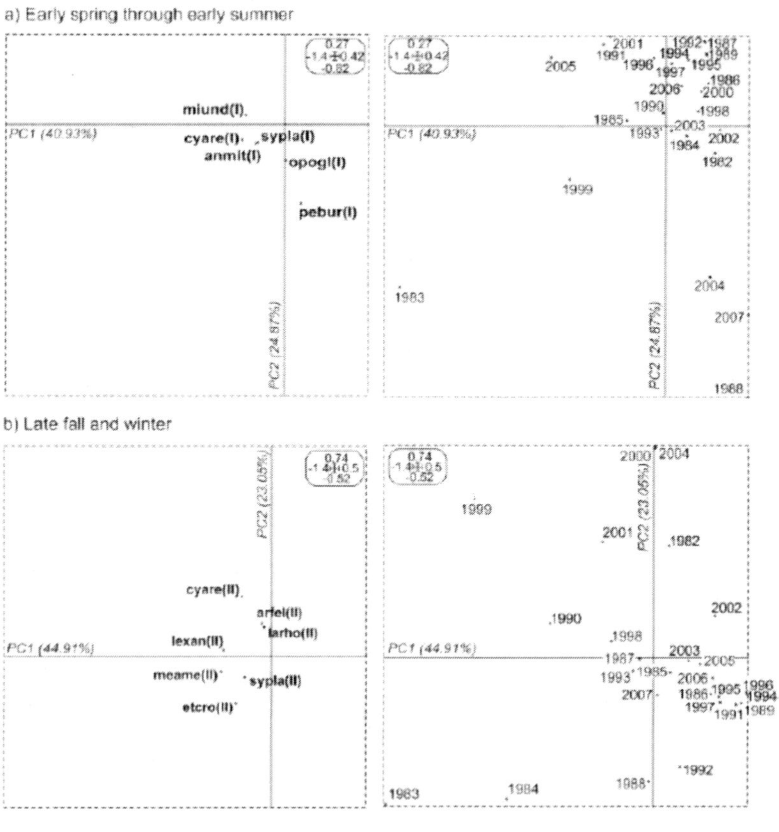

c) Summer and fall

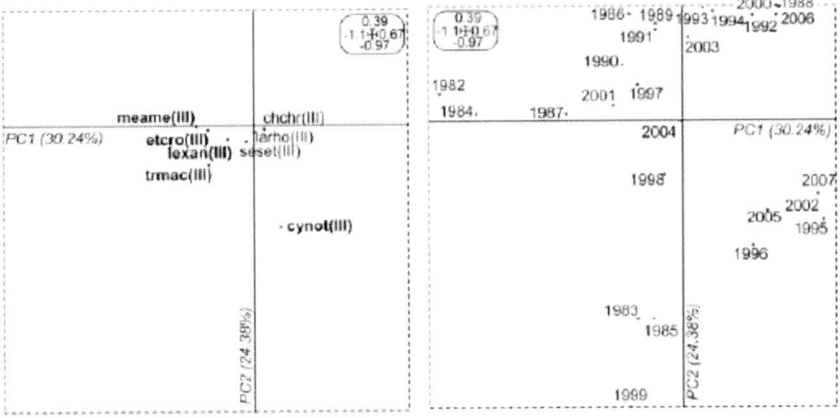


**FIGURE 4.** Principal components analyses conducted on log10($x$ + 1) transformed standardized annual juvenile abundance indices of **(a)** six fish species characterized by juvenile seasonal peaks in early spring through early summer (seasonal group I), **(b)** seven fish species characterized by juvenile seasonal peaks in late fall and winter (seasonal group II), and **(c)** eight fish species characterized by juvenile seasonal peaks in summer and fall (seasonal group III). Covariances between species and projections of years on the principal components 1 and 2 (PC1–PC2) are represented in the left and right columns, respectively. Bold labels indicate species that were retained for co-inertia analysis of environment–juvenile abundance relationships (i.e., species with total contributions>5% on PC1–PC2; Table 3). Scales are given in the rounded boxes. Fish species codes and seasonal groups are defined in Table 2. Six species were represented in more than one seasonal group as a result of large juvenile abundances throughout several seasons (Figure 3; Table 2): sand seatrout and blackcheek tonguefish in groups I and II; and fringed flounder, pinfish, spot, and southern kingfish in groups II and III.

thread herring in 1988, 2004, and 2007; for Atlantic croakers in 1985 and 2005; and for sand seatrout, bay anchovy, and blackcheek tonguefish in 1983 and 1999. For the six species from group I, JAIs were generally lower for 10 out of the 26 years of the time series (years grouped in the top-right part of the PC1–PC2 plane; Figure 4a). These six species all presented total contributions greater than 5% on the PC1–PC2 factorial plane (Table 3).

The two first PCs of the PCA conducted on fish group II (species for which juvenile abundances peaked in late fall and winter; Figure 4b) explained 68.0% of interannual variability in juvenile abundances. Sand seatrout and spot presented relativelyhigh JAIs in 1990, 1999, and 2001 and lower JAIs

**TABLE 3**. Absolute contributions (%) of environmental variables and fish species on the two first principal components (PC1, PC2, and sum of PC1–PC2) of the normed and centered principal components analyses (PCAs), respectively. For each data set, three PCAs were conducted: one on early spring–early summer values (group I), one on late fall–winter values (group II), and one on summer–fall values (group III). Projections of variables–species and years on the PC1–PC2 plane are depicted in Figures 4 and 5. Codes of environmental variables are defined in Table 1. Fish species codes and groups are defined in Table 2. See Methodsfor details on the selection of variables and species retained for the co-inertia analysis.

| | PCA group I | | | PCA group II | | | PCA group III | | |
|---|---|---|---|---|---|---|---|---|---|
| Variable or species | PC1 | PC2 | Sum | PC1 | PC2 | Sum | PC1 | PC2 | Sum |
| **Environmental Variables** | | | | | | | | | |
| soi | 22.91 | 3.85 | 26.76 | 13.46 | 15.33 | 28.79 | 3.93 | 0.54 | 4.47 |
| nao | 8.34 | 0.34 | 8.68 | 1.80 | 19.63 | 21.43 | 9.98 | 0.64 | 10.62 |
| AT | 16.58 | 19.77 | 36.35 | 25.27 | 12.56 | 37.83 | 14.92 | 27.54 | 42.46 |
| WT | 8.65 | 35.04 | 43.69 | 24.52 | 9.96 | 34.48 | 25.63 | 5.48 | 31.11 |
| WS | 3.78 | 20.09 | 23.87 | 6.78 | 6.79 | 13.57 | 11.03 | 27.68 | 38.71 |
| uW | 0.10 | 3.24 | 3.34 | 0.50 | 11.68 | 12.18 | 4.68 | 0.32 | 5.00 |
| vW | 9.43 | 0.02 | 9.45 | 10.08 | 0.00 | 10.08 | 0.77 | 11.47 | 12.24 |
| AP | 25.04 | 2.55 | 27.59 | 8.69 | 11.51 | 20.20 | 28.99 | 4.00 | 32.99 |
| RD | 5.13 | 15.06 | 20.19 | 8.86 | 12.49 | 21.35 | 0.05 | 22.28 | 22.33 |
| **Fish Species** | | | | | | | | | |
| anmit | 16.47 | 4.61 | 21.08 | | | | | | |
| arfel | | | | 1.13 | 14.31 | 15.44 | | | |
| chchr | | | | | | | 1.01 | 0.14 | 1.15 |
| cyare | 34.76 | 2.83 | 37.59 | 10.97 | 44.43 | 55.40 | | | |
| cynot | | | | | | | 7.12 | 83.10 | 90.22 |
| etcro | | | | 16.39 | 23.27 | 39.66 | 23.32 | 0.20 | 23.52 |
| larho | | | | 0.61 | 11.31 | 11.92 | 0.00 | 0.40 | 0.40 |
| lexan | | | | 29.48 | 0.73 | 30.21 | 7.98 | 1.60 | 9.58 |
| meame | | | | 32.37 | 1.90 | 34.27 | 37.05 | 0.02 | 37.07 |
| miund | 28.69 | 0.95 | 29.64 | | | | | | |
| opogl | 0.05 | 15.89 | 15.94 | | | | | | |
| pebur | 5.62 | 71.68 | 77.30 | | | | | | |
| seset | | | | | | | 0.72 | 2.09 | 2.81 |
| sypla | 14.38 | 4.01 | 18.39 | 9.02 | 4.02 | 13.04 | | | |
| trmac | | | | | | | 22.76 | 12.42 | 35.18 |

in 1992. Southern kingfish, fringed flounder, and blackcheek tonguefish had higher JAIs in 1983, 1984, and 1988 and lower JAIs in 1982 and 2002. Hardhead catfish and pinfish presented relatively high JAIs in 2000 and 2004 (Figure 4b). The seven species from group II were generally characterized by low JAIs for 13 out of the 26 years of the time series (years grouped on the bottomleft part of the PC1–PC2 plane; Figure 4b). All seven species presented total contributions greater than 5% on the PC1–PC2 plane (Table 3).

The two first PCs of the PCA conducted on fish group III (species for which juvenile abundances peaked in summer and fall; Figure 4c) explained 54.6% of interannual variability in juvenile abundances. Silver seatrout presented relatively high JAIs in 1995, 1996, 2000, 2002, 2005, and 2007

(Figure4c). Hogchokers and spot had high JAIs in 1983, 1985, 1998, and 1999 (Figure 4c). Fringed flounder and southern kingfish JAIs were also generally higher in 1982, 1984, and 1987 (Figure 4c). Atlantic bumpers, pinfish, and Atlantic moonfish had minor contributions to interannual patterns in JAIs during this season, their contributions being less than 5% on the PC1–PC2 plane (Table 3). As a result, these three species 420 CARASSOU ET L. were ignored for the COIA of environment–juvenile abundance relationships.

## Interannual Variations in Environmental Conditions

The two first PCs of the PCA conducted on environmental averages from group I (Figure 5a) explained 48.0% of interannual variability in environmental conditions during early spring through early summer. Six variables had total contributions greater than 20% of the PC1–PC2 plane (Table 3). Years 1990, 1999, 2000, and 2006 were generally characterized by relatively higher water and air temperatures (Figure 5a). Higher river discharge and wind speeds were observed in 1991, 1995, 1997, 1998, and 2003, and the contrary was true in 1985, 1986, 1996, and 2007 (Figure 5a).High values of SOI and atmospheric pressure characterized 1982 and 1989, and the contrary was true in 1983 and 1987 (Figure 5a). In total, four variables were selected as representative of environmental interannual variability in early spring through early summer based on a compromise between their correlations and relative contributions (Figure 5a; Table 3) and consistency with variables selected for the two other seasonal groups (see below): SOI, water temperature, river discharge, and wind speed (Figure 5a).

The two first PCs of the PCA conducted on environmental averages from group II (Figure 5b) explained 50.9% of interannual variability in environmental conditions during late fall and winter. Six variables had total contributions greater than20% on the PC1–PC2 plane (Table 3). Years 1986, 1989, 1992, and 1997 were characterized by high values of wind speed and river discharge and low values of SOI and atmospheric pressure, whereas the contrary was true for 1985 and 1999 (Figure 5b). The NAO presented relatively high values in 1994 and 2004 and relatively low values in 1995, 2000, and 2002 (Figure 5b). Relatively high water and air temperatures were observed in 1988, 1990, 1998, and 2007, and the contrary was true in 1983, 1987, 1991, 1993, 2003, and 2006 (Figure 5b). Four variables were selected as representative of environmental interannual variability in late fall and winter based on a compromise between their correlations and relative contributions (Figure 5b; Table 3): NAO, water temperature, river discharge, and wind speed (Figure 5b).

The two first PCs of the PCA conducted on environmental averages from group III (Figure 5c) explained 45.7% of interannual variability in environmental conditions during summer and fall. Five variables had total contributions greater than 20% on the PC1–PC2 plane (Table 3). Wind speed appeared relatively high in 1988, 1995, 1999, 2003, and 2005 and relatively low in 1982, 1984, 1985, 1986, 1990, 1993, and 2000 (Figure 5c). Water and air temperatures were relatively high in 1998, 2002, 2004, 2006, and 2007 and were relatively low in 1987, 1991, and 1992 (Figure 5c). Particularly low river discharge occurred in 1982 and 1986, and high river discharge occurred in 1997 (Figure 5c). Four variables were selected as representative of environmental interannual variability in summer and fall based on a compromise between their correlations and relative contributions (Figure 5c; Table 3): atmospheric pressure, water temperature, river discharge, and wind speed (Figure 5c).

Of noticeable interest is also the observation that for two out of the three seasonal groups, river discharge appeared highly negatively correlated with SOI (in early spring–early summer and in late fall–winter; Figure 5a, b). Similarly, a slight positive correlation was observed between NAO and water temperature in late fall and winter, and a strong negative correlation was observed between NAO and water temperature in summer and fall (Figure 5b, c).

## Environment–Juvenile Abundance Relationships

The three first axes of the COIA allowed 78.3% of the common structure between environmental and fish data sets to be visualized (Figure 6a, b). The significance level from the Monte Carlo permutation test was 0.027. However, the total inertia was 0.80, indicating a relatively low statistical agreement between the two data sets. Nevertheless, relationships emphasized by the COIA appeared to be of meaningful ecological sense. The common structure between fish and environmental data sets was mainly driven by river discharge during early spring through early summer, wind speed during late fall and winter, and water temperature during summer and fall (Figure 6a, b; Table 4). For brevity, only the clearest relationships linking environmental factors and species presentingtotal contributions greater than 10% on the three first axes of the co-inertia will be commented on further (Figure 6; Table 4).

Juvenile abundances of sand seatrout during early spring through early summer were positively associated with river discharge during this season (Figure 6a). Juvenile abundances of Atlantic croakers in early spring through early summer were negatively associated with river discharge and NAO, and to a lesser extent with water temperature, during late fall and winter (Figure 6a, b). Juvenile abundances of Gulf butterfish during early spring through early summer were positively associated with SOI and water temperature and negatively associated with river discharge during this season (Figure 6a, b). Juvenile abundances of sand seatrout during late fall and winter were positively associated with water temperature and NAO during late fall–winter and were negatively associated with river discharge during late fall–winter and early spring–early summer (Figure 6a, b).

Juvenile abundances of fringed flounder during late fall and winter were negatively associated with water temperature and positively associated with river discharge during early spring through early summer (Figure 6a, b). Juvenile abundances of southern kingfish during late fall and winter were negatively associated with wind speed during this season, and were positively associated with river discharge during early spring through early summer (Figure 6a, b). Juvenile abundances of silver seatrout during summer and fall were positively associated with water temperature and wind speed and negatively associated with atmospheric pressure

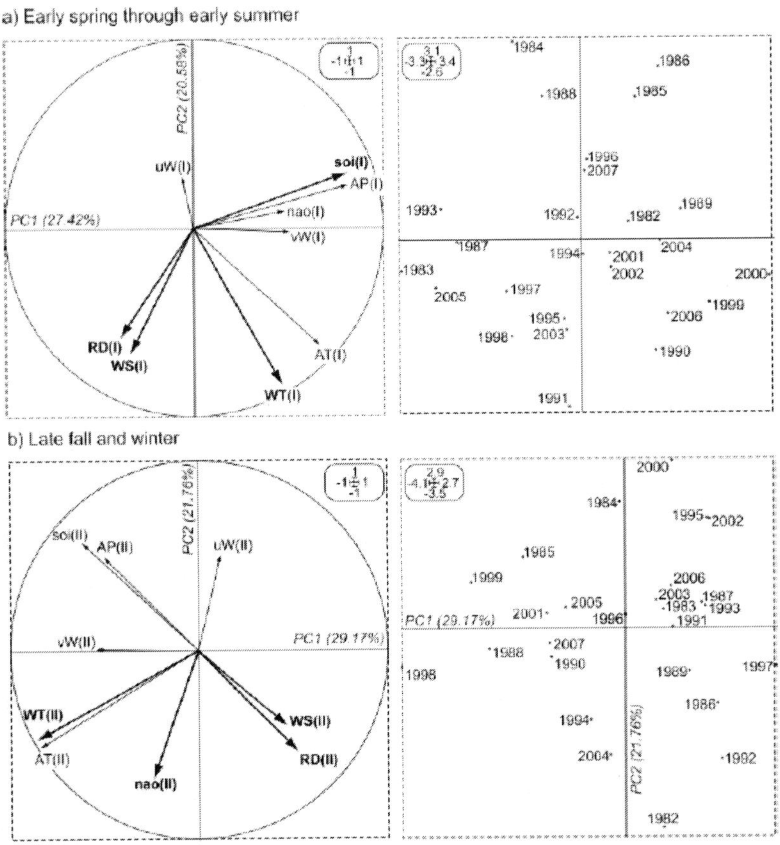

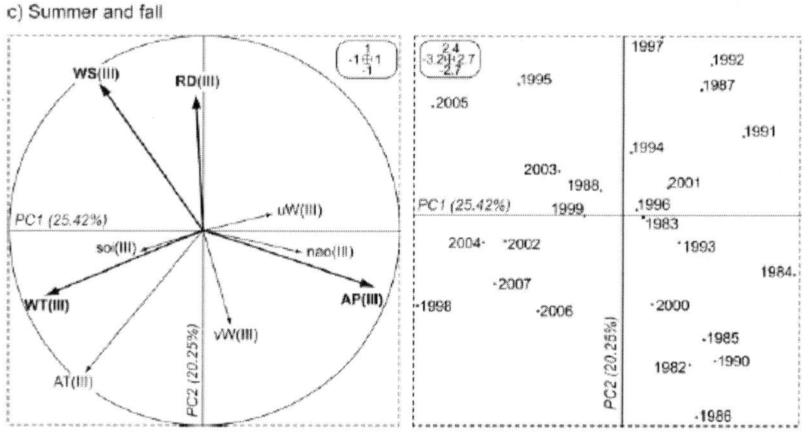


**FIGURE 5.** Principal components analyses conducted on normalized environmental factors averaged over **(a)** early spring through early summer (i.e., February–July; seasonal group I), **(b)** late fall and winter (November–February; seasonal group II), and **(c)** summer and fall (i.e., June–November; seasonal group III). Correlations between environmental factors and projections of years on the principal components 1 and 2 (PC1–PC2) are represented in the left and right columns, respectively. Bold labels indicate variables that were retained for co-inertia analysis of environment–juvenile abundance relationships (i.e., independent variables with a total contribution > 20% on PC1–PC2; Table 4). Scales are given in the rounded boxes. Codes for environmental factors are defined in Table 1.

during this season (Figure 6a, b). Juvenile abundances of fringed flounder during summer and fall were positively associated with atmospheric pressure and negatively associated with wind speed during this season, and were positively associated with river discharge during early spring through early summer (Figure6a, b). Juvenile abundances of spot during summer and fall were positively associated with river discharge during early spring through early summer (Figure 6a). Juvenile abundances of southern kingfish during summer and fall were positively associated with atmospheric pressure and negatively associated with wind speed during this season, and were positively associated with river discharge and negatively associated with wind speed during early spring through early summer (Figure 6a, b). Finally, juvenile abundances of hogchokers during summer and fall were positively associated with atmospheric pressure and negatively associated with wind speed during this season, and were positively associated with river discharge and negatively associated with water temperature and SOI during early spring through early summer (Figure 6a, b). Overall, river discharge during early spring through early

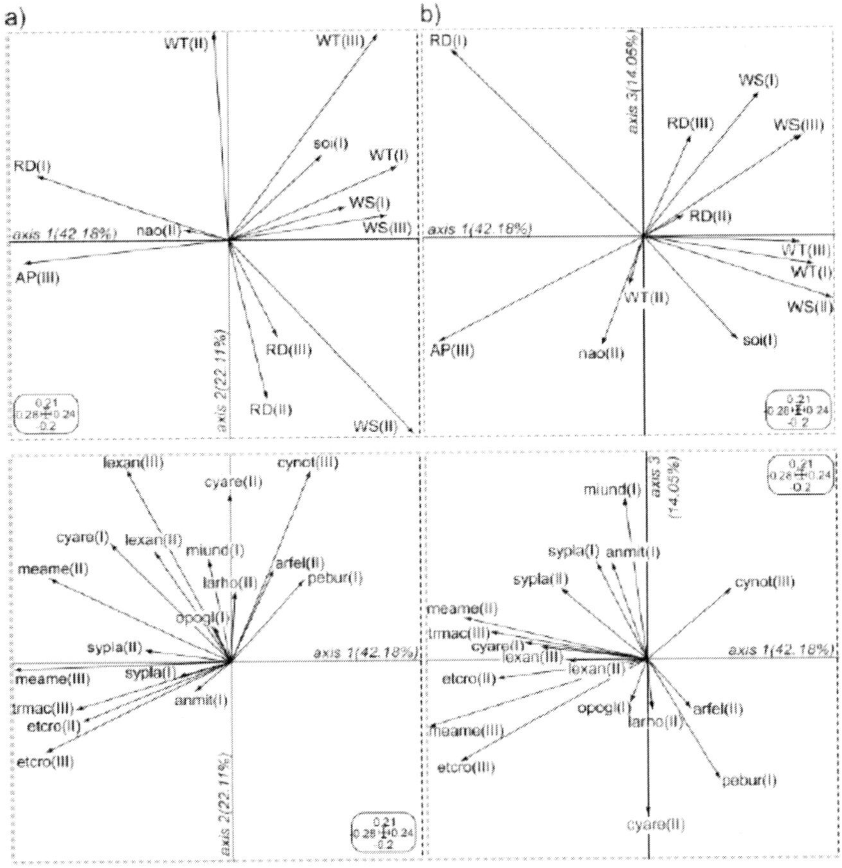


**FIGURE 6**. Co-inertia analysis (see Methods) coupling 12 normalized environmental factors (four variables/seasonal group) and log10($x$ + 1) transformed standardized juvenile abundance indices of 18 fish species (six species from seasonal group I, early spring–early summer; seven species from seasonal group II, late fall–winter; and five species from seasonal group III, summer–fall; Table 2) on **(a)** axes 1 and 2 of the co-inertia and **(b)** axes 1 and 3 of the co-inertia. Projections of environmental variables and fish species are given in the top and bottom panels, respectively. Total inertia is 0.80, and the percentage of variance explained is 64.3% on axes 1 and 2 and 56.2% on axes 1 and 3. Significance level ($P$) of the fixed-$D$ Monte Carlo permutation test with 1,000 permutations was 0.027. Fish species codes and seasonal groups are defined in Table 2; environmental factor abbreviations are defined in Table 1.

summer appeared to be related to juvenile fish abundances of many species from several seasonal groups and also presented the highest total contributions of all environmental variables on the three first axes of the co-inertia (sum of axes 1, 2, and 3 = 50.4%; Table 4).

**TABLE 4.** Absolute contributions (%) of fish species and environmental variables on the first three axes of the co-inertia analysis. Projections of environmental variables and fish species on axes 1–2 and axes 1–3 of the co-inertia are depicted in Figure 6a and 6b. Codes of environmental variables are given in Table 1; fish species codes are defined in Table 2 (seasonal groups: I = early spring–early summer; II = late fall–winter; III = summer–fall).

| Species or variable | Axis 1 | Axis 2 | Axis 3 |
|---|---|---|---|
| | **Fish Species** | | |
| anmit(I) | 0.50 | 0.39 | 7.63 |
| cyare(I) | 6.50 | 7.77 | 0.28 |
| miund(I) | 0.21 | 5.47 | 22.43 |
| opogl(I) | 0.10 | 0.55 | 1.37 |
| pebur(I) | 2.29 | 3.52 | 12.21 |
| sypla(I) | 1.07 | 0.09 | 7.86 |
| arfel(II) | 0.77 | 4.42 | 1.91 |
| cyare(II) | 0.00 | 15.52 | 20.49 |
| etcro(II) | 10.09 | 1.91 | 0.28 |
| larho(II) | 0.00 | 2.47 | 1.82 |
| lexan(II) | 2.56 | 6.59 | 0.00 |
| meame(II) | 15.23 | 3.96 | 1.60 |
| sypla(II) | 3.18 | 0.07 | 4.18 |
| cynot(III) | 3.08 | 20.95 | 4.07 |
| etcro(III) | 16.09 | 4.45 | 8.87 |
| lexan(III) | 4.90 | 20.54 | 0.13 |
| meame(III) | 22.36 | 0.02 | 4.09 |
| trmac(III) | 10.99 | 1.21 | 0.69 |
| | **Environmental Variables** | | |
| soi(I) | 3.83 | 3.77 | 9.01 |
| WT(I) | 12.87 | 2.91 | 0.67 |
| WS(I) | 6.00 | 0.54 | 17.77 |
| RD(I) | 16.96 | 2.39 | 31.02 |
| nao(II) | 0.75 | 0.05 | 9.60 |
| WT(II) | 0.08 | 24.75 | 1.68 |
| WS(II) | 16.33 | 22.07 | 3.57 |
| RD(II) | 0.61 | 13.63 | 0.32 |
| WT(III) | 10.89 | 24.41 | 0.03 |
| WS(III) | 11.20 | 0.30 | 8.61 |
| AP(III) | 19.44 | 0.25 | 9.28 |
| RD(III) | 0.98 | 4.88 | 8.38 |

# DISCUSSION

Whereas a variety of local environmental variables were identified as potential controls of juvenile fish abundances in our study, three of them appeared to affect juvenile dynamics of **TABLE 4.** Absolute contributions (%) of fish species and environmental variables on the first three axes of the co-inertia analysis. Projections of environmental variables and fish species on axes 1–2 and axes 1–3 of the co-inertia are depicted in Figure 6a and 6b. Codes of environmental variables are given in Table 1; fish species codes are defined in Table 2 (seasonal groups: I = early spring–early summer; II = late fall–winter; III = summer–fall).

many species: river discharge, wind speed, and water temperature. Climatic indices that were representative of large-scale, oceanwide climatic conditions (e.g., SOI and NAO) were also found to play a role.

## River Discharge

River discharge appeared to have a major effect on interannual variations in juvenile fish abundances in the Mobile Bay area. This result is consistent with previous studies reporting the influence of river discharge on larval or juvenile fish abundances (or both) along the East Coast (Crecco et al. 1986; Rulifson and Manooch 1990; North and Houde 2003) and West Coast (Turner and Chadwick 1972; Kimmerer et al. 2001) of the United States. However, the studies cited above were conducted on anadromous fish species (striped bass *Morone saxatilis*, white perch *Morone americana*, and American shad *Alosa sapidissima*) that directly utilize riverine habitats during the larval or juvenile stages. Our results indicate that juvenile abundances of nonanadromous coastal marine fish species are also affected by river discharge variability in waters off Mobile Bay, as has been suggested for Gulf menhaden *Brevoortia patronus*, king mackerel *Scomberomorus cavalla*, and Atlanticbumpers in the Mississippi River delta (Grimes and Finucane 1991; Govoni 1997). In Europe, variability in recruitment of European anchovy *Engraulis encrasicolus* (in the Mediterranean Sea, the Bay of Biscay, and the Black Sea), European whiting *Merlangius merlangus* (in the Black Sea), and European bass *Dicentrarchus labrax* (in Portugal) has also been linked to variability in river discharge within their respective water bodies (Daskalov 1999, 2003; Lloret et al. 2004; Planque and Buffaz 2008; Vinagre et al. 2009).

A likely explanation for the observed effect of river discharge on juvenile abundances of these nonanadromous species may be the stimulation of coastalmarine primary and secondary production by terrestrial nutrient inputs (Day et al. 1989), which in turn affects the abundance and distribution of zooplanktonic prey of larval and juvenile fish (Kimmerer 2002; Martino and Houde 2010). Moreover, the inputs of the freshwater layer in coastal surface waters affect water column stratification, which also influences the distribution of fish larvae and juveniles and their zooplankton food in the estuaries (Mann 1993; Kimmerer 2002; Martino and Houde 2010). The influence of river discharge on juvenile fish abundances may thus result either from variable feeding conditions for larval and juvenile fish or from variability in salinity habitats associated with water column stratification. In the present study, among species for which juvenile abundances peaked in summer and fall (e.g., the fringed flounder, spot, southern kingfish, and hogchoker) or in late fall and winter (e.g., the Atlantic croaker, sand seatrout, fringed flounder, and southern kingfish), many were associated with river discharge during early spring through early summer, consistent with a probable effect of river discharge on larval and early juvenile feeding conditions.

The present study also demonstrated a correlation between river discharge and large-scale climatic conditions (SOI), illustrating the sensitivity of watershed processes to global climate variation (Nohara et al. 2006).More specifically, the fluctuations of El Ni˜no–La Ni˜na phases, reflected by SOI in the present study, have been shown to affect precipitation patterns in the study region, which in turn strongly affect river discharge interannual variability (Srivastava et al. 2010). Our results from the north-central Gulf of Mexico, along with those from other marine systems, thus suggest that large-scale variations in discharge arising from climate variability as well as from human watershed regulations (such as dams) can strongly affect coastal marine fish population dynamics.

## Wind

The JAIs of four species—namely southern kingfish during late fall–winter and summer–fall and silver seatrout, fringed flounder, and hogchokers during summer and fall—were found to relate with wind speed in the study area. Variations in juvenile abundances of marine fishes from awide diversity of coastal systems, including the U.S. East Coast (Hare and Cowen 1996; Luettich et al. 1999), South Africa (Hutchings et al. 1998), Denmark (Nielsen et al. 1998), and Japan (Nakata et al. 2000), have been linked to wind-driven mechanisms of larval transport. However, in contrast to larval fish, juvenile fish examined in the present study were well developed and able to control their swimming direction. Wind is also known to affect vertical mixing of the water column, thereby influencing primary and secondary production processes and leading to variable feeding conditions for juveniles (Mann 1993). Turbulence generated by wind stress also affects encounter rates of juvenile fishwith their zooplanktonic prey (Mann 1993).

In addition, wind stress can also be an important driver of coastal circulation, thus affecting the distribution of juvenile zooplanktonic prey (North and Houde 2003).However,most of the relationships found between wind speed and juvenile abundances in the present study were negative and were generally found to occur with no seasonal lags (e.g., juvenile abundances of silver seatrout and fringed flounder during late fall and winter were negatively related to wind speed during this season). This tends to suggest that strong winds resulted in smaller numbers of juveniles collected at the three sampling stations. Consequently, it remains difficult to disentangle an effect of wind stress on species catchability from its potential influence on water column structure and associated larval transport, larval and juvenile feeding, or both.

## Temperature

In the present study, variations in juvenile abundances of many species were related to water temperature variations (e.g., the Atlantic croaker and Gulf butterfish during early spring through early summer; sand seatrout and fringed flounder during late fall and winter; and silver seatrout and hogchoker during summer and fall; Figure 6a, b). Temperature effects on juvenile fish abundances have been widely documented in a variety of marine ecosystems but primarily for fishes in northern latitudes, such as Atlantic cod *Gadus morhua* (North Atlantic), European anchovy and European whiting (Black Sea), walleye pollock *Theragra chalcogramma* (Alaska), European sprat *Sprattus sprattus* (Baltic Sea), and others (Daskalov 1999; Planque and Fr´edou 1999; MacKenzie and K¨oster 2004; Ciannelli et al. 2005). Temperature variations were also shown to be correlated with NAO in late fall–winter and summer–fall in the present study (Figure 5b, c). This is consistent with observations from the U.S. East Coast by Hare and Able (2007), who proposed that winter severity, and in particular minimum temperature (which tracked NAO variability well), was the main environmental control of winter-spawned Atlantic croaker recruitment success. In the present study, the relationship between Atlantic croaker abundance during early spring through early summer and water temperature during late fall and winter also reflected a similar relationship, tracking variations of NAO during late fall and winter (Figure 6b). However, this relationship was found to be negative, such that juvenile abundances of Atlantic croakers during early spring through earlysummer appeared to increase when water temperature and NAO during fall and winter decreased. This inconsistency between Hare and Able's (2007) result and our observations could be accounted for by the fact that Atlantic croakers are at the edge of the species' biogeographical range in the northeast continental shelf of the United States, where Hare and Able (2007) conducted their study. Conversely, the northern Gulf of Mexico is not only characterized by the highest relative probability of occurrence for this species (Froese and Pauly 2010) but also by much less severe winters, which could explain a different relationship between this species and temperature conditions than those observed elsewhere. This is consistent with the hypothesis that correlations between environmental factors and juvenile fish abundances vary as a function of relative locations within the biogeographical ranges of fish species (Myers 1998).

Other species' abundances were related to water temperature in our study: positive relationships were observed for Gulf butterfish in early spring through early summer, for silver seatrout in summer and fall, and for sand seatrout during late fall and winter (Figure 6a, b). Conversely, fringed flounder during late fall and winter and hogchokers during summer and fall were negatively associated with water temperature during early spring through early summer (Figure 6a, b). Gulf butterfish spawn from fall through spring in the study area (Pattillo et al. 1997; Hernandez et al. 2010a, 2010b). Growth and survival of larvae and early juveniles of this species could thus be favored by warm conditions during early spring through early summer, explaining the positive relationship we observed with water temperature during early spring through early summer for this species. Conversely, fringed flounder and hogchokers spawn in spring and summer in the region (Pattillo et al. 1997; Hernandez et al. 2010a, 2010b). Larval and early juvenile survival for these two species could thus be favored in years with cool water temperatures in spring and summer, contributing to higher juvenile abundances of fringed flounder in late fall and winter and of hogchokers in summer and fall. Such effects of water temperature on larval and early juvenile growth and survival during spring and summer (contributing to variable juvenile abundances in summer, late fall, and winter) were also reported for larval striped bass along the U.S. East Coast (Secor and Houde 1995; Rutherford et al. 1997). **Other Potential Drivers**

Although the multivariate approach used in this study identified the major environmental variables involved in shaping multispecies patterns in juvenile fish abundances, the total inertia remained relatively low (0.80), indicating that other biological and environmental factors that have not been considered in the present study play an important role. For example, dissolved oxygen concentrations (particularly the occurrence of hypoxic water masses during summer) have been shown to contribute to the dynamic of coastal marine communities in the Mississippi River delta (Rabalais and Turner 1998). The higher probability of occurrence of hypoxic events during warm years may also further help in explaining some of the negative relationships between juvenile abundances and water temperature observed in the present study (e.g., for the fringed flounder and hogchoker). uch hypoxic events indeed occur every summer inMobile

Bay (Park et al. 2007) and the Mississippi–Alabama shelf (Brunner et al. 2006), and they may affect juvenile abundances of fish species whose larvae or juveniles (or both) are found during summer in the study area. Similarly, seasonal jellyfish invasions reported in coastal Alabama waters, which are also more frequent during warm conditions, could also influence juvenile abundance patterns for some coastal fish species (Graham et al. 2003). Biological variables, such as spawning stock biomass, may further help to increase the amount of explained variability in juvenile abundances for some of the species examined (Clark and Hare 2002; Ottersen et al. 2006). Sand seatrout, Atlantic croakers, and southern kingfish, for example, are heavily harvested in our region (NOAA2006), and all three species are bycatch of the shrimp industry (Diamond et al. 2000; Steele et al. 2002). The influence of general harvest pressure and shrimp fishery effort on juvenile abundance and survival of these fishes in our region is an interesting area of research and should be given particular attention in a separate analysis. Finally, single-species analyses need to be conducted to account for the potential additive effects of several environmental and biological factors on fish juvenile dynamics. Such single-species analyses should be conducted at first on species for which otoliths records are available and can be used to build the size-at-age relationships requested for the development of age-specific models (e.g., Atlantic croaker).

To conclude, this study provides the first investigation of environmental influences on juvenile fish dynamics by using a multispecies and multivariable approach in the north-centralGulf of Mexico, outside of the Mississippi River delta. It confirms the importance of river discharge for fish dynamics in the region and for nonanadromous coastal marine fish species in general. Consistent with previous studies, juvenile fish abundances in river-dominated, productive coastal ecosystems appear to track some of the large-scale climate patterns represented by synthetic indices such as the SOI or NAO and by local environmental conditions, among which river discharge is prevailing. This result has strong implications for management since it emphasizes the strong associations between watershed processes and the production of adjacent coastal fisheries.

## ACKNOWLEDGMENTS

This study was funded by the Fisheries Oceanography of CoastalAlabama program at theDauphin Island Sea Laboratory, supported by ADCNR.We thank Marcus Drymon (Dauphin Island Sea Laboratory) for help in the acquisition and processing of the FAMP data used in this study, and we are grateful to the technical personnel from MRD for field and laboratory work.

# REFERENCES


1. Able, K. W., and M. P. Fahay. 2010. Ecology of estuarine fishes: temperate waters of the westernNorth Atlantic. Johns Hopkins University Press, Baltimore, Maryland.
2. Brander, K. M. 2007. Global fish production and climate change. Proceedings of the National Academy of Sciences of the USA 104:19709–19714.
3. Browder, J. A. 1993.Apilotmodel of the Gulf ofMexico continental shelf. Pages 279–284 *in* V. Christensen and D. Pauly, editors. Trophic models of aquatic ecosystems. International Center for Living Aquatic Resources Management (ICLARM) Conference Proceedings 26, ICLARM, Manila.
4. Brunel, T., and J. Boucher. 2007. Long-term trends in fish recruitment in the north-east Atlantic related to climate change. Fisheries Oceanography 16:336–349.
5. Brunner, C. A., J. M. Beall, S. J. Bentley, and Y. Furukawa. 2006. Hypoxia hotspots in the Mississippi Bight. Journal of Foraminiferal Research 36:95–107.
6. Ciannelli, L., K. M. Bailey, K. S. Chan, A. Belgrano, and N. C. Stenseth. 2005. Climate change causing phase transitions of walleye pollock (*Theragra chalcogramma*) recruitment dynamics. Proceedings of the Royal Society of London B 272:1735–1743.
7. Clark, W. G., and S. R. Hare. 2002. Effects of climate, stock size, and region on recruitment and growth of Pacific halibut. North American Journal of Fisheries Management 22:852–862. Crecco, V., T. Savoy, and W. Whitworth. 1986. Effects of density-dependent and climatic factors on American shad, *Alosa sapidissima*, recruitment: a predictive approach. Canadian Journal of Fisheries and Aquatic Sciences 43:457–463.
8. Cushing, D. H. 1996. Towards a science of recruitment in fish populations, excellence in ecology series, book 7. International Ecology Institute, Oldendorf/Luhe, Germany.
9. Daskalov, G. M. 1999. Relating fish recruitment to stock biomass and physical environment in the Black Sea using generalized additive models. Fisheries Research 41:1–23.
10. Daskalov, G. M. 2003. Long-term changes in fish abundance and environmental indices in the Black Sea. Marine Ecology Progress Series 255:259–270.
11. Day, J. W., Jr., C. A. S. Hall, W. M. Kemp, and A. Y´a˜nez-Arancibia. 1989. Estuarine ecology. Wiley, New York.
12. Diamond, S. L., L. G. Cowell, and L. B. Crowder. 2000. Population effects of shrimp trawl bycatch on Atlantic croaker. Canadian Journal of Fisheries and Aquatic Sciences 57:2010–2021.
13. Ditty, J. G., and F. M. Truesdale. 1983. Comparative larval development of *Peprilus burti*, *P. triacanthus* and *P. paru* (Pisces: Stromateidae) from the western North Atlantic. Copeia 1983:397–406.
14. Dol´edec, S., and D. Chessel. 1994. Co-inertia analysis: an alternative method for studying species-environment relationships. Freshwater Biology 31:277–294.
15. Dray, S., D. Chessel, and J. Thioulouse. 2003. Co-inertia analysis and the linking of ecological data tables. Ecology 84:3078–3089.
16. Fahay, M. P. 2007. Early stages of fishes in the western North Atlantic Ocean (Davis Strait, southern Greenland and Flemish Cap to Cape Hatteras), volumes 1 and 2. Northwestern Atlantic Fisheries Organization, Dartmouth, Nova Scotia.
17. Froese, R., and D. Pauly, editors. 2010. FishBase: world wide web electronic publication. Available: www.fishbase.org./ (December 2010).
18. Fuiman, L. A., and R. G. Werner, editors. 2002. Fishery science: the unique contributions of early life stages. Blackwell Scientific Publications, Oxford, UK.
19. Gallaway, B. L., and K. Strawn. 1974. Seasonal abundance and distribution of marine fishes at a hot-water discharge in Galveston Bay, Texas. Contributions in Marine Science 18:71–137.

20. Govoni, J. J. 1997. The association of the population recruitment of gulf menhaden, *Brevoortia patronus*, with Mississippi River discharge. Journal of Marine Systems 12:101–108.
21. Graham,W.M., D. L. Martin, D. L. Felder, V. L. Asper, and H. M. Perry. 2003. Ecological and economic implications of a tropical jellyfish invader in the Gulf of Mexico. Biological Invasions 5:53–69.
22. Grimes, C. B. 2001. Fishery production and the Mississippi River discharge. Fisheries 26(8):17–26.
23. Grimes, C. B., and J. H. Finucane. 1991. Spatial distribution and abundance of larval and juvenile fish, chlorophyll and macrozooplankton around the Mississippi River discharge plume, and the role of the plume in fish recruitment.Marine Ecology Progress Series 75:109–119.
24. Hare, J. A., and K. W. Able. 2007. Mechanistic links between climate and fisheries along the east coast of the United States: explaining population outbursts of Atlantic croaker (*Micropogonias undulatus*). Fisheries Oceanography 16:31–45.
25. Hare, J. A., and R.K. Cowen. 1996. Transportmechanisms of larval and juvenile bluefish (*Pomatomus saltatrix*) from South Atlantic Bight spawning grounds to Middle Atlantic Bight nursery habitats. Limnology and Oceanography 41:1264–1280.
26. Hernandez, F. J., Jr., S. P. Powers, andW.M.Graham. 2010a. Seasonal variability in ichthyoplankton abundance and assemblage composition in the northern Gulf of Mexico. U.S. National Marine Fisheries Service Fishery Bulletin 108:193–207.
27. Hernandez, F. J., Jr., S. P. Powers, andW.M. Graham. 2010b. Detailed examination of ichthyoplankton seasonality from a high-resolution time series in the northern Gulf of Mexico during 2004–2006. Transactions of the American Fisheries Society 139:1511–1525.
28. Hollowed, A. B., S. R. Hare, and W. S. Wooster. 2001. Pacific basin climate variability and patterns of northeast Pacific marine fish production. Progress in Oceanography 49:257–282.
29. Houde, E. D. 1997. Patterns and consequences of selective processes in teleost early life histories. Pages 173–196 *in* R. C. Chambers and E. D. Trippel, editors. Early life history and recruitment of fish populations. Chapman and Hall, London.Hutchings, L.,M. Barange, S. F. Bloomer, A. J. Boyd, R. J.M. Crawford, J. A.
30. Huggett, M. Kerstan, M. Korr^ubel, J. A. A. de Oliveira, S. J. Painting, A. J. Richardson, L. J. Shannon, F. H. Sch¨ulein, C. D. van der Lingen, and H. M. Verheye. 1998. Multiple factors affecting South African anchovy recruitment in the spawning, transport and nursery areas. South African Journal ofMarine Science 19:211–225.
31. Kimmerer, W. J. 2002. Effects of freshwater flow on abundance of estuarine organisms: physical effects or trophic linkages? Marine Ecology Progress Series 243:39–55.
32. Kimmerer,W., J. Cowan, L.Miller, and K. Rose. 2001. Analysis of an estuarine striped bass population: effects of environmental conditions during early life. Estuaries and Coasts 24:557–575.
33. Lankford, T. E., Jr., and T. E. Targett. 2001. Low-temperature tolerance of age-0 Atlantic croakers: recruitment implications for U.S. mid-Atlantic estuaries. Transactions of the American Fisheries Society 130:236–249.
34. Legendre, P., and L. Legendre. 1998. Numerical ecology. Elsevier, Amsterdam. Lehodey, P., J. Alheit, M. Barange, T. Baumgartner, G. Beaugrand, K.Drinkwater, J. M. Fromentin, S. R. Hare, G. Ottersen, R. I. Perry, C. Roy, C. D. van der Lingen, and F. Werner. 2006. Climate variability, fish, and fisheries. Journal of Climate 19:5009–5030.
35. Lloret, J., I. Palomera, J. Salat, and I. Sole. 2004. Impact of freshwater input and wind on landings of anchovy (*Engraulis encrasicolus*) and sardine (*Sardinapilchardus*) in shelf waters surrounding the Ebre (Ebro) River delta (northwestern Mediterranean). Fisheries Oceanography 13:102–110.

36. Luettich, R. A. J., J. L. Hench, C. W. Fulcher, F. E. Werner, B. O. Blanton, and J. H. Churchill. 1999. Barotropic tidal and wind-driven larval transport in the vicinity of a barrier island. Fisheries Oceanography 8:190–209.
37. MacKenzie, B. R., and F. W. K¨oster. 2004. Fish production and climate: spratin the Baltic Sea. Ecology 85:784–794.
38. Mann, K. H. 1993. Physical oceanography, food chains, and fish stocks: a review. ICES Journal of Marine Science 50:105–119.
39. Martin, F. D., andG. E.Drewry. 1978. Development of fishes of the Mid-Atlantic Bight. an atlas of egg, larval, and juvenile stages, volume 6: Stromateidae through Ogcocephalidae. U.S. Fish and Wildlife Service Biological Services Report FWS/OBS-78/12.
40. Martino, E. J., and E. D. Houde. 2010. Recruitment of striped bass in Chesapeake Bay: spatial and temporal environmental variability and availability of zooplankton prey. Marine Ecology Progress Series 409:213–228.
41. Miller, B. S., and A.W. Kendall Jr., editors. 2009. Early life history of marine fishes. University of California Press, Berkeley.
42. Myers, R. A. 1998. When do environment-recruitment correlations work? Reviews in Fish Biology and Fisheries 8:285–305.
43. Nakata, H., M. Fujihara, Y. Suenaga, T. Nagasawa, and T. Fujii. 2000. Effect of wind blows on the transport and settlement of brown sole (*Pleuronectes herzensteini*) larvae in a shelf region of the Sea of Japan: numerical experiments with an Euler-Lagrangian model. Journal of Sea Research 44:91–100.
44. Nielsen, E., O. Bagge, and B. R. MacKenzie. 1998. Wind-induced transport of plaice (*Pleuronectes platessa*) early life-history stages in the Skagerrak-Kattegat. Journal of Sea Research 39:11–28.
45. NOAA (National Oceanic and Atmospheric Administration). 2006. Alabama tables of commercial and recreational fisheries. NOAA NMFS. Available: www.st.nmfs.noaa.gov/st5/publication/econ/2006/Gulf ALTables Econ.pdf. (January 2011).
46. NOAA (National Oceanic and Atmospheric Administration). 2010a. National Weather Service, Climate Prediction Center. NOAA NWS CPC. Available: http://www.cpc.noaa.gov. (December 2010).
47. NOAA (National Oceanic and Atmospheric Administration). 2010b. NationalData Buoy Center. NOAA NDBC. Available: www.ndbc.noaa.gov. (December 2010).
48. Nohara, D., A. Kitoh, M. Hosaka, and T. Oki. 2006. Impact of climate change on river discharge projected by multimodel ensemble. Journal of Hydrometeorology 7:1076–1089.
49. North, E. W., and E. D. Houde. 2003. Linking ETM physics, zooplankton prey, and fish early-life histories to striped bass *Morone saxatilis* and white perch *M. americana* recruitment. Marine Ecology Progress Series 260:219–236.
50. Ottersen, G., D. Ø. Hjermann, and N. C. Stenseth. 2006. Changes in spawning stock structure strengthen the link between climate and recruitment ina heavily fished cod (*Gadus morhua*) stock. Fisheries Oceanography 15:230–243.
51. Park, K., C. K. Kim, and W. W. Schroeder. 2007. Temporal variability in summertime bottom hypoxia in shallow areas of Mobile Bay, Alabama. Estuaries and Coasts 30:54–65.
52. Pattillo,M. E., T. E. Czapla, D. M. Nelson, andM. E.Monaco. 1997. Distribution and abundance of fishes and invertebrates in Gulf of Mexico estuaries, volume 2—species life history summaries. Estuarine Living Marine Resources Report 11, NOAA/NOS Strategic Environmental Assessments Division, Silver Spring, Maryland.
53. Planque, B., and L. Buffaz. 2008. Quantile regression models for fish recruitment-environment relationships: four case studies. Marine Ecology Progress Series 357:213–223
54. .Planque, B., and T. Fr´edou. 1999. Temperature and the recruitment of Atlantic cod (*Gadus morhua*). Canadian Journal of Fisheries and Aquatic Sciences 56:2069–2077.

55. Rabalais, N. N., and R. E. Turner, editors. 1998. Coastal hypoxia: consequencesfor living resources and ecosystems. American Geophysical Union, Coastaland Estuarine Studies, AGU Books Board, Washington, D.C.
56. Rabalais, N. N., R. E. Turner, D. Justi´c, Q. Dortch,W. J.Wiseman, and B. K. S. Sen Gupta. 1996. Nutrient changes in the Mississippi River and system responses on the adjacent continental shelf. Estuaries and Coasts 19:386–407.
57. Richards, W. J., editor. 2006. Early stages of Atlantic fishes: an identificationguide for the western central North Atlantic, volumes 1 and 2. CRC Press, Boca Raton, Florida.
58. Rotunno, T., and R. C. Cowen. 1997. Temporal and spatial spawning patterns of the Atlantic butterfish, *Peprilus triacanthus*, in the south and middle Atlantic bights. U.S. National Marine Fisheries Service Fishery Bulletin 95: 785–799.
59. Rulifson, R. A., and C. S. Manooch. 1990. Recruitment of juvenile striped bass in the Roanoke River, North Carolina, as related to reservoir discharge. North American Journal of Fisheries Management 10: 397–407.
60. Rutherford, E. S., E. D. Houde, and R. M. Nyman. 1997. Relationship of larvalstage growth and mortality to recruitment of striped bass, *Morone saxatilis*, in Chesapeake Bay. Estuaries 20:174–198.
61. Schroeder, W. W. 1978. Riverine influence on estuaries: a case study. Pages 347–364 *in* M. L.Wiley, editor. Estuarine interactions. Academic Press, New York.
62. Secor, D. H., and E. D. Houde. 1995. Temperature effects on the timing of striped bass egg production, larval viability, and recruitment potential in the Patuxent River (Chesapeake Bay). Estuaries 18:527–544.
63. Srivastava, P., V. Sharda, K. Ingram, M. Chelliah, and L. Kalin. 2010. Quantification of El Ni˜no Southern Oscillation (ENSO) impact on precipitation and stream flows for improved management of water resources in Alabama. Proceedings of the Alabama Mississippi Bays and Bayous Symposium, Mobile. Available: www.mobilebaynep.com/wp-content/uploads/2010/07/ 2010ProceedingsFinal.pdf. (January 2011).
64. Steele, P., T.M. Bert, K. H. Johnston, and S. Lerett. 2002. Efficiency of bycatchreduction devices in small otter trawls used in the Florida shrimp fishery. U.S. National Marine Fisheries Service Fishery Bulletin 100:338–350.
65. Switzer, T. S. 2003. The ecology of two estuarine-dependent tonguefishes, the blackcheek tonguefish (*Symphurus plagiusa*) and the offshore tonguefish (*S. civittatum*), in coastal Louisiana. Doctoral dissertation. Louisiana State University, Baton Rouge.
66. Thioulouse, J., D. Chessel, S. Dol´edec, J. M. Olivier, F. Goreaud, and R. Pelissier. 2001. ADE-4: ecological data analysis: exploratory and Euclidean methods in environmental sciences, version 2001. Universit´e Lyon 1, Lyon, France.
67. Thioulouse, J. D., M. Simier, and D. Chessel. 2004. Simultaneous analysis of a sequence of paired ecological tables. Ecology 85:272–283.
68. Turner, J. L., and H. K. Chadwick. 1972. Distribution and abundance of youngof- the-year striped bass, *Morone saxatilis*, in relation to river flow in theSacramento-San Joaquin estuary. Transactions of the American Fisheries Society 101:442–452.
69. USGS (United States Geological Survey). 2010a. Alabama River at Clairborne Lock and Dam near Monroeville, Alabama. National Water Information System, USGS 02428400. Available: waterdata.usgs.gov/usa/ nwis/uv?site no=02428400. (December 2010).
70. USGS (United States Geological Survey). 2010b. Tombigbee River at Coffeeville Lock and Dam near Coffeeville, Alabama. National Water Information System, USGS 02469761. Available: waterdata.usgs.gov/usa/ nwis/uv?site no=02469761. (December 2010).
71. Vinagre, C., F. D. Santos, H. N. Cabral, and M. J. Costa. 2009. Impact of climate and hydrology on juvenile fish recruitment towards estuarine nursery grounds in the context of climate change. Estuarine Coastal and Shelf Science 85:479–486.

72. Wood, R. J., and H. M. Austin. 2009. Synchronous multidecadal fish recruitment patterns in Chesapeake Bay, USA. Canadian Journal of Fisheries and Aquatic Sciences 66:496–508.

# CHAPTER 15

# Morphological variations of wild populations of Nile tilapia (Oreochromis niloticus) living in extreme environmental conditions in the Kenyan Rift-Valley

**Titus Chemandwa Ndiwa & Dorothy Wanja Nyingi & Julien Claude & Jean-François Agnèse**

*Department of Clinical Studies, College of Agriculture and Veterinary Sciences, University of Nairobi, P. O Box 29053-00625, Nairobi, France*

*Kenya Wetlands Biodiversity Research Team, National Museums of Kenya, P.O. Box 40658-00100, Nairobi, Kenya*

*Département Forme, UMR IRD 226 CNRS 5554, Institut des Science de l'Evolution, Université de Montpellier 2, Montpellier, France*

## ABSTRACT

In this study, we use geometric morphometric method to establish morphological differences between natural populations of Nile tilapia from two extreme environmental conditions (high temperature and salinity) in Kenya, and compare them to two populations from regions experiencing less extreme conditions. To determine genetic influence on morphology, we correlated genetic data with morphological data. The study observed significant morphological differences between all studied populations, including three closely related hot spring populations (Bogoria, Chelaba and

Turtle Springs), and two populations with similar genetic background inhabiting saline environments in Lake Turkana basin (Turkana and Crocodile Lake populations). In addition, allometric growth patterns of the seven populations differed significantly, demonstrating that developmental changes (plastic or adaptive) operated between these populations. Positive correlation between morphometric and genetic data confirmed the influence of genetic factors on morphology. All observed differences were attributed either to genetic and/or environmental factors, which seemed to play a major role in influencing morphology of wild Nile tilapia populations. We recommend further studies to be carried out under controlled conditions to confirm the role of temperature, pH and salinity in morphological diversification of Nile tilapia.

**Keywords:** Morphometrics Salinity Temperature Nile tilapia Rift-Valley Kenya

## INTRODUCTION

Cichlids are tropical freshwater fishes well known for their high rates of speciation often resulting in rapid radiation. They exhibit remarkably high levels of genetic and morphological diversity, which affect their morphology, ecology, behaviour and genomes (Nelson 1994; Barlow 2000; Chakrabarty 2005). A large extent of cichlid diversity is observed in three main East African Lakes (Victoria, Tanganyika and Malawi) where 1800 to 2000 species (60 % of all cichlid species worldwide) are present in species flocks (Greenwood 1984; Turner et al. 2001). These cichlid species flocks are primarily monophyletic groups of closely related species inhabiting the same ecosystem, and evolved from a single ancestral species (Greenwood 1974). Within these three East African Lakes, some cichlid species went through rapid morphological diversification and speciation, leading to formation of many new species within a few thousand generations (Sage et al. 1984; Meyer et al. 1990; Owen et al. 1990). In Lake Victoria for instance, the radiation of monophyletic haplochromine species occurred within a geologically short time of about 750,000 years and resulted to the emergence of approximately 500 cichlid species (Greenwood 1984; Abila et al. 2004; Kerschbaumer and Sturmbauer 2011). Additionally, evolutionary older cichlid assemblages of Lake Tanganyika have shown the highest degree of morphological differentiations (Salzburger et al. 2002).

Under the ecological theory of diversification (Schluter 2000), a strong relationship exists between speciation events and colonization of new environments, or niche diversification, which may in turn induce changes in organism's behaviour, morphology and physiology (Price et al. 2003).

Generally, selection pressures that exist in new environments favour divergence of populations from their ancestral forms. A classic example is provided by the case of the three spined sticklebacks (*Gasterosteus aculeatus*) in Canada, where two sympatric forms (limnetic and benthic) emerged as a consequence of ecological divergence (Schluter and McPhail 1993). This example among others (Meyer 1987; Day et al. 1994; Imre et al. 2002) has provided evidence of morphological diversifications as a result of environmental variations. However, Gagnaire et al. (2013) proposed that ecological divergence may actually involve multiple facets of species divergence on which divergent selections may operate, hence the need for an integrative approach to study the process of speciation. Some of the factors proposed to influence divergence include variations in quantitative phenotypic traits, regulation in gene expression and differential transmission of particular allelic combinations (Gagnaire et al. 2013).

The Nile tilapia (*Oreochromis niloticus*), the subject of our study, is a teleost fish of the family Cichlidae, has its native range in Africa and is widely known for its importance in aquaculture. This species has been recognized for its ability to tolerate a wide range of salinity, dissolved oxygen and temperature (Farmer and Beamish 1969; Avella et al. 1993), high fecundity and rapid growth rate (Welcomme 1967) and an omnivorous mode of feeding (El-Sayed 1999). Due to these attributes, the species has the ability to adapt rapidly to new environments, showing a wide range of biological responses to different environmental conditions both in culture and in nature (Schofield et al. 2011; Grammer et al. 2012). As a result of this plasticity (or of the high degree of physiological tolerance found in the Nile Tilapia), the species has been able to successfully establish itself in extreme environmental conditions such as temperate winter conditions (Peterson et al. 2005; Grammer et al. 2012), hot springs (Trewavas 1983; Nyingi et al. 2009; Ndiwa et al. 2014) and saline waters (Schofield et al. 2011). Studies by Collyer et al. (2005) in pupfish (*Cyprinodon tularosa*) noted that the geographical spread of species across a broad range of ecological conditions is accompanied by equally diverse morphological variations, which are strongly correlated to the environmental conditions.

Within the Kenyan Rift Valley, Nile tilapia has colonized numerous newly formed habitats within a system characterised by a number of drainages comprising a series of rivers, springs and lakes. This complex system is a host to four out of the seven described sub-species of *Oreochromis niloticus* by Trewavas (1983): *O. n. baringoensis* from Lake Baringo; *O. n. sugutae* from River Suguta; *O. n. vulcani* from Lake Turkana. Recently Nyingi et al. (2009) discovered a natural population from Lake Bogoria Hotel genetically differentiated from but nevertheless close to the Lake Baringo population. The purpose of the current study is to investigate whether populations of Nile Tilapia have undergone significant morphological diversifications, and whether these differences can be related to environmental factors (either as the result of phenotypic plasticity or adaptive evolution).

According to Collyer et al. (2005, 2007), variations in salinity have the potential to affect fish body shape. These authors reported a general decrease in body circumference to overall body size at high salinity on pupfish (*Cyprinodon Tularosa*). Similarly, high water temperatures can influence both the growth rate and development rate of an organism (Atkinson 1995). In Zebra fish (*Danio rerio*) for instances, raising of juveniles under four different developmental temperatures resulted to highly variable phenotypes (Sfakianakis et al. 2011). A similar study by Georgakopoulou et al. (2007) on European seabass (*Dicentrarchus labrax*, L) showed that samples reared under lower temperatures (15 °C) were more slender than those reared under higher temperature (20 °C). Even though no studies have been carried out to establish the effect of water pH on fish shape, extreme changes in water pH have been reported to cause fish deaths by altering other aspects of water chemistry, for instance increase in pH increases toxicity of ammonia (Schofield 1976; Wurts and Durborow 1992).

Taking into consideration the local genetic variations observed in three recently discovered populations of *O. niloticus* within the hot springs (Chelaba, Lake Bogoria Hotel and Turtle) of Loboi Swamp (Nyingi et al. 2009; Ndiwa et al. 2014), we hypothesized that these fish populations may also differ in shape due to the influence of high salinity and high temperature. In addition to the Loboi Swamp fish, another population of *O. niloticus* found in the northern part of the Kenyan Rift-Valley also experiences extreme environmental conditions. The Crocodile Lake, one of the three Lakes located on the Central Island of Lake Turkana is saline, with electrical conductivity as high as 10,590 μS/cm and alkaline with a pH of around 10.0 (Avery 2010). Depending on the level of Lake Turkana, Crocodile Lake can either be submerged or emerged. During high water level the lake has been known to disappear (Hay 1976; Cerling 1979) while in low water levels, it remains isolated. The last time the two lakes were reported to have connected was in 1902 (Hopson 1982). So far only three species of fish have been reported to survive in Crocodile Lake namely: *Clarias gariepinus*, *Haplochromis sp.*, and *O. niloticus*.

Both the Loboi and Crater Lake populations have adapted to extreme conditions, high temperature and high salinity levels respectively, while other closely related populations (Lakes Baringo, Turkana and Victoria) do not experience such conditions. These populations thus present an opportunity to observe how extreme conditions affect the morphology of the fish. To achieve this goal, we compared individuals from seven populations, four with extreme conditions (hot springs; Lake Bogoria Hotel Spring, Chelaba Spring, Turtle Spring and saline; Crocodile Lake populations), and three that experience more favourable environmental conditions (Lakes Baringo, Turkana and Victoria). Six of the studied Nile tilapia populations occur in their natural habitats except the Lake Victoria population, which was introduced into Lake Victoria from Lake Albert in Uganda (Fryer and Isles 1972; Trewavas 1983).

The study provides an insight into morphological responses of these fish populations to extreme temperature and salinity. Occurrence of phenotypic differences between the populations studied will confirm that the environment or/and the genetic background play an important role in evolution of cichlids, while lack of phenotypic differences will imply that the high degree of tolerance in Nile Tilapia is related to its propensity to invade new environment, and in keeping a very canalized phenotype.

# MATERIALS AND METHODS

Specimens from three hot spring populations of Loboi swamp drainage system (Chelaba, Lake Bogoria Hotel and Turtle) and four other lakes (Baringo, Turkana, Crocodile and Victoria) were studied (Table 1; Fig. 1). The three hot springs are characterised by elevated water temperatures (approximately 36 °C), low dissolved oxygen concentration (3.5–3.9 $mgl^{-1}$) and slight acidity pH (6.4–6.9) (Ashley et al. 2004). The springs drain into Loboi swamp of about 1.5 $km^2$, which is covered by dense vegetation of *Typha domingensis* and *Cyperus papyrus*. As the water exits the swamp, it is poorly aerated (Dissolved Oxygen (DO) = 0.2–1.1 $mgl^{-1}$) and has a higher pH (7.4–8.3) (Ashley et al. 2004). Comparatively, Crocodile Lake is a crater formed island lake within the Lake Turkana. It is characterised by relatively higher water pH (10) and almost three times higher salinity (conductivity, 10,590 μS/cm) than Lake Turkana, which has pH and a conductivity of 9.3 and 3420 μS/cm respectively (Avery 2010). Lake Baringo, a fresh water Lake located about 52 km from the Loboi swamp has DO concentration ranging from 5.89–6.7 mg/L, temperature range of 22.1–31.8 °C, pH ~ 8.5 and conductivity of about 580 μS/cm (Omondi et al. 2014). Lake Victoria, also fresh water body is characterized by water temperatures ranging from 23.8 to 31.5 °C, pH 6.2–9.2, and conductivity of approximately 170 μS/cm (Ochumba 1990; Crul 1995).

**Table 1** Names of the seven populations of *O. niloticus* studied, their respective sample sizes (N), coordinates of study sites, habitat characteristics, surface water temperature range, conductivity (salinity) and size of the water body

| **Population** | **Number** | **Co-ordinates** | **Habitat** | **Temperature** | **Conductivity** | **Surface area** |
|---|---|---|---|---|---|---|
| Bogoria Hotel Spring | 41 | 0°21′44″ N, 36°03′04″ E | Hotspring | 32–36 °C | <100 µS/cm | 1.5 $Km^2$ |
| Chelaba Spring | 30 | 0°21′30″ N, 36°02′58″ E | Hotspring | 32–36 °C | <100 µS/cm | 1.5 $Km^2$ |
| Turtle Spring | 24 | 0°21′44″ N, 36°02′41″ E | Hotspring | 32–36 °C | <100 µS/cm | 1.5 $Km^2$ |
| Crocodile Lake | 28 | 3°29′34″ N, 36°01′54″ E | Saline | 25–30 °C | 10,590 µS/cm | <1 $Km^2$ |
| Lake Turkana | 38 | 3°30′46″ N, 35°54′53″ E | Saline | 26–27 °C | 3420 µS/cm | 6405 $Km^2$ |
| Lake Baringo | 43 | 0°36′43″ N, 36°01′31″ E | Normal | 22–32 °C | 580 µS/cm | 130 $Km^2$ |
| Lake Victoria | 33 | 0°38′17″ N, 33°58′35″ E | Normal | 23–32 °C | 170 µS/cm | 68,800 $Km^2$ |

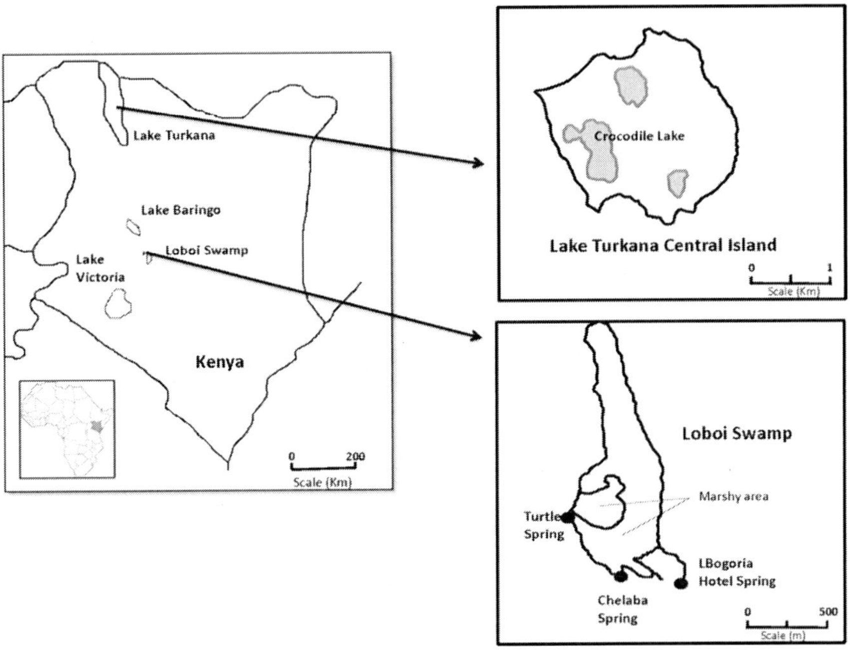


**Figure. 1** Map showing the location of all the six natural populations and one introduced population of *O. niloticus*: Lake Turkana at Lake Turkana drainage system, Crocodile Lake in the Central Island of Lake Turkana, Lake Baringo at Lake Baringo drainage system, Turtle Spring, Chelaba Spring and Bogoria Hotel Spring in the Loboi Swamp, and Lake Victoria (introduced) at *L. Victoria* drainage system

Fish specimens were collected from the hot springs (Chelaba, Lake Bogoria Hotel and Turtle) and Lakes Baringo, Victoria, Turkana and Crocodile using seine nets and scoop nets. Fish were anaesthetised and killed using an overdose of MS-222 before fixing in 4 % formalin solution for seven days. The formalin was washed and voucher specimens preserved in 70 % ethanol, and are curated as reference material at the National Museums of Kenya. The standard lengths (distance from the tip of the snout to the base of caudal fin) of the fish were measured using a digital Vernier calliper in order to estimate their sizes.

Body shape of the samples was quantified using landmark based geometric morphometric method (Rohlf and Marcus 1993). Images of preserved samples of fish were obtained by using a flatbed scanner (Herler et al. 2007) and imported into the program tpsUtil version 1.53 (Rohlf 2012). A total of 14 landmarks (Fig. 2) were chosen based on their capacity to capture overall body shape, and their coordinates (x,y) digitized on the left hand side of each sample using the program tpsDig version 2.16 (Rohlf 2010). The coordinates were saved in a tps format data file and used for further morphometric analysis. A total of 237 adult fish specimens from seven wild population of Nile tilapia were analysed. Maturity of the specimens was ascertained by observing their gonads. Their standard length sizes ranged from 40 to 207 mm.

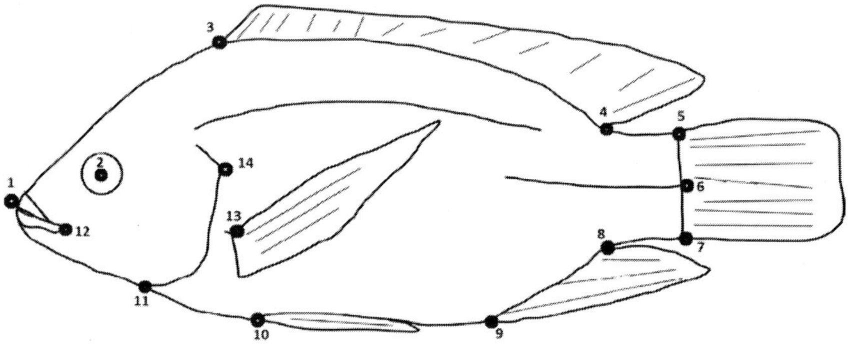


**Figure. 2** Position of 14 different landmarks used in the study: (*1*) anterior tip of the snout with mouth closed, (*2*) centre of the eye, (*3* & *4*) anterior and posterior insertions of the dorsal fin, (*5*) dorsal origin of caudal fin, (*6*) point halfway between dorsal and ventral origin of caudal fin, (*7*) ventral origin of caudal fin, (*8*) posterior insertion of anal fin, (*9*) anterior origin of anal fin, (*10*) anterior origin of pelvic fin, (*11*) juncture of the ventral edge of operculum with the ventral outline, (*12*) most posterior end of the mouth, (*13*) anterior origin of pectoral fin and (*14*) most posterior end of operculum

## Data analysis

The data file was imported into R software program and analysed using Momocs (Bonhomme et al. 2014), Geormoph (Adams and Otarola-Castillo 2013; Dryden 2013) and Ape packages (Padis et al. 2004; Padis 2012). Coordinates were scaled to unit centroid size, translated, rotated and optimally superimposed through the generalized partial Procrustes Analysis. Procrustes coordinates were projected into Eucledian tangent shape space for further analysis (Dryden and Mardia 1998; Claude 2008). In order to establish morphological variations between the different populations of fish, we used principal component analysis (PCA) of the superimposed coordinates of the landmarks. Variations were visualised by plotting multivariate ordination of specimens on the first two principal component (PC) axes (PC1 and PC2). Contribution of original variables on corresponding eigenvectors or PC loadings were analysed to establish shape features that had the greatest influence on each PC.

Nested MANCOVA models were conducted on the principal components (PC) of shape data in order to assess the influence of locality, and environmental factors. Habitat and locality were considered as explanatory variables and centroid size as a covariate in the model. Size was added as a covariate to understand the effect of habitat and locality on allometric growth patterns.

Response ~ Size + Habitat + Habitat:Size + Habitat:locality + Habitat:Locality:Size

To represent and appraise morphological differences between the localities, a linear discriminant analysis was performed on the dataset of the Procrustes coordinates.

The Fst values for 5 populations of Nile tilapia (Chelaba, Bogoria, Turtle, Turkana, Baringo and Victoria) used in the current study were obtained from our previous studies (Ndiwa et al. 2014). Alternatively, the Fst values for Crocodile Lake and Lake Victoria populations were obtained from unpublished data. All our Fst values were calculated from microsatellite molecular data using Genepop software (Rousset 2008). A mantel test analysis (Mantel 1967) was performed through permutation to correlate the Fst Values and Mahalanobis distance. This analysis aimed at confirming the influence of genetic factors on the morphology. Lastly, minimum spanning network based on Mahalanobis distance was constructed on R based on Ape package (Padis et al. 2004; Padis 2012) in order to visualize phenotypic relationship between the seven wild populations of Nile tilapia.

# RESULTS

Comparison of body sizes of the fish samples from the seven populations of Nile tilapia (Fig. 3) indicated minimal variations between specimens from Lake Baringo, Chelaba Spring, Crocodile Lake and Turtle Spring. However, greater body size differences were observed between the latter populations and populations from Bogoria Spring, Lake Turkana and Lake Victoria. Body size comparison among individuals of same population showed least variations within Crocodile Lake (Fig. 3).

The first shape PC accounted for 23 % of the observed variations. This PC was mostly related with the variation of the head, caudal peduncle and anal fin (Fig. 4). The PC2 on the other hand accounted for 14.5 % of variation mostly within the depth of the body and caudal peduncle (Fig. 4a).

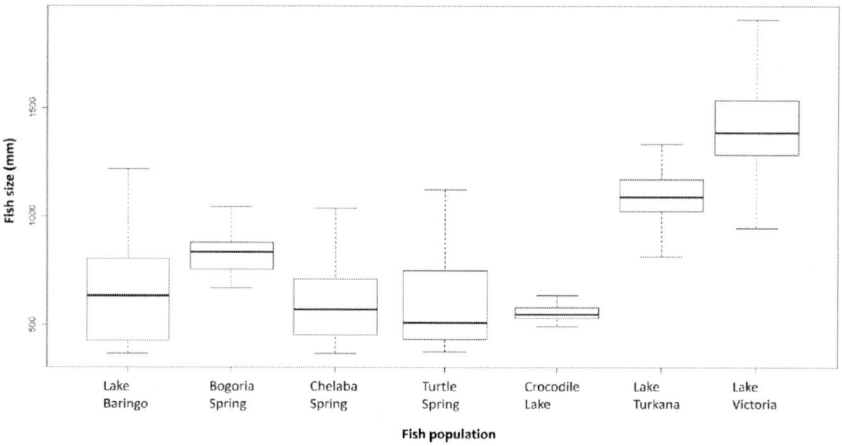


**Figure. 3** Size distribution of the fish samples collected from the seven localities

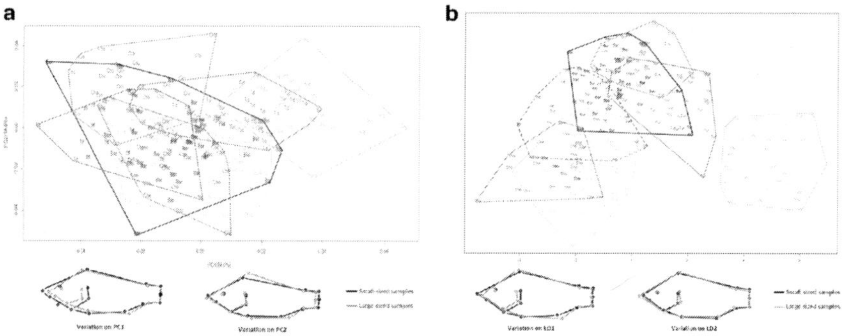


**Figure. 4 a** Differentiation of the seven populations of Nile tilapia based on Principal Component Analysis (PCA). On fish shape images (PC 1, 2), Grey = mature (large samples); Black = small (young) samples. In the scatterplot, cro = crocodile; Tur = Turkana; Che = Chelaba; Bog = Bogoria; Trt = Turtle; Bar = Baringo; Vic = Victoria **b**Differentiation of the seven populations of Nile tilapia using Linear Discriminant Analysis (LDA). On fish shape images (LDA 1, 2), Grey = mature (large samples); Black = small (young) samples. In the scatterplot, cro = crocodile; Tur = Turkana; Che = Chelaba; Bog = Bogoria; Trt = Turtle; Bar = Baringo; Vic = Victoria

Even though there were overlaps between most of the populations under study, significant morphological differences were observed between all populations. A strong separation was observed between the hot spring populations and both Crocodile Lake, and Lake Turkana populations based on PC1. However, PC2 showed greater differentiation between Lakes Turkana and Crocodile populations than the latter (Fig. 4a).

Surprisingly, we observed that populations of Lakes Turkana and Victoria were similar morphologically. These results were unexpected since the two systems have different ecological characteristics.

## Differences between populations and habitats

Both differences in habitats and localities significantly explained the shape variations between the seven populations of Nile tilapia studied (Table 2). Linear discriminant analysis based on the sampling locations correctly assigned 84.8 % of the individuals to their respective localities. All samples from Lake Victoria were assigned to their real group (Table 3). However, most of individuals from the hot springs populations that were not correctly assigned to their respective localities were either assigned to another hot spring population(s), or to L. Baringo (Table 3). Assignment of individuals from other localities is shown on Table 3.

**Table 2** Results of multi-variate analysis of covariance (MANCOVA) performed on body shape considering all the samples

| Factor | df | Pillai | approx. F | num. df | den. df | *P*-value |
|---|---|---|---|---|---|---|
| Size | 1 | 0.8282 | 40.167 | 24 | 200 | <0.0001 |
| Habitat | 2 | 1.2667 | 14.464 | 48 | 402 | <0.0001 |
| Habitat:Size | 2 | 0.3991 | 2.088 | 48 | 402 | <0.0001 |
| Habitat:Locality | 4 | 2.0469 | 8.864 | 96 | 812 | <0.0001 |
| Habitat:Locality:Size | 4 | 0.5402 | 1.321 | 96 | 812 | 0.0268 |

**Table 3** Linear discriminant analysis (LDA) assignment of individuals to different localities and groupings according to habitat types, percentages of assignation are in brackets

| Origin of the fish | False assignation | Habitat grouping |
|---|---|---|
| Lake Bogoria Hotel Spring (82.5 %) | Chelaba Spring (10 %) | Hotspring - Hotspring |
| | Lake Baringo (5 %) | Hotspring - Normal |
| | Turtle Spring (2.5 %) | Hotspring - Hotspring |
| Chelaba Spring (78.13 %) | Bogoria Spring | Hotspring - |

| | | |
|---|---|---|
| | (12.5 %) | Hotspring |
| | Turtle Spring (9.37 %) | Hotspring - Hotspring |
| Turtle Spring (76.19 %) | Chelaba Spring (4.76 %) | Hotspring - Hotspring |
| | Crocodile Lake (4.76 %) | Hotspring - Saline |
| | Lake Baringo (14.29 %) | Hotspring - Normal |
| Lake Baringo (91.43 %) | Bogoria Spring (8.57 %) | Normal - Hotspring |
| Lake Turkana (94.59 %) | Crocodile Lake (5.41 %) | Saline - Saline |
| Crocodile Lake (83.87 %) | Turtle Spring (3.23 %) | Saline - Hotspring |
| | Lake Turkana (3.23 %) | Saline -Saline |
| | Lake Baringo (9.67 %) | Saline - Normal |
| Lake Victoria (97.06 %) | Lake Baringo (2.94 %) | Normal - Normal |

Linear discriminant analysis (LDA) clustered the seven populations of *O. niloticus* into two distinct groups. The first cluster consisted entirely of individuals from Lake Victoria, while the second cluster comprised the other six populations (Fig. 4b).

However, partial separations were observed among the hot spring populations and also among Lakes Turkana and Crocodile populations (Fig. 4b). Shape variations related with both LDA1 and LDA2 were attributed to differences within the head and the caudal region.

Because we observed a significant interaction of locality with size, we compared small sized individuals to large sized individuals from each locality in order to establish their differences and similarity in growth. Generally, we observed a decrease in head size and length of the mouth with increase in size in all the six populations except Crocodile Lake, where no marked differences were observed between the heads of the small and large sized samples. We also observed differences in body depth between the small and large sized samples (Fig. 5).

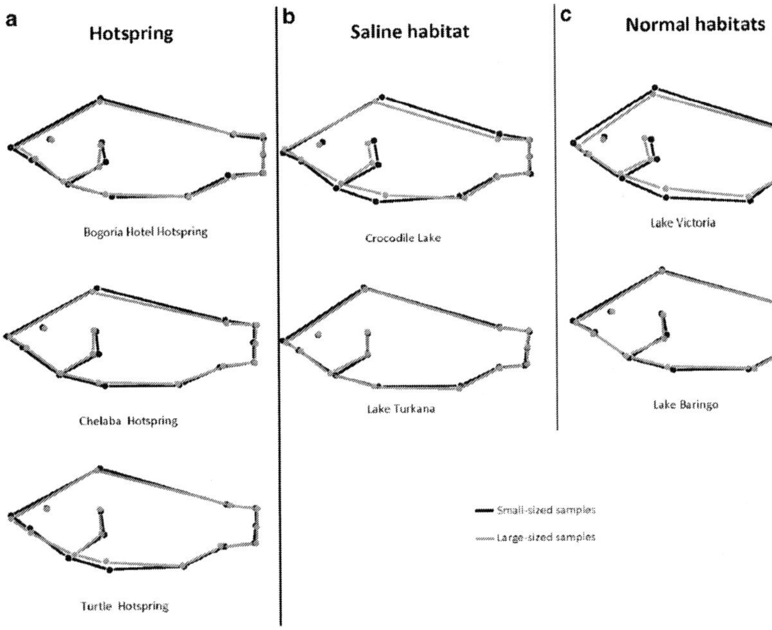


**Figure. 5** Growth patterns in the seven populations of *Oreochromis niloticus*. *Grey colour*represents the shape of the large samples while the *black colour* represents the shape of the small sized individuals

The closely related hot spring populations displayed different patterns of growth irrespective of their close spatial relationship. In Bogoria spring, small sized individuals were characterized by more elongated mouths compared to the other two spring populations. On the other hand, the heads of small sized individuals from Chelaba Spring showed slight elongation with greater dorsal displacement of their mouths. We also observed posterior displacement of the anal fin as individuals increase in size for Bogoria and Chelaba populations. Finally, large sized samples from Chelaba and Turtle Springs had elongated caudal peduncle length compared to small sized individuals, while no differences were observed for Lake Bogoria Hotel Spring population (Fig. 5).

Growth patterns of the Lake Turkana system (Turkana and Crocodile) populations differed significantly among themselves, and among other *O. niloticus* populations. For instance, great change in shape was observed between the small sized samples and the large sized samples involving drastic decrease of the depth of the body and caudal peduncle. Dorsal displacement of the mouth and anterior displacement of the eye, dorsal fin and operculum were also noted. Smaller differences were observed between small and large samples in Lake Turkana; anterior elongation of the mouth for small sized individual, anterior displacement of the lower jaw and pelvic fin of mature individuals, and slight elongation of the caudal peduncle (Fig. 5).

Within Lakes Victoria and Baringo populations, marked differences were observed between small and large sized samples. Comparison between their growth patterns showed that Lake Victoria samples had deeper caudal peduncles that increased in length with maturity of the fish, while Lake Baringo samples had slender caudal peduncles, which also increased in length with fish growth (Fig. 5).

Additionally, correlation of Fst values and Mahalanobis distance based on Mantel test analysis showed existence of significant relationship ($P < 0.05$) (Fig. 6).

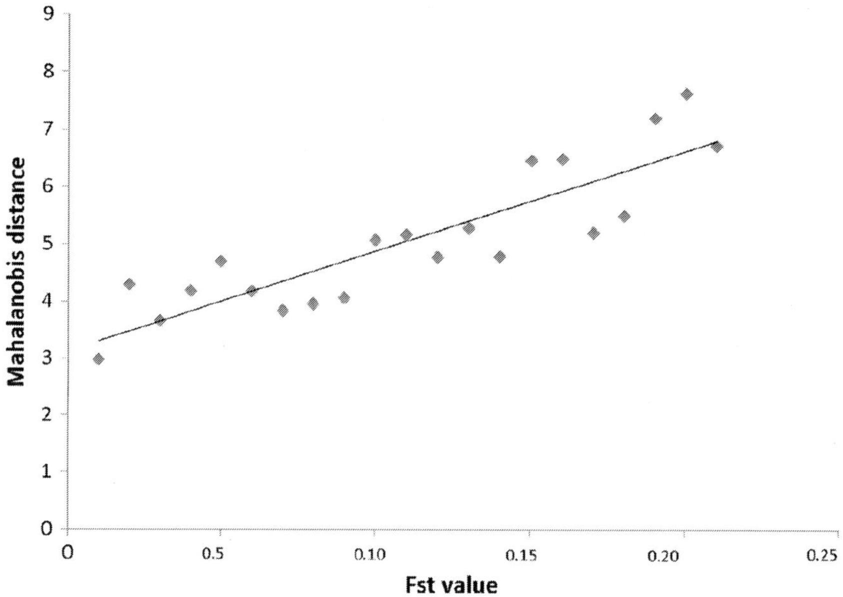


**Figure. 6** Correlation of Fst values against Mahalanobis distance showing positive correlation

Minimum spanning network computed from Mahalanobis distances between localities showed close proximity between the three hot spring populations. Additionally, Lake Turkana population showed close relationship to Lake Victoria population, while Crocodile Lake and Lake Turkana populations were separated (Fig. 7).

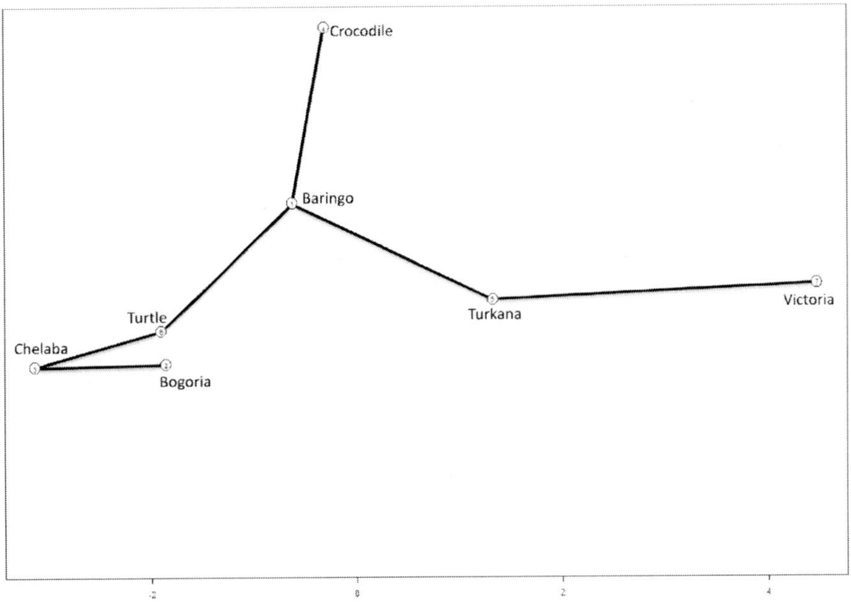


**Figure. 7** Minimum spanning network showing overall similarity in body shape among the *O. niloticus* populations

# DISCUSSION

Morphological analysis revealed that specimens studied were first differentiated by their sizes. Fishes from Lakes Turkana and Victoria were the largest with a mean standard length sizes of 122.8 ± 4.38 mm (mean ± standard error) and 163.3 ± 4.40 mm, respectively. The smallest specimens were from Turtle Spring and Crocodile Lake, and exhibited mean standard length sizes of 76.6 ± 6.11 mm and 70.1 ± 1.72 mm, respectively. Differences were also observed in terms of variability of sizes; some samples had higher size range (Lake Victoria) than others (Crocodile Lake). These observed differences may partly represent artefacts due to sampling, and may not completely reflect the actual differences of fish sizes in different localities despite the knowledge that Lakes Turkana and Victoria host large specimens of *O. niloticus* (Trewavas 1983) due to their larger sizes and greater depths, compared to biotopes investigated in the present study.

The PC analysis revealed a large amount of shape differences whereby variations of the head, caudal peduncle and anal fin accounted for 23 % (Fig. 4a). These variations allowed discrimination of the Lake Victoria specimens from those of other Lakes (Baringo, Turkana and the hot springs). Within these latter populations, fishes from Lake Turkana were placed between the two groups of fishes.

Differentiation of the Lake Victoria population from all other population based on linear discriminant analysis was even clearer (Fig. 4b). There was also a clear separation between the hot springs specimens on one hand, and Lake Turkana and Crocodile Lake specimens on the other hand. Specimens from Lake Baringo appeared to fall between these two groups of specimens. All observed variations on both LDA1 and LDA2 were related to differences within the head and the caudal region. These results are congruent with the genetic variation observed by Agnèse et al. (1997), who demonstrated that fishes from Lake Baringo and Lake Turkana are genetically close. Later on, Nyingi et al. (2009) and Ndiwa et al. (2014) observed that fishes from the hot springs are even more closely related to fishes from Lake Baringo. If one assumes that fishes from Crocodile Lake should have a genetic background identical to the fishes of Lake Turkana, and that fishes from Lake Victoria introduced from Lake Albert (Fryer and Isles 1972; Trewavas 1983) should be less genetically related to the other fish studied, then morphological differentiation seems to be congruent with genetic differentiation. Indeed, our studies based on mantel test confirmed the role of genetic factors in influencing morphology. In addition, links between morphological and genetical differentiation have been reported by Stewart and Albertson (2010), who integrated both genetic and geometric morphometric approaches in studying scale eating cichlids of Lake Tanganyika, which had been reported to show frequency balanced polymorphism in the left-handedness and right-handedness of the mouth, an adaptation previously reported by Takahashi et al. (2007). Their morphometric results strongly supported the genetic basis of jaw laterality.

In our study, we also observed different allometric patterns of growth in all the seven populations of Nile tilapia. Considering the fact that most of the localities differ greatly in their environmental conditions, the observed allometries may be partly attributed to environmental differences, which might have acted on developmental stages, and shaped the differences observed between *O. niloticus* populations. Whether differences are adaptive or the by-product of phenotypic plasticity is still speculative at that stage. However, common garden experiments are necessary in order to establish the role of various environmental factors on fish morphology.

Shape changes found in our study could be functional and environmentally related. Indeed, our observations are congruent to findings by Ramler et al. (2014) who investigated the effect of different temperatures on body size and shape on stickleback (*Gastrerosteus aculeatus*). The authors reported significant differences between individuals across different temperature groups. Similarly, study on zebrafish (*Danio rerio*) juveniles reported occurrence of highly variable phenotypes induced by diverse thermal conditions during the early ontogenic stages (Sfakianakis et al. 2011). However, results obtained by Fujimura and Okada (2008) comparing (during their developmental stages) the lower jaw bones of a cichlid fish (*Haplochromis chilotes*) from Lake Victoria and another Riverine cichlid (*O. niloticus*) contradicted our findings. Irrespective of the habitat differences, the authors noted that most shape changes had a similar pattern during the growth of the two species showing that the development of these fish were relatively robust.

When looking at closely genetically related populations, one can observe a certain amount of morphological differentiation. The first case concerned specimens from the three hot springs. Specimens from Bogoria and Chelaba or Bogoria and Turtle springs were unevenly distributed along the first and second axis of the LDA analysis respectively (Fig. 4b). Differences between Chelaba and Bogoria Springs were mainly due to variations in body length and body depth, while differences between Bogoria/Chelaba and Turtle were as a result of variations in head length and body depth.

If we consider that the genetic background of these three populations is the same (Ndiwa et al. 2014), then these observed morphological differentiations could be attributed to phenotypic plasticity of the species. Very likely, environmental characteristics of these hot springs are slightly different. Bogoria Spring is the biggest of the three hot springs and runs a few hundred meters before entering the swamp. The two other springs have a smaller flow rate and enter in the swamp after only few meters. Nevertheless, as only a few genes can strongly act on morphological differentiation, one cannot exclude that morphological differences observed between the three hot spring populations was due to genetic differences. Ndiwa et al. (2014) observed that gene flow between these populations is limited by the swamp, which is covered by a dense vegetation of *Typha domingensis* and *Cyperus papyrus*. Decomposition of organic matter, and shading provided by these vegetation is responsible for low oxygen concentration and low temperatures, which act as barriers to free movement of the fish from one place to another. Similarly, Ndiwa et al. (2014) also observed that these fish are slightly introgressed by *O. leucostictus* genes from escapees of this species originating from aquacultural ponds within the drainage. The account of possible introgression of alien genes acting on morphological differentiation cannot be excluded.

The second case of genetically related populations in our study concerns fish from Lakes Turkana and Crocodile. There was a clear morphological differentiation between these two groups of fish as shown on Fig. 4b.

Specimens from Crocodile Lake were generally smaller than those from Lake Turkana, and differed in body depth, head size and caudal length.

These two populations are genetically very close (unpublished results) since Crocodile Lake population originated from Lake Turkana about one century ago. Most likely, the observed differences can be as a result of phenotypic plasticity. These lakes offer two quite different environments to the fish. Lake Turkana is a large alkaline (pH 9.3) and slightly saline lake (conductivity 3420 μS/cm) hosting numerous fish species (48 following Hopson 1982). Comparatively, Crocodile Lake is a small size lake (about 0.32 $km^2$), with high alkalinity (pH 10) and salinity (conductivity 10,590 μS/cm). Only three species of fish are present, two Cichlids (*Aplochromis* sp., *O. niloticus*) and one Clariid (*Clarias gariepinus*).

In conclusion, our study observed that populations of Nile tilapia inhabiting different localities displayed differentiated body shapes. The study also noted correlation between genetic distance and morphological distance. Despite these findings, further laboratory studies are necessary in order to confirm the importance of genetic and environmental factors in influencing morphological development. This can be achieved by breeding genetically identical siblings in different environmental conditions: first under low versus high salinity, and secondly under low versus high temperature.

## ACKNOWLEDGMENTS

We thank Mr. Joseph Gathua (National Museums of Kenya, NMK), Mr. Gilbert Kosgei (Kenya Wetlands Biodiversity Research Group, KENWEB) and Dr. Judith Nyunja (Kenya Wildlife Service, KWS) for their support during collection of samples. We are also grateful to KENWEB, NMK, Institut Recherche pour le Développement (IRD) and French Embassy, Nairobi, Kenya for their financial support during this study.

## REFERENCES

1. Abila R, Barluenga M, Engelken J, Meyer A, Salzburger W (2004) Population-structure and genetic diversity in a haplochromine cichlid fish (corrected) of a satellite lake of Lake Victoria. Mol Ecol 13(9):2589–2602
2. Adams DC, Otarola-Castillo E (2013) Geomorph: an R package for the collection and analysis of geometric morphometric shape data. Methods Ecol Evol 4:393–399
3. Agnèse JF, Adépo-Gouréne B, Abban EK, Fermon Y (1997) Genetic differentiation among natural populations of the Nile tilapia *Oreochromis niloticus* (Teleostei, Cichlidae). Heredity 79:88–96

4. Ashley GM, Maitima MJ, Muasya AM, Owen RB, Driese SG, Hover VC, Renaut RW, Goman MF, Mathai S, Blatt SH (2004) Sedimentation and recent history of a freshwater wetland in a semi-arid environment: Loboi Swamp, Kenya, East Africa. Sedimentology 51:1301–1321
5. Atkinson D (1995) Effects of temperature on the size of aquatic ectotherms: exceptions to the general rule. J Therm Biol 20:61–74
6. Avella M, Berhault J, Bornancin M (1993) Salinity tolerance of two tropical fishes, Oreochromis aureus and *O niloticus* I Biochemical and morphological changes in the gill epithelium. J Fish Biol 42:243–254
7. Avery S (2010) Hydrological impacts of Ethiopias Omo basin on Kenya's Lake Turkana water levels and fisheries. Final Report prepared for the African Development Bank, Tunis
8. Barlow G (2000) The cichlid fishes: the natures grand experiment in evolution. Persuus publishing Cambridge, Massachusetts
9. Bonhomme V, Picq S, Gaucherel C, Claude J (2014) Momocs: outline analysis using R. J Stat Softw 56(13)
10. Cerling TE (1979) Palaeochemistry of Plio-Pleistocene Lake Turkana, Kenya. Palaeogeogr Palaeoclimatol Palaeoecol 27:247–285
11. Chakrabarty P (2005) Testing conjectures about morphological diversity in cihlids of Lakes Malawi and Tanganyika. Copeia 2:359–373
12. Claude J (2008) Morphometrics with R. Springer, New York
13. Collyer ML, Novak JM, Stockwell CA (2005) Morphological divergence of native and recently established populations of white sand pupfish (*Cyprinodon Tularosa*). Copeia 2005(1):1–11
14. Collyer ML, Stockwell CA, Adams DC, Reiser MH (2007) Phenotypic plasticity and contemporary evolution in introduced populations: evidence from translocated populations of sand pupfish (*Cyprinodon Tularosa*). Ecol Res 22:902–910
15. Crul RCM (1995) Limnology and hydrology of Lake Victoria, Comprehensive and Comparative Study of Great Lakes. UNESCO/IHP-IV Project M-5.1.
16. Day T, Pritchard J, Schluter D (1994) A comparison of two sticklebacks. Evolution 48:1723–1734
17. Dryden IE (2013) Shapes: Statistical shape analysis. R package version 1.1–6. Retrieved on march 2006 from http://www.maths.nottingham.ac.uk/~ild/shapes
18. Dryden IE, Mardia KV (1998) Statistical Shape Analysis. Wiley, Chichester
19. El-Sayed AFM (1999) Alternative dietary protein sources for farmed tilapia, *Oreochromis* spp. Aquaculture 179:149–168
20. Farmer GJ, Beamish FWH (1969) Oxygen consumption of *Tilapia nilotica* in relation to swimming speed and salinity. Z Fish Rex Bd Canada 26(11):2807–2821
21. Fryer G, Isles TD (1972) The cichlid fishes of the Great Lakes of Africa. Oliver and Boyd, Edinburgh
22. Fujimura K, Okada N (2008) Shaping of the lower jaw bone during growth of Nile tilapia *Oreochromis niloticus* and Lake Victoria cichlid *Haplochromis chilotes*: a geometric morphometric approach. Develop Growth Differ 50:653–663
23. Gagnaire PA, Normandeau E, Pavey SA, Bernatchez L (2013) Mapping phenotypic, expression and transmission ratio distortion QTL using RAD markers in the Lake Whitefish (*Coregonus clupeaformis*). Mol Ecol 22:3036–3048
24. Georgakopoulou E, Sfakianakis DG, Kouttouki S, Divanach P, Kentouri M, Koumoundouros G (2007) The influence of temperature during early life on phenotypic expression at later ontogenetic stages in sea bass. J Fish Biol 70(1):278–291
25. Grammer GL, Slack WT, Peterson MS, Dugo MA (2012) Nile tilapia *Oreochromis niloticus*(Linnaeus, 1758) establishment in temperate Mississippi, USA: multi-year survival confirmed by otolith ages. Aquat Invasions 7(3):367–376
26. Greenwood PH (1974) The cichlid fishes of Lake Victoria, East Africa: the biology and evolution of a species flock. Bull Br Mus Nat Hist 6:1–134

27. Greenwood PH (1984) African cichlids and evolutionary theories. In: Echelle AA, Kornfield I (eds) evolution of fish species flocks. University of Maine at Orono press, Maine
28. Hay RL (1976) Geology of Olduvai gorge. University of Carlifornia Press, Bakely
29. Herler J, Lipej L, Makovec T (2007) A simple technique for simple imaging of live and preserved small fish specimens. Cybium 31:39–44
30. Hopson AJ (1982) Lake Turkana; A report on the findings of the Lake Turkana Project 1972 1975, vol 16. Overseas Development Administration, London
31. Imre I, McLaughlin RL, Noakes DLG (2002) Phenotypic plasticity in brook charr: changes in caudal fin induced by water flow. J Fish Biol 61(5):1171–1181
32. Kerschbaumer M, Sturmbauer C (2011) The utility of geometric morphometrics to elucidate pathways of cichlid fish evolution. Int J Evol Biol 2011:8. doi:10.4061/2011/290245
33. Mantel N (1967) The detection of disease clustering and a generalized regression approach. Cancer Res 27(2):209–220
34. Meyer A (1987) Phenotypic plasticity and heterochrony in *Cichlasoma managuense* (Pisces, Cichlidae) and their implications for speciation in cichlid fishes. Evolution 41:1357–1369
35. Meyer A, Kocher TD, Basasibwaki P, Wilson AC (1990) Monophyletic origin of Lake Victoria cichlid fishes suggested by mitochondrial DNA sequences. Nature 347:550–553
36. Ndiwa CT, Nyingi D, Agnèse JF (2014) An important natural genetic resource of *Oreochromis niloticus* (Linnaeus, 1758) threatened by aquaculture activities in Loboi drainage, Kenya. PLoS One 9(9):e106972
37. Nelson JS (1994) Fishes of the world, 3rd edn. Wiley, New York
38. Nyingi D, De-Vos L, Aman R, Agnèse JF (2009) Genetic characterization of an unknown and endangered native population of the Nile tilapia *Oreochromis niloticus* (Linnaeus, 1758) (Cichlidae; Teleostei) in the Loboi Swamp (Kenya). Aquaculture 297:57–63
39. Ochumba PBO (1990) Massive fish kills within the Nyanza Gulf of Lake Victoria. Kenya Hydrobiol 208:93–99
40. Omondi R, Kembenya E, Nyamweya C, Ouma H, Machua SK, Ogari Z (2014) Recent limnological changes and their implication on fisheries in Lake Baringo, Kenya. J Ecol Nat Environ 6(5):154–163
41. Owen RB, Crossley R, Johnson TC, Tweddle D, Kornfield I, Davison S, Eccles DH, Engtrom DE (1990) Major low levels of Lake Malawi and their implecation for speciation rates in cichlid fishes. Proc R Lond B 240:519–553
42. Padis E (2012) Analysis of phylogenetics and evolution with R, Second edn. Springer, New York
43. Padis E, Claude J, Strimmer K (2004) APE: analysis of phylogenetics and evolution in R language. Bioinformatics 20:289–290
44. Peterson MS, Slack WT, Woodley CM (2005) The occurrence of nonindigenous Nile Tilapia, *Oreochromis niloticus* (Linnaeus) in coastal Mississippi: ties to aquaculture and thermal effluent. Wetlands 25:112–121
45. Price TD, Qvarnstrom A, Irwin DE (2003) The role of phenotypic plasticity in driving genetic evolution. Proc R Soc Lond B 270:1433–1440
46. Ramler D, Mitteroecker P, Shama LNS, Wegner KM, Ahnelt H (2014) Nonlinear effects of temperature on body form and developmental canalization in the threespine stickleback. J Evol Biol 27:497–507
47. Rohlf FJ (2010) *tpsDig, digitize landmarks and outlines, version 2.16*. Department of Ecology and Evolution, State University of New York at Stony Brook
48. Rohlf FJ (2012) *tpsUtil, file utility program, version 1.53*. Department of Ecology and Evolution, State University of New York at Stony Brook
49. Rohlf FJ, Marcus LF (1993) A revolution in morphometrics. Trends Ecol Evol 8:129–132

50. Rousset F (2008) Genepop 007: a complete re-implementation of the genepop software for Windows and Linux. Mol Ecol Resour 8:103–106
51. Sage RD, Loiselle PV, Basasibwaki P, Wilson AC (1984) Molecular versus morphological change among cichlid fishes of Lake Victoria. In: Echelle AA, Kornfield I (eds) Evolution of fish species flocks. University of Maine at Orono press, Orono
52. Salzburger W, Meyer A, Baric S, Verheyen E, Sturmbauer C (2002) Phylogeny of the Lake Tanganyika Cichlid species flock and its relationship to the central and East African Haplochromine cichlid fish faunas. Syst Biol 51(1):113–135
53. Schluter D (2000) The ecology of adaptive radiation. Oxford University Press, Oxford
54. Schluter D, McPhail JD (1993) Character displacement and replicate adaptive radiation. Trends Ecol Evol 8:197–200
55. Schofield CL (1976) Acid precipitation: effects on fish. Ambio 5(5–6):228–230
56. Schofield PJ, Peterson MS, Lowe MR, Brown-Peterson NJ, Slack WT (2011) Survival, growth and reproduction of nonindigenous Nile tilapia (Oreochromis niloticus (Linnaeus, 1758). I Physiological capabilities to various temperatures and salinities. Mar Freshw Res 62:439–449
57. Sfakianakis DG, Leris I, Laggis A, Kentouri M (2011) The effect of rearing temperature on body shape and meristic characters in zebra fish (*Danio rerio*) juveniles. Environ Biol Fish. doi:10.1007/s10641-011-9833-z
58. Stewart TA, Albertson RC (2010) Evolution of unique predatory feeding apparatus: functional anatomy, development and a genetic locus for jaw laterality in Lake Tanganyika scale-eating cichlids. BMC Biol. doi:10.1186/1741-7007-8-8
59. Takahashi R, Watanabe K, Nishida M, Hori M (2007) Evolution of feeding specialization in Tanganyikan scale-eating cichlids: a molecular phylogenetic approach. BMC Evol Biol 7:195. doi:10.1186/1471-2148-7-195
60. Trewavas E (1983) Tilapiine fishes of the genera *Sarotherodon*, Oreochromis and Danakilia. British Museum (Natural History), London
61. Turner GF, Seehausen O, Knight ME, Allender C, Robinson RL (2001) How many species of cichlid fishes are there in African lakes? Mol Ecol 10(3):793–806
62. Welcomme RL (1967) Observations on the biology of Introduced species of Tilapia in Lake Victoria. Revue Zool Bot Afr 76(3–4):249–279
63. Wurts WA, Durborow RM (1992) Interaction of pH, carbon dioxide, alkalinity and hardness in fish ponds. SRAC Publication No 464

# CHAPTER 16

# Movements of Brown Smoothhounds, *Mustelus Henlei*, in Tomales Bay, California

**Brent R. Campos, Maxfield A. Fish, Gardner Jones, Rebecca W. Riley, Peter J. Allen, Peter A. Klimley, Joseph J. Cech Jr & John T. Kelly**

*Department of Wildlife, Fish, and Conservation Biology, University of California, Davis, One Shields Avenue, Davis, CA 95616, USA*

*Department of Wildlife, Humboldt State University, 1 Harpst St, Arcata, CA 95521, USA*

*Department of Zoology, University of Manitoba Winnipeg, Z320 Duff Roblin Building, Winnipeg, MB R3T 2N2, Canada*

*Department of Natural Resources Conservation, University of Massachusetts, 204 Holdsworth Hall, Amherst, MA 01003, USA*

## ABSTRACT

Ultrasonic telemetry was used to analyze the effects of environmental variables on movement directions and movement rates of brown smoothhounds, *Mustelus henlei*, in Tomales Bay, California. Ultrasonic transmitters were surgically implanted in the peritoneal cavities of one male and five female brown smoothhounds and tracked during the period of 29 June to 15 July 2004. Coarse-scale tracking consisted of locating all tagged individuals multiple times during a single session, while fine-scale tracking consisted of following a single individual continuously during a session. Coarse-scale tracking suggested movement toward the inner bay with incoming and high tides and toward the outer bay with outgoing and low tides (P = 0.01), whereas the diel cycle had no apparent effect on their movement directions. Mean shark movement rate was 0.09 m $s^{-1}$ (range: 0.01–0.34 m $s^{-1}$), with diel and tidal cycles both having significant effects on their rates of movement (P = 0.02 and $P < 0.01$), respectively. We tracked two female sharks on a fine scale over three tracking sessions in July 2004. Both individuals exhibited higher rates of movement during the night compared to the day ($P < 0.01$). While one shark's rate of movement was not significantly affected by tidal stage, the other's was ($P < 0.001$).

**Keywords:** Shark, *Mustelus henlei,* Telemetry, Movement, Behavior.

## INTRODUCTION

The brown smoothhound (*Mustelus henlei*) is an epibenthic elasmobranch ranging in shallow waters of the east Pacific Ocean from the Gulf of California to Coos Bay, Oregon, in the northern hemisphere, and from Ecuador to Peru in the southern hemisphere (Love 1996). Brown smoothhounds are found on a diversity of substrates along the continental slope, but show a preference for shallow muddy or sandy bays, such as Tomales Bay in northern California, utilizing them for feeding and refuge from larger predators (Love 1996). Throughout their range, Tomales Bay being no exception, brown smoothhounds are known to feed heavily on benthic organisms including crabs, shrimp, and polychaete worms (Haeseker and Cech 1993; Love 1996). Brown smoothhounds remain in bays throughout most of the year, but may leave them during the winter as the waters cool and rains reduce the salinity (Love 1996; Hopkins and Cech 2003).

Prior to this study, two other elasmobranch species had been tracked in Tomales Bay. Coarse-scale, population-based tracking studies indicated that time of day was the most significant influence on the movements of bat ray (*Myliobatus californica*) [Matern et al. 2000], and that movement direction in leopard sharks (*Triakis semifasciata*) was related to tide (Ackerman et al. 2000). Fine-scale, individual-based tracking (Klimley et al. 2005) demonstrated that bat rays exhibit two modes of travel: rapid and directional, and slow and non-directional. While some of the bat rays in Klimley et al. (2005) behaved consistently with the behavioral thermoregulation hypothesis presented by Matern et al. (2000), some did not, emphasizing the importance of being cognizant of individual behavior when interpreting the results of population-based telemetry studies. There have been no past studies of the movement patterns of the brown smoothhound.

Our objective was to relate the directionality and rate of movement of brown smoothhounds to environmental variables using both coarse-scale and fine-scale tracking. Coarse-scale tracking consisted of locating multiple individuals sequentially during an environmental-based sampling period (e.g. day or night, high or low tide), while fine-scale tracking consisted of constantly following a single individual for an extended period while recording the same environmental variables. We reasoned that the movement patterns of brown smoothhounds would resemble those of sympatric leopard sharks because of the overlap in their morphology (Setka and Cech 1994), habitat selection (Love 1996), and food habits (Haeseker and Cech 1993; Webber and Cech 1998). As observed in leopard sharks (Ackerman et al. 2000), we hypothesized that the brown smoothhounds would: (1) move during incoming and high tides towards the inner bay where food would be more available in the submersed littoral zone of the inner bay, and (2) that their rate of movement would be greater during nighttime than during daytime.

# MATERIALS AND METHODS

## Study site

We tracked smoothhound sharks in Tomales Bay, located north of San Francisco Bay in Marin County, California. It is a 20-km-long straight bay, averaging 1.4 km in width and 3 m in depth (Fig. 1). The cumulative area of the bay is ca. 24 km$^2$. The bay has been divided into an inner and outer bay based on physical and chemical characteristics (Smith et al. 1991; Hopkins and Cech 2003). Longitudinal gradients of the physical and chemical properties of the water in Tomales Bay are large relative to lateral and vertical gradients, and vertical stratification is weak (Smith et al. 1991). The water in the shallow inner bay (12 km) has large seasonal and daily shifts in temperature, whereas the deeper outer bay (8 km) is more influenced by oceanic conditions and therefore the water temperature remains more constant (Matern et al. 2000). Exchange of water between the inner and outer bay during the summer is minimal, and the inner bay tends to be 2–6°C warmer and 1–3 ppt more saline than the outer bay at this time of year (Smith et al. 1991). A deep channel exists on the west side of the outer bay and is flanked by a sandbar, whereas the inner bay consists of extensive expanses of intertidal mud flats (Smith et al. 1991).

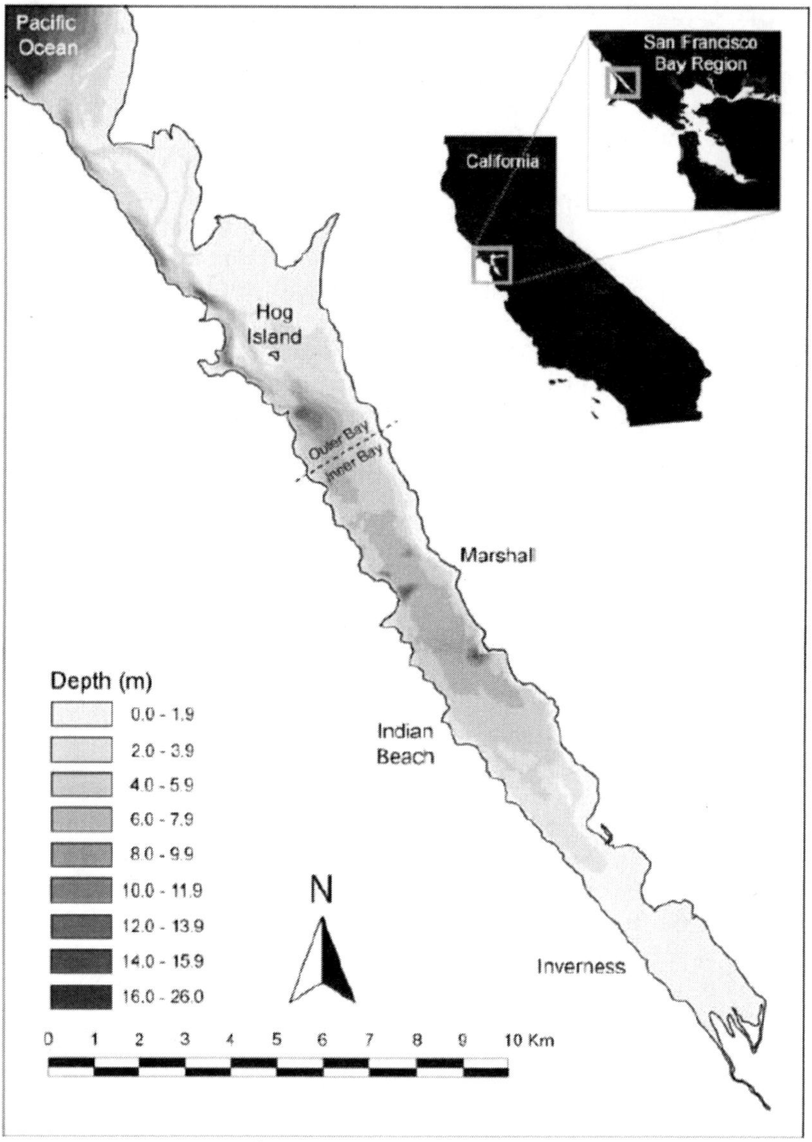


**Figure. 1** Bathymetric map of Tomales Bay and its position in California

## Fish collection

We collected six adult brown smoothhounds, one male and five females, on 24–30 June 2004 (Table 1). The sharks were captured using gillnets (100 m x 3 m, 6 cm stretch mesh) set for 20–60 min perpendicular to the western shoreline of the inner and outer regions of Tomales Bay. We immediately placed captured brown smothhounds into a cooler on the boat with oxygenated sea water until transmitter implantation (5 min-1.5 h).

**Table 1** Capture information for brown smoothhounds tracked in Tomales Bay. (TL = total length, M = male, F = female)

| Date | Shark | Location Captured | Sex | TL (cm) | Weight (kg) | Time of Catch | Time of Release |
|---|---|---|---|---|---|---|---|
| 24 Jun 2004 | BS1 | N. Indian Beach | F | 87 | 2.35 | 14:45 | 15:17 |
| 24 Jun 2004 | BS2 | S. Indian Beach | F | 95 | 3.03 | 14:45 | 16:55 |
| 25 Jun 2004 | BS3 | N. Indian Beach | F | 86 | 2.41 | 15:10 | 15:40 |
| 30 Jun 2004 | BS4 | Hog Island | M | 79 | 1.45 | 7:55 | 8:40 |
| 30 Jun 2004 | BS5 | S. Indian Beach | F | 74 | 1.22 | 12:20 | 13:30 |
| 30 Jun 2004 | BS6 | S. Indian Beach | F | 86 | 2.07 | 12:20 | 14:19 |

## Transmitter implantation

Each shark was anesthetized by immersion in water saturated with $CO_2$, made by mixing 40 ml concentrated acetic acid and 170 g sodium bicarbonate into 20 L of sea water (Peake 1988). Immediately after gill ventilation movements ceased (3–5 min), the shark was removed from the cooler, placed in a supine position on a portable operating table, and gills were irrigated with sea water (alternating, 1-L beakers of water with the anesthetic and sea water). We made a 2.5-cm, axial incision adjacent to the ventral midline and inserted an ultrasonic transmitter (Lotek, CAFT16-1), which was 54-mm long and 16-mm in diameter and weighed 18.9 g in water, in the peritoneal cavity and closed the incision with silk sutures and cyanoacrylate glue. After the surgery (duration <10 min), we recorded the shark's mass, length, and sex (Table 1). Once the shark recovered from anesthesia, it was immediately released at the capture site. Three days were allowed for the individual to recover from the stress of capture and anesthesia before we began tracking.

## Telemetry: coarse-scale

Coarse-scale tracking was conducted over 2.5 weeks from 29 June to 15 July 2004 during day and night and all tidal stages (incoming, high, outgoing, and low). Each shark's transmitter emitted a unique signal, which was detected by a submerged directional hydrophone (Lotek, LHP-1) attached to the boat gunwale. The signal was converted to a radio frequency by an upconverter (Lotek, UUCN-150) and decoded by a radio receiver (Lotek, SRX-400). The receiver recorded which shark was being tracked and digitally stored its identity and its geographic coordinates obtained from a global positioning system (GPS; Garmin, GPSMAP 76S; 5 m accuracy). The range of transmitter detection was ca. 500 m, depending on ambient noise and the bay's surface conditions. When searching for sharks, transects were made 500 m offshore along one side of the bay, submerging the hydrophone to search for the transmitters' signal at 500-m distance intervals. Upon detecting a transmitter's signal, we proceeded toward its source until directly overhead, indicated by the strong and non-directional nature of the transmitter's signal (Winter 1996). We then recorded the transmitter's identity, its position, time of day, and measured five hydrographic variables: depth (with a sounding line), substrate (using indicator clay on a sounding-line weight), and salinity, temperature, and dissolved oxygen at the surface and at ca. 3.3 m depth with a meter and probe (YSI, Model 85). We then left the shark and resumed searching for other individuals. We attempted relocation of each shark multiple times during each sampling session, allowing frequent calculations of movement directions and rates.

## Data analysis: coarse-scale

Brown smoothhound positions obtained during coarse-scale tracking were projected onto a bathymetric map of Tomales Bay, using a 5 m grid digital elevation model. The digital chart was obtained from the GIS Service Center of the Information Technology Division of the California Department of Fish and Game, located in Sacramento, California. The map was displayed on a PC using a geographic information system (GIS) (ESRI, ArcView 3.2), following Klimley et al. (2005).

Tidal stages were determined using tide tables and were split into four stages: incoming, high, outgoing and low. High and low tides were defined as 1 h before and after the extreme high or low. The diel cycle was split into day and night based on the times of sunrise and sunset according to the U.S. Naval Observatory for Point Reyes Station, California. The distance to the ocean (the mouth of Tomales Bay) was calculated for every shark location using the Geofunc Microsoft Excel add-in. Using SigmaStat 3.5 statistical software (Systat Software, Inc.), two-way ANOVAs were employed to analyze the effects of the diel and tidal cycles on shark distance-to-ocean (ie. movement direction), followed by a Tukey pairwise multiple comparison test when significant differences were detected. To boost our sample size per treatment for the distance-to-ocean analyses we grouped tidal-stage shark positions into high/incoming and low/outgoing stages.

Shark movement rates were calculated by dividing distance traveled between successive positions by the time interval between position readings. The movement rate data was analyzed in cm $sec^{-1}$ after a natural-log transformation, but reported in the text as m $sec^{-1}$ to aid comprehension. Two-way ANOVAs were then used to analyze differences in movement rates between tidal stages and between diel stages, followed by a Tukey pairwise multiple comparison test when significant differences were detected.

## Telemetry: fine-scale

In fine-scale tracking, we continuously tracked a single shark (BS2 or BS3 depending on the tracking session) for two 20–21 h intervals and one 4-h interval over a period from 1–13 July 2004. Tracking intervals were chosen to include as many combinations of tidal and diel stages as possible. After locating our target shark for the tracking session, we positioned our boat directly above it and followed it continuously, recording the time of day and the geographical coordinates every 6–15 min with a few exceptions (i.e. the shark could not be detected or we returned to the dock for refueling/crew exchange). During fine-scale tracking sessions, the same five hydrographic variables were measured every hour.

## Data analysis: fine-scale

Brown smoothhound positions recorded during fine-scale tracking were imported into a GIS (ESRI, ArcView 3.2) in the same manner described for coarse-scale analysis. In order to describe the general patterns of fine-scale brown smoothhound movements, consecutive points were selected from each track and classified by polygraphic shape: points in no apparent order, but separated, were classified as continuous, non-uniform swimming (CS) movement; points that were linear were classified as straight-line swimming (SS) movement; points that were very close to one another ($<10$ m $h^{-1}$) were classified as non-movement (NM). We used software (Oriana 3) to calculate circular statistics that quantified the directionality of the polygraphic shapes in each tracking session, indicated by the coefficient of angular concentration (Rayleigh's 'r'). An r value of 1 indicates maximum directionality, whereas 0 indicates omni-directionality. Furthermore, Rayleigh's uniformity test was used to determine whether the distributions of directional movements were significantly non-uniform. Because of our low, fine-scale tracking sample size (N =2), we did not analyze the effects of the tidal and diel cycles on the distribution of BS2 and BS3 as we did with our coarse-scale tracking data.

Our methods for computing fine-scale rates of movement and the statistical analyses of these data were the same as described above for the coarse-scale analysis, except that the fine-scale data could not be transformed to fit a normal distribution. Thus, a Kruskal-Wallis ANOVA on ranks was used, followed by Dunn's pairwise multiple comparison test when significant differences were detected. The data from tracks BS3-A and -B were combined before analysis.

# RESULTS

## Coarse-scale tracking

We gathered a total of 121 locations for all six smoothhounds during coarse-scale tracking. Whereas there was no apparent pattern to the sharks' movement directions based on the diel cycle (Fig. 2a), a pattern did exist for the tidal cycle (Fig. 2b). Mean shark positions were located significantly further from the mouth of the bay during high/incoming tides than during low/outgoing tides (P = 0.014) (Table 3). Though, BS6 was the only individual to exhibit this directional movement at a significant level in Tukey multiple pairwise comparisons (P = 0.01; Table 4).

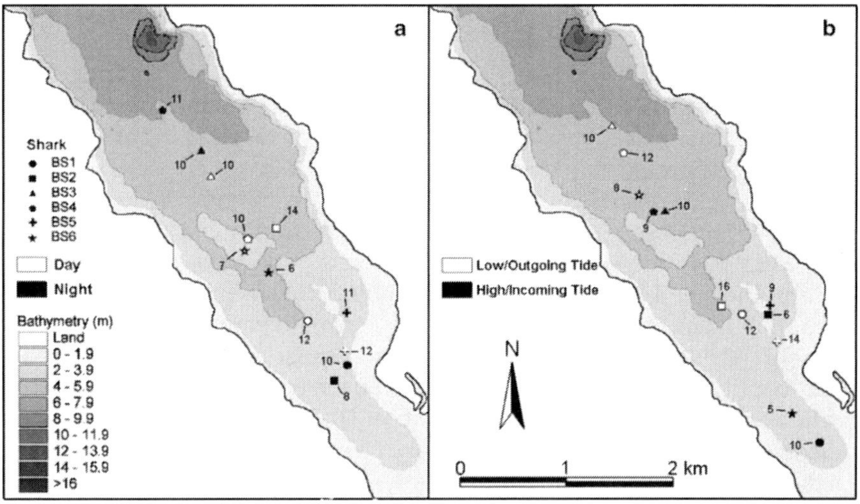


**Figure. 2** Mean locations for individual sharks through the diel (on the left; part a) and tidal (on the right; part b) cycles overlain on a bathymetric map of a section of Tomales Bay. The sample size for each symbol is indicated on the map

**Table 2** Two-way ANOVA results from analyses of coarse-scale tracking of brown smoothhound movements in Tomales Bay. Rate of movement (ROM) data were converted to cm $s^{-1}$then natural-log transformed for the analyses. Means reported are least square means. Bold text indicates a statistically significant result.

| Tracking Period | N | Mean ROM ($m\ s^{-1}$) | Least Squares Mean [ln(ROM $cm\ s^{-1}$)] | F-statistic | P-value |
|---|---|---|---|---|---|
| Day | 39 | 0.073 | 1.54 | 5.331 | **0.024** |
| Night | 51 | 0.104 | 2.00 | | |
| High Tide | 12 | 0.090 | 2.00 | 5.28 | **0.003** |
| Incoming Tide | 12 | 0.044 | 1.30 | | |
| Low Tide | 17 | 0.052 | 1.28 | | |
| Outgoing Tide | 47 | 0.117 | 2.19 | | |
| Entire | 88 | 0.091 | | | |

**Table 3** Two-way ANOVA results from analyses of coarse-scale tracking of brown smoothhound movements in Tomales Bay. Rate of movement (ROM) data were converted to cm $s^{-1}$then natural-log transformed for the analyses. Means reported are least square means. Bold text indicates a statistically significant result.

| Tracking Period | N | Mean Distance to Ocean (km) | Least Squares Mean (km) | F-statistic | P-value |
|---|---|---|---|---|---|
| Day | 66 | 14.65 | 14.62 | 0.002 | 0.962 |
| Night | 55 | 14.56 | 14.61 | | |
| High/Incoming Tide | 44 | 14.93 | 14.93 | 6.174 | **0.014** |
| Low/Outgoing Tide | 77 | 14.42 | 14.33 | | |
| High Tide | 17 | 15.17 | Not Tested | | |
| Incoming Tide | 28 | 14.80 | | | |
| Low Tide | 17 | 14.72 | | | |
| Outgoing Tide | 59 | 14.32 | | | |
| Entire | 121 | 14.61 | | | |

The sharks' movement rates ranged over 0.01–0.34 m $sec^{-1}$, with an overall mean of 0.09 m $sec^{-1}$. The sharks exhibited significantly different rates of movement in night (mean = 0.10 m $sec^{-1}$) vs. day (mean = 0.07 m) (P = 0.024) and across the tidal cycle (P = 0.003) (Table 2). However, the only significant difference in movement rates between tidal stages occurred between outgoing and low tides (P = 0.008) and outgoing and incoming tides (P = 0.029) (Table 5). Movement rates between all other tidal combinations were statistically indistinguishable (Table 5). Due to the time gap between positions during coarse-scale tracking, these data represent minimal movement rates.

**Table 4** Tukey's test results for the analysis of brown smoothhound positions in the Tomales Bay relative to the tidal cycle. Means reported are least square means. Bold text indicates a statistically significant result. Overall P-value = 0.014. (H/I = high/incoming tides, L/O = low/outgoing tides)

| Shark | Mean Distance To Ocean (km) | | q-statistic | P-value |
|---|---|---|---|---|
| | H/I | L/O | | |
| BS1 | 15.84 | 14.94 | 2.388 | 0.094 |
| BS2 | 15.56 | 15.03 | 1.06 | 0.455 |
| BS3 | 14.14 | 13.14 | 2.538 | 0.076 |
| BS4 | 13.78 | 13.27 | 1.26 | 0.375 |
| BS5 | 15.44 | 15.62 | 0.493 | 0.728 |
| BS6 | 15.85 | 13.98 | 3.692 | **0.01** |

**Table 5** Tukey's test results for the coarse-scale rate of movement analysis through the tidal cycle. Overall P-value = 0.003. Bold text indicates a statistically significant result. (O = outgoing tide, L = low tide, H = high tide, I = incoming tide)

| Comparison | q-statistic | P-value |
|---|---|---|
| O vs. L | 4.676 | **0.008** |
| O vs. I | 4.028 | **0.029** |
| O vs. H | 0.857 | 0.930 |
| H vs. L | 2.783 | 0.211 |
| H vs. I | 2.504 | 0.297 |
| I vs. L | 0.089 | 1.000 |

Hydrographic gradients we measured were small throughout the study site for the duration of our study. Dissolved oxygen levels (mean: 8.5 mg $O_2$ $L^{-1}$) were quite homogeneous throughout the bay with the lowest reading (7.5 mg $O_2$ $L^{-1}$) being well above stressful levels (Davis 1975; Lai et al. 1990). Similarly, temperature (mean = 19.2°C, range = 17.7–19.9°C), salinity (mean = 34.2 ppt, range = 33.8–34.4 ppt), and substrate composition (95% of measurements were mud) all showed minimal variation within the bay. However, because we only measured these variables in waters used by the sharks and not what was available but not used, we cannot draw any correlations between the movements of the sharks and these hydrographic variables.

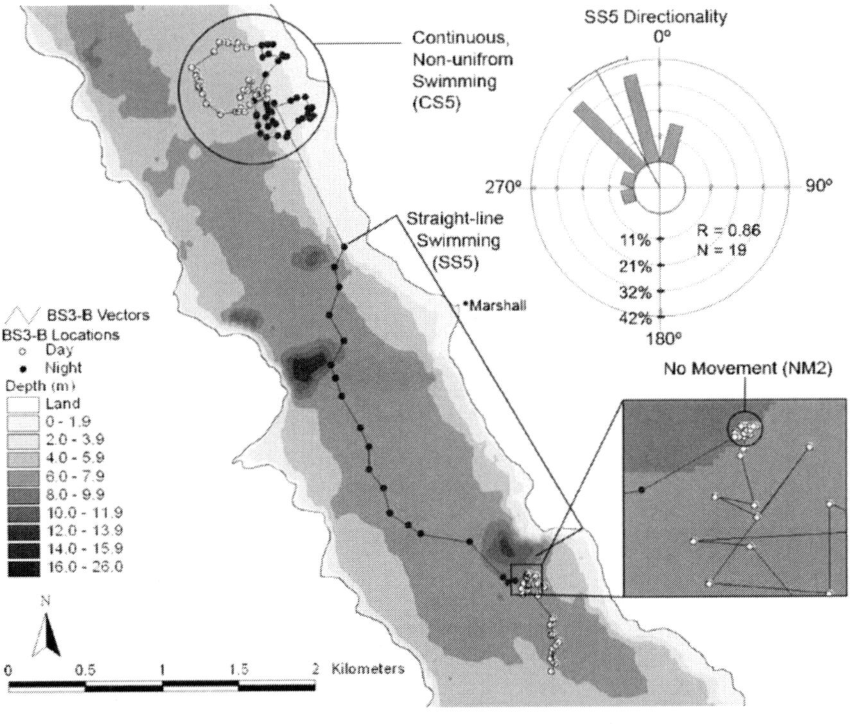


**Figure. 3** A fine-scale track of brown smoothhound BS3 in Tomales Bay. Dark circles designate nighttime positions; white circles daytime positions. This figure illustrates all three swimming classifications exhibited during our fine-scale tracking: 1) continuous, non-uniform swimming (CS), 2) straight-line swimming (SS), and 3) no movement (NM). The rectangular inset shows an expanded view of NM2 when the shark appeared to be motionless. The shark first exhibited Straight-line Swimming (SS5) at sunset during the change of the tide to outgoing. The inlaid circular histogram indicates strong directionality during SS5. The shark performed continuous, non-uniform swimming (CS5) from night to day

**Table 6** Directionality of movements for individual brown smoothhounds tracked on a fine scale based on classification of movement polygons. Bold text indicates a statistically significant result. (A = Day 1, B = Day 2, CS = continuous, non-uniform swimming, SS = straight-line swimming, NM = non-movement, r = Rayleigh's coefficient of angular concentration, P = probability of uniform distribution)

| Shark and Track | Tracking Period | Time Period | N | r | P |
|---|---|---|---|---|---|
| BS2 | Entire | 13:24–9:13 | 113 | | |
| | SS1 | 13:38–14:42 | 7 | 0.82 | **0.005** |
| | CS1 | 14:53–16:50 | 12 | 0.14 | 0.79 |
| | CS2 | 16:57–1:54 | 22 | 0.08 | 0.88 |
| | SS2 | 4:51–6:32 | 6 | 0.61 | 0.11 |
| | NM1 | 7:33–9:13 | 9 | 0.19 | 0.73 |
| BS3-A | Entire | 10:15–14:11 | 29 | | |
| | CS3 | 10:21–11:51 | 14 | 0.11 | 0.84 |
| | SS3 | 11:59–11:41 | 14 | 0.3 | 0.30 |
| BS3-B | Entire | 14:54–11:46 | 201 | | |
| | SS4 | 15:01–16:19 | 12 | 0.53 | **0.029** |
| | CS4 | 16:38–19:10 | 29 | 0.13 | 0.61 |
| | NM2 | 19:16–20:37 | 13 | 0.19 | 0.65 |
| | SS5 | 20:47–23:28 | 19 | 0.86 | **< 0.001** |
| | CS5 | 1:37–11:46 | 80 | 0.05 | 0.85 |

## Fine-scale tracking

Two brown smoothhounds were tracked on a fine scale over three tracking periods. BS3 was tracked from 10:15 to 14:11 on 1 July 2004 and from 14:54 on 7 July to 11:46 on 8 July 2004. BS2 was tracked from 13:24 on 12 July to 09:13 on 13 July 2004. Both sharks displayed the three patterns previously defined: CS, SS, NM. Figure 3 shows the fine-scale track of BS3 in its entirety, exhibiting examples of all three patterns of movement. We determined the degree of directionality to the segments of the tracks that we segregated into the three categories (Table 6). As expected, no directionality was indicated for the segments of the track exhibiting little movement while more directionality was exhibited while actively swimming. Shark BS2 exhibited no significant directionality to NM1, CS1, and CS2 (Table 6). The same shark showed a significant degree of directionality during SS1 and near-significant directionality during SS2. Similarly, the r-values for SS1 and SS2 were considerably higher than those of NM1, CS1, CS2. BS3 showed no significance in directionality for NM2,

CS3, CS4, CS5, or SS3. However, BS3 did show significant directionality during SS4 and SS5. Furthermore, SS4 and SS5 had greater r-values than CS4 or CS5.

Both BS2 and BS3 displayed movement rates drastically higher for nighttime than during daytime, but only BS2 exhibited differences between tidal stages (Table 7). For BS2, movement rates averaged 0.18 and 0.12 m $sec^{-1}$ for nighttime and daytime, respectively. For BS3, movement rates averaged 0.20 m $sec^{-1}$ and 0.10 m $sec^{-1}$ for nighttime and daytime, respectively. BS2's mean movement rates during outgoing and low tides were near equal at 0.19 m $sec^{-1}$ and 0.19 m $sec^{-1}$, and both were significantly higher than those during incoming tide (O vs. I: Diff of ranks = 28.97, Q = 4.834, $P < 0.05$; L vs I: Diff of ranks = 27.81, Q = 3.805, $P < 0.05$; O vs. L: Diff. of ranks = 1.16, Q = 0.18, $P > 0.05$). Only one location of BS2 was obtained during high tide, thus, high tide was excluded in the analysis of its movement rates. Unlike BS2, BS3's rates of movement did not differ among tidal stages. Only two locations of BS3 were obtained during low tide, thus, low tide was excluded in the analysis of its rates of movement.

**Table 7** Kruskal-Wallis ANOVA on ranks results for the fine-scale tracking of sharks BS2 and BS3 in Tomales Bay. Due to low sample sizes, high tide was excluded in the analysis of movement rates of BS2 and low tide was excluded in the analysis of movement rates of BS3. Bold text indicates a statistically significant result. (A = Day 1, B = Day 2, ROM = rate of movement)

| Shark | Tracking Period | N | Mean ROM (m/s) | Median ROM (m/s) | H-statistic | P-value |
|---|---|---|---|---|---|---|
| BS2 | Day | 41 | 0.12 | 0.08 | 6.875 | **0.009** |
| | Night | 36 | 0.18 | 0.16 | | |
| | High Tide | 1 | 0.01 | 0.01 | | |
| | Incoming Tide | 20 | 0.05 | 0.01 | 25.36 | < **0.001** |
| | Low Tide | 16 | 0.19 | 0.14 | | |
| | Outgoing Tide | 40 | 0.19 | 0.16 | | |
| BS3-A&B | Day | 106 | 0.10 | 0.08 | 38.176 | < **0.001** |
| | Night | 61 | 0.20 | 0.18 | | |
| | High Tide | 30 | 0.13 | 0.08 | 1.144 | 0.726 |
| | Incoming Tide | 97 | 0.12 | 0.11 | | |
| | Outgoing Tide | 38 | 0.16 | 0.11 | | |
| | Low Tide | 2 | 0.14 | 0.14 | | |

BS2 was located over substrates consisting of mud (85% of samples), gravel/algae (2%), mud/algae (5%), gravel (5%), algae (2%) and sand (1%) in water column depths of 2.6–4.7 m. At the surface, the salinity recorded over the shark was relatively constant at 34 ppt, while dissolved oxygen ranged over 89.2–99.5% of air saturation, and temperature ranged over 19.3–20.1°C. At 3.3 m depth, salinity stayed constant at 34 ppt, while dissolved oxygen saturation fluctuated over 88.9–99.8%, and temperature ranged over19.4–20.1°C.

BS3 spent 95% of its time over a muddy substrate and 5% of its time over a sand/clay substrate in depths ranging from 2.0 to 9.6 m. At the surface, salinity was 34 ppt, dissolved oxygen fluctuated over 95.4–101.7% of air saturation, and temperature ranged over 16.5–19.6°C. At 3.3 m below the surface, salinity was 34 ppt, dissolved oxygen fluctuated over 95.4–107.0%, and temperature ranged over 16.5–19.4°C.

## DISCUSSION

Our results from these two, distinct methods of tracking elucidated the complex movement patterns of brown smoothhounds in Tomales Bay, California. For coarse-scale tracking data, the tidal cycle appeared to be the dominant factor affecting their distribution, whereas both the diel and tidal cycles affected their movement rates. At the individual level, (i.e. fine-scale tracking) the diel cycle affected the movement rates of both BS2 and BS3, while the tidal cycle influenced the movement rate of only BS2. Their movements were contained within waters with weak temperature, salinity, and dissolved oxygen gradients typical of the inner bay in summer (Smith et al. 1991). Tracking studies of other elasmobranchs in Tomales Bay also endeavored to relate movements to diel and tidal cycles. Matern et al. (2000) found a distinct periodicity to bat rays' movements between the inner bay and the outer bay, hypothesized to be related to the large temperature gradient between the two areas, but their movement rates did not correlate with either diel or tidal cycles. In contrast, Ackerman et al. (2000) found that leopard sharks showed strong movements with tides and increased their movement rates at night compared to day. As hypothesized, the brown smoothhounds displayed similar movement patterns to those of leopard sharks. Interestingly, brown smoothhound and leopard sharks often associate with the same substrates (Love 1996; Ackerman et al. 2000), have similar diet compositions (Haeseker and Cech 1993; Webber and Cech 1998), and share almost identical blood and muscle characteristics (Setka and Cech 1994).

Sharks were located farthest toward the inner bay during high and incoming tides. These movements into freshly flooded mud flats on incoming and high tides are consistent with what is known about their foraging ecology. Brown smoothhounds are known to feed heavily on benthic organisms including crabs, shrimp, and polychaete worms (Haeseker and Cech 1993; Love 1996), many of which are widely available in the tidal mud flats when they are flooded. As the sharks moved inward and outward they generally swam in shallow water, stayed near shore, and may have been foraging. Alternatively their behavior also may reflect predator avoidance (e.g., minimizing contact with larger sharks). Sevengill sharks are present in the bay and they grow to be twice as long as brown smoothhounds, which have been recorded as their prey (Love 1996).

The similarities in the body design and feeding habits of smoothhounds to other sharks that move with tides supports our findings that brown smoothhound sharks also may move in relation to tidal stage. In temperate, estuarine habitats on the eastern U.S. coast, the dusky shark and the sandbar shark moved predominately in the direction of tidal currents (Huish and Benedict 1978; Medved and Marshall 1983). Ackerman et al. (2000) explained the tidally influenced movement on leopard sharks as a means of reducing energy expenditure. Our findings suggest that brown smoothhounds' movements in Tomales Bay are associated with tide to some degree, but not to the degree of the leopard shark. The movement rates through the tidal cycle observed in this study are more difficult to interpret.

From the coarse-scale data, smoothhounds exhibited a maximum rate of movement on an outgoing tide, mid-range rates of movement during high tide, and minimum rates of movement during incoming and low tide. The observed behavior may perhaps be explained by their foraging ecology. The mid-range rates of movement exhibited during high tides may reflect active foraging behavior over mud flats. During an outgoing tide it may be beneficial to a shark to rapidly move off these flats to avoid being stranded. At low tide brown smoothhounds have generally retreated to deeper water where they may await their next foraging opportunity while digesting their last meal, thus exhibiting slower rates of movement. Looking at the individual level, both sharks we tracked exhibited some patterns not clear at the population level. Although fine-scale tracking likely yielded our most accurate movement rates, our sample sizes across the tidal spectrum were inadequate to fully interpret those results. However, across the diel cycle, our sample size doubled per treatment as treatment variables halved versus the tidal cycle (i.e. day and night), permitting comparison to the coarse-scale results.

As we expected, at a coarse-scale, our brown smoothhounds showed a significant increase in their movement rates at night versus day. At a fine scale, the difference in movement rate across the diel cycle was even more distinct. The sharks showed a marked increase in their movement rates at night, with BS2 moving 50% faster and BS3 moving 100% faster than during the day. These rates were nearly equivalent to those found in leopard sharks by Ackerman et al (2000). Based on our field observations during both tracking modes, the higher resolution of the individual tracking method seems to be the more accurate and precise measure of the smoothhounds' actual movement rates over the diel cycle.

## Movement polygons

Among the movement patterns obtained from fine-scale tracking, CS was the most common. Our primary hypothesis for the motive behind this movement pattern was foraging. In our field observations, and when plotted on the bathymetric map, continuous non-uniform swimming was generally displayed just off the shoreline and on mudflats where smoothhounds are known to feed (Haeseker and Cech 1993).

Straight-line (SS) swimming was defined as a period of movement in which individuals showed directionality by a high 'r' coefficient of concentration. Using fine-scale tracking, Klimley (1993) described highly directional swimming in scalloped hammerhead sharks *(Sphyrna lewini)* in the Sea of Cortez and in bat rays *(Myliobatis californica)* in Tomales Bay (Klimley et al. 2005). Regarding the cues that brown smoothhounds use for directing its movement, Tomales Bay's high turbidity precludes effective use of vision for directionality, suggesting cues other than vision are at play. Brown smoothhound SS occurred more often during a moving tide, as opposed to a slack tide, in Tomales Bay (SS1, SS2, SS5), perhaps suggesting the use of rheotactic orientation and passive transport. However, Klimley (1993) showed that the heading of a flow-oriented shark would change unless maintained by forward momentum. Klimley et al. (2005) suggested that bat rays swimming in straight-line manner may have been guided by the strong linear magnetic field that runs parallel with the length of Tomales Bay, which was created by a large geological (San Andreas) fault. It is plausible that brown smoothhounds use the same magnetic cues.

The combination of late maturity, low fecundity, and slow growth in most shark species has made them especially vulnerable to over-exploitation, and shark populations around the world are declining at a rapid rate due to overfishing (Musick 1999). In Tomales Bay, female brown smoothhounds are known to birth young throughout the summer (Hopkins and Cech 2003). A better understanding of the brown smoothhounds' movements provides natural resource managers with more information to aid in managing and protecting California's small commercial fishery for brown smoothhounds. While the small scale of this study precludes a detailed assessment of the environmental variables influencing the movements of brown smoothhounds, it is apparent that variables such as tide and diel cycle influence the distribution of the species within inshore habitats. Further large-scale studies are necessary to tease apart the interactions of brown smoothhounds with their environment and to assess the possible impact of anthropogenic activities.

## ACKNOWLEDGEMENTS

We thank the staff of the UC Davis Bodega Marine Laboratory (especially S. McAffee) for logistical support; D. Kratville, N. Quesenberry, J. Graham, S. Feltman, P. Lin, E. Chen, R. McGraw, and J. Phillips for technical assistance; V. deVlaming for manuscript review; and the Department of Wildlife, Fish and Conservation Biology, the Lloyd Swift Scholarship Fund, and the University of California Agricultural Experiment Station (grant no. 3455-H to JJC) for financial support.

## REFERENCES

1. Ackerman JT, Kondratieff MC, Matern SA, Cech JJ Jr (2000) Tidal influence on spatial dynamics of leopard sharks, *Triakis semifasciata*, in Tomales Bay, California. Environ Biol Fishes 58:33–43 doi:10.1023/A:1007657019696
2. Davis JC (1975) Minimal dissolved oxygen requirements of aquatic life with emphasis on Canadian species: a review. J Fish Res Board Can 32:2295–2332
3. Haeseker SL, Cech JJ Jr (1993) Food habits of the brown smoothhound shark (*Mustelus henlei*) from two sites in Tomales Bay. Calif Fish Game 79:89–95
4. Hopkins TE, Cech JJ Jr (2003) The influence of environmental variables on the distribution and abundance of three elasmobranches in Tomales Bay, California. Environ Biol Fishes 66:279–291 doi:10.1023/A:1023907121605
5. Huish MT, Benedict C (1978) Sonic tracking of dusky sharks in the Cape Fear River, North Carolina. J Elisha Mitchell Sci Soc 93:21–26
6. Klimley AP (1993) Highly directional swimming by scalloped hammerhead sharks, *Sphyrna lewini*, and subsurface irradiance, temperature, bathymetry, and geomagnetic field. Mar Biol (Berl) 117:1–22 doi:10.1007/BF00346421
7. Klimley AP, Kelly JT, Kihslinger RL (2005) Directional and non-directional movements of bat rays, *Myliobatis californica*, in Tomales Bay, California. Environ Biol Fishes 74:79–88 doi:10.1007/s10641-005-3997-3
8. Lai NC, Graham JT, Burnett L (1990) Blood respiratory properties and the effect of swimming on blood gas transport in the leopard shark *Triakis semifasciata*. J Exp Biol 151:161–641
9. Love M (1996) Probably more than you want to know about the fishes of the Pacific coast. Really Big Press, Santa Barbara
10. Matern SA, Cech JJ Jr, Hopkins TE (2000) Diel movements of bat rays, *Myliobatis californica*, in Tomales Bay, California: Evidence for behavioral thermoregulation? Environ Biol Fishes 58:173–182 doi:10.1023/A:1007625212099
11. Medved RJ, Marshall JA (1983) Short-term movements of young sandbar sharks, *Carcharhinus plumbeus* (Pisces, Carcharhinidae). Bull Mar Sci 33:87–93
12. Musick JA (1999) Ecology and conservation of long-lived marine animals. Am Fish Sym 23:1–10

13. Peake S (1998) Sodium bicarbonate and clove oil as potential aneshetics for nonsalmonid fishes. N Am J Fish Manage 18:919–924 doi:10.1577/1548-8675(1998)018<0919:SBACOA>2.0.CO;2
14. Setka JD, Cech JJ Jr (1994) Blood and muscle characteristics of leopard shark (*Triakis semifasciata*) and brown smoothhound (*Mustelus henlei*). Calif Fish Game 80:89–96
15. Smith SV, Hollibaugh JT, Dollar SJ, Vink S (1991) Tomales Bay metabolism: C-N-P stoichiometry and ecosystem heterotrophy at the land-sea interface. Estuaries 33:223–257
16. Webber JD, Cech JJ Jr (1998) Nondestructive diet analysis of the leopard shark from two sites in Tomales Bay, California. Calif Fish Game 84:18–24
17. Winter JD (1996) Advances in underwater biotelemetry. In: Murphy BR, Willis DW (eds) Fisheries techniques, second edition. American Fisheries Society, Bethesda, p 732

# CHAPTER 17

# Otolith Elemental Signatures Reflect Residency in Coastal Water Masses

**Mary M. Nishimoto Libe Washburn Robert R. Warner Milton S. Love Georges L. Paradis**

*Marine Science Institute, University of California, Santa Barbara, CA 93106, USA*

*Institute for Computational Earth System Science and Department of Geography, University of California, Santa Barbara, CA 93106, USA*

*Ecology, Evolution, and Marine Biology, University of California, Santa Barbara, CA 93106, USA*

## ABSTRACT

We examined variability in otolith chemistry of wild caught fish in relation to in situ temperature and salinity within the California Current System. Barium, magnesium, and iron from the most recent growth zone in otoliths differentiated pelagic juvenile shortbelly rockfish (*Sebastes jordani*) residing in water masses with distinct temperature and salinity properties from central and southern California spanning nearly 500 km of coastline. The 3-element signature also discriminated fish that resided in different water masses that were associated with mesoscale cyclonic eddy circulation in the Santa Barbara Channel. Variability in otolith chemistry reflected the spatial patterns of both horizontal gradients and vertical gradients in water mass properties related to circulation. Although we found that the concentrations of particular elements in otoliths were correlated to ambient temperature or salinity, we suggest that these parameters are more useful as an identifying signature of distinct water masses associated with unique otolith signatures rather than as factors directly affecting otolith chemistry. Other factors varying among the water masses or among the fish populations residing in the water masses may also affect otolith chemistry. We discuss how oceanographic phenomena associated with the 1997–1998 El Niño and the persistent, recirculating eddy in the Channel may have affected coastal ocean conditions and variation in otolith chemistry of fish in the study area.

**Keywords:** Otolith microchemistry Coastal circulation Water masses Early life history

## INTRODUCTION

Elemental and isotopic assays of fish otoliths have been directed at questions concerning the tracking and mixing of marine and freshwater populations. Inferences about fish homing behavior (Secor et al. 1995; Thorrold et al. 2001), connectivity estimates (Miller and Shanks 2004; Standish et al. 2008), and larval dispersal histories (Swearer et al. 1999; FitzGerald et al. 2004; Hamilton et al. 2008) have been based on otolith chemistry as geographic tags. These applications of otolith chemistry do not necessarily require the determination of how element concentrations, salinity, and temperature in the environment are related to element concentrations in the otolith; what is required is that the chemical assays of fish otoliths discriminate individuals from different locations. However, at least two important limitations in utilizing otolith chemistry to assess the movement of individuals and the mixing of populations among locations stem from not knowing how environmental variability relates to the spatial and temporal variability in otolith chemistry.

The first limitation is the reliability of site-specific otolith elemental signatures over time. Sufficient consistency in the spatial variability of otolith elemental signatures allows discrimination among habitat sites within a year; however, many studies that have compared otolith chemical composition among locations over two or more years have found significant interannual variability (Milton et al. 1997; Dove and Kingsford 1998; Patterson et al. 1999; Campana et al. 2000; Gillanders 2002; Hamer et al. 2003; Miller and Shanks 2004; Brown 2006). This temporal inconsistency is especially problematic if the goal, for example, is to examine a population over time to estimate connectivity via migration or larval dispersal among spatially distinct habitats.

The second limitation is that individuals of unknown residency can only be assigned to the set of sampled locations that define the spatial variability of the signatures. There is an unknown degree of error in assigning fish of unknown residency to a specific site of a set of sampled locations, because fish residing in unsampled locations might share the same elemental signature as fish from the set of sampled locations. This can occur even if the spatial variability of the otolith elemental signatures defined by a set of sampled locations is consistent over time and the elemental signatures are good discriminators of the sites sampled.

Presently, inferences are limited to the geographic scale of sampling, the localities sampled, and the time of sampling within individual studies, because of the uncertainty of changing environmental conditions that affect the variability in otolith chemistry over time and space. We turn our attention from the utility of using otolith chemistry as a location-based discriminator, and focus on the prospect of utilizing otolith chemistry to identify water mass residency. A motivation for this study is that if unique elemental signatures are associated with distinct water masses, then a reference or atlas identifying these relationships might be developed. Otolith chemistry and oceanography might prove useful for reconstructing the transport histories of individuals and estimating the connectivity of populations across broad regions and over time.

The central goal of this study was to test the hypothesis that variability among otolith trace element signatures of wild-caught fish was associated with in situ water masses. To identify distinct water masses, we examined temperature and salinity measured during pelagic juvenile fish sampling off central California in May 1998 (Sakuma et al. 2000) and off the southern California coast in June 1998 (Nishimoto and Washburn 2002) spanning a distance of nearly 500 km (Fig. 1). The otolith signatures of shortbelly rockfish (*Sebastes jordani*) from these surveys were defined by a suite of elements assayed from the outer otolith material grown in the recent days preceding capture. Otolith sampling included fish found concentrated in a persistent cyclonic eddy in the western Santa Barbara Channel (Nishimoto and Washburn 2002; Figs. 1b and 2). We examined the variability in otolith chemistry in relation to temperature and salinity across the study area and the three-dimensional circulation of the eddy.

## MATERIALS AND METHODS

### Water mass sampling and otolith chemistry

Similar methods were used in the central and southern California surveys (detailed in Wyllie Echeverria et al. 1990; Nishimoto and Washburn 2002; Sakuma et al. 2000) to collect the otoliths and oceanographic data for this study. Fish were collected at night with a modified Cobb mid-water trawl

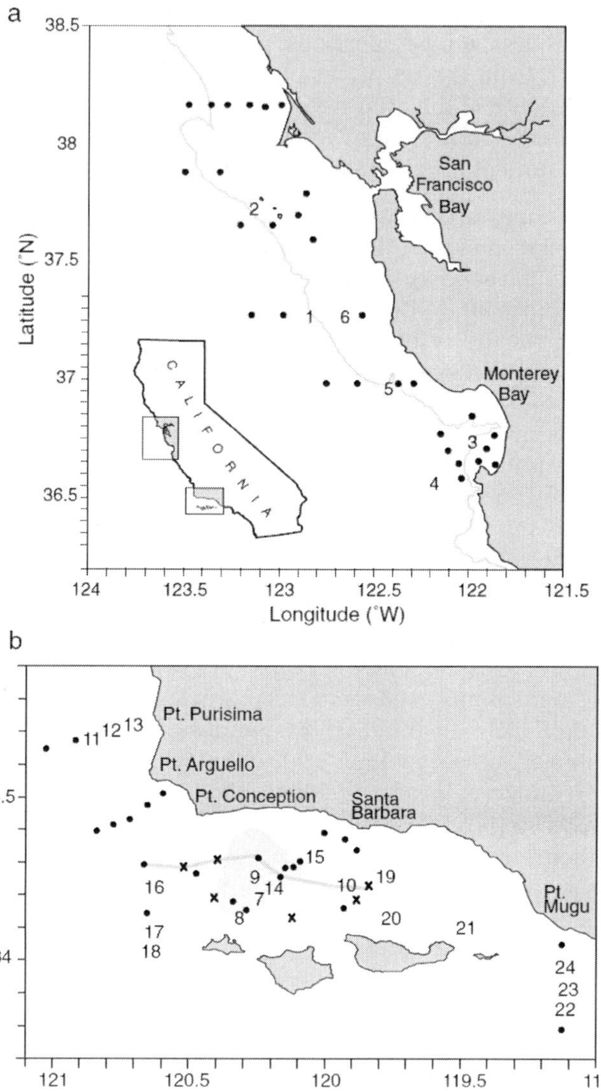


**Figure. 1** The study area. Midwater trawling and oceanographic surveys in late spring 1998 were conducted off (**a**) central California and (**b**) southern California. Shortbelly rockfish specimens and temperature-salinity profiles were collected in central California from stations numbered 1–6, and in southern California from stations 7–24. The eddy (shaded gray area) in the western Santa Barbara Channel was defined by station dynamic height $\phi < 0.37$ (5 dbar/200 dbar) (Nishimoto and Washburn 2002). The closed circles mark stations where fish were absent or no otoliths were sampled

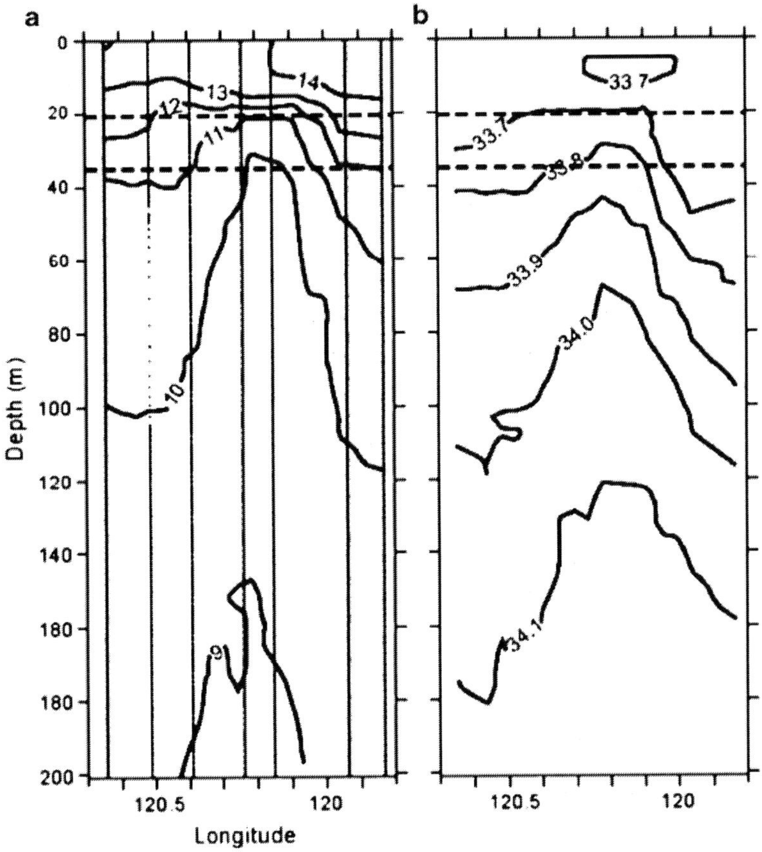


**Figure 2** East-west vertical sections of: (**a**) potential temperature, T (°C); and (**b**) salinity, S (psu), along the section indicated by the line in Fig. 1. Vertical dotted lines in (**a**) indicate T-S profile locations. The cyclonic flow of the eddy produced the large uplift of isotherms and isohalines. Two horizontal dashed lines bound the 14-m trawling stratum. The average headrope depth for all hauls in the Santa Barbara Channel survey was 20 m (from Nishimoto and Washburn (2002))

with a 9 mm codend towed at depth for 15 min at ~5 km $h^{-1}$ covering ~1.5 km (Nishimoto and Washburn 2002). The opening of the net used in both surveys was approximately 10 m wide and 14 m high when trawling at a headrope depth of 20 m. The depth interval for each haul was estimated as the distance from the mean depth, $d_0$, of the headrope (rope across the top of the net opening) to 14 m below $d_0$ (Table 1). Vertical profiles of potential temperature T and salinity S (averaged into 1-m depth bins) were obtained either immediately before or after each haul to at least 200 m or a few meters above shallower bottom depths. A water mass where fish were sampled was defined by the profile within the depth interval fished.

**Table 1** Station code; survey date; net headrope depth $d_0$, m; number of otoliths sampled by LA-ICPMS, N; number of fish measured, (N); size range of fish, mm SL, +/− 1 sd

| Station | Date | $d_0$ | N | (N) | SL |
|---|---|---|---|---|---|
| 1 | 5/11/98 | 75 | 1 | (1) | 12.8 |
| 2 | 5/12/98 | 30 | 3 | (3) | 19.4 ± 4.5 |
| 3 | 5/15/98 | 36 | 1 | (1) | 15.2 |
| 4 | 5/15/98 | 77 | 1 | (1) | 21.7 |
| 5 | 5/17/98 | 36 | 1 | (1) | 18.9 |
| 6 | 5/27/98 | 22 | 4 | (3) | 30.2 ± 4.2 |
| 7 | 6/2/08 | 19 | 8 | (8) | 32.7 ± 9.1 |
| 8 | 6/2/08 | 19 | 1 | (1) | 35.9 |
| 9 | 6/3/98 | 19 | 3 | (3) | 39.0 ± 7.3 |
| 10 | 6/4/08 | 19 | 3 | (3) | 31.4 ± 2.3 |
| 11 | 6/7/98 | 19 | 7 | (7) | 16.3 ± 2.0 |
| 12 | 6/7/98 | 19 | 5 | (5) | 16.8 ± 2.5 |
| 13 | 6/7/98 | 19 | 2 | (2) | 24.6 ± 3.5 |
| 14 | 6/9/98 | 19 | 9 | (8) | 24.1 ± 4.4 |
| 15 | 6/9/98 | 19 | 2 | (2) | 35.6 ± 1.7 |
| 16 | 6/12/98 | 19 | 6 | (6) | 20.3 ± 2.9 |
| 17 | 6/12/98 | 19 | 1 | (1) | 17.8 |
| 18 | 6/12/98 | 19 | 2 | (2) | 16.9 ± 0.6 |
| 19 | 6/13/98 | 19 | 2 | (2) | 26.7 ± 3.5 |
| 20 | 6/13/98 | 19 | 2 | (2) | 38.4 ± 0.4 |
| 21 | 6/13/98 | 19 | 1 | (1) | 47.2 |
| 22 | 6/14/98 | 28 | 1 | (1) | 22.4 |
| 23 | 6/14/98 | 24 | 1 | (1) | 27.4 |
| 24 | 6/14/98 | 21 | 1 | (1) | 20.0 |

Otoliths from 68 specimens of shortbelly rockfish sampled from the central and southern California collections were assayed using laser ablation inductively coupled plasma mass spectrometry (LA-ICPMS) (Fig. 1, Table 1). The fish ranged from 12.8 mm to 47.2 mm SL. Shortbelly rockfish were the most abundant of the pelagic juvenile rockfishes collected in the survey. Adults of this active, schooling species range from southern British Columbia to southern Baja California, and larvae are found as much as 400 km from shore (Love et al. 2002). Fifty-seven specimens represent all areas defined by Nishimoto and Washburn (2002) where shortbelly rockfish were collected during 3–15 June 1998 in southern California (Fig. 1b, Table 1). Eleven otolith samples from central California were collected during 11–27 May 1998 from the Farallon Islands west of San Francisco to Cypress Point, Monterey (stations 1–6 in Fig. 1a and Table 1). The otoliths of one to four specimens per station were assayed (Table 1).

The otolith samples from the two surveys were handled somewhat differently. Specimens from the 1998 southern California survey were subsampled in 2000 from an ETOH-preserved archive, measured, and the sagittae extracted. The fish had been frozen at sea, then thawed and archived within several months after collection. The otoliths were rinsed in deionized water, air dried and stored in plastic bags for several months until the time when each otolith was affixed to a plastic slide using epoxy resin (Epo-Thin, Buehler). The fish collected in the central California survey were frozen at sea. Within several months, the specimens were thawed and measured, and the otoliths extracted, rinsed, dried and affixed to glass slides using a clear nail polish epoxy (Sally Hansen's Hard-as-Nails brand).

All otoliths were polished down to about 15 μm above the center of otolith nucleus using a lapping wheel and 9, 3, and 1 μm 3 M diamond polishing films. After polishing, decontamination steps described by Ruttenberg et al. (2005) and Warner et al. (2005) were performed in a clean laboratory on individually isolated otoliths.

Otolith material produced within 1 week preceding capture was analyzed on a Finnigan MAT Element 2-sector field ICPMS as described in Warner et al. (2005) and Ruttenberg et al. (2005). A VG UV microprobe Nd:YAG (Neodymium-doped Yttrium Aluminum Garnet) 266 nm laser ablation system was outfitted with a helium aerosol carrier system to transfer the sample from the enclosed sample chamber to the ICPMS as described in Zacherl et al. (2003). The laser was set at 0.1 mJ at 3 Hz. The laser was used to ablate three sample spots in the same growth zone near the edge of the anterior rostrum of each otolith with care taken to avoid epoxy that overlaid one or two outermost increments. For the southern California otoliths, run in May 2002, the laser aperture was set to 1 μm which emitted a beam that ablated a pit diameter of about 20 μm. For the central

California otoliths, run in September 2003, the laser aperture was set to 2 μm which ablated a pit diameter of about 30 μm.

Two samples were collected at each of the three spots to assay a suite of isotopes: Calcium 48 ($^{48}Ca$), Strontium 86 ($^{86}Sr$) Barium 138 ($^{138}Ba$) and Lead 208 ($^{208}Pb$) were determined by using low resolution mode ($R = 300$); Magnesium 24 ($^{24}Mg$), Calcium 48 ($^{48}Ca$), Manganese 55 ($^{55}Mn$), Iron 56 ($^{56}Fe$), and Zinc 64 or Zn 66 ($^{64}Zn$ for SBC samples and $^{66}Zn$ for CC samples) were determined in medium resolution mode ($R = 4{,}000$). Eight laser pulses were emitted on a targeted spot to collect each sample. Each targeted spot was pre-ablated with two laser pulses as a precautionary cleaning measure. The isotope intensities of each sample were blank-corrected by subtracting isotope intensities of a 1% nitric acid ($HNO_3$) instrument blank preceding the sample sequence. The plasma conditions were kept similar for both ablated material and solution-based standards by aspirating a 1% nitric acid solution during the ablation process and keeping the gas flows constant. The abundance of an element was expressed as a ratio relative to the amount of calcium to control for the amount of material analyzed per sample spot. A solution standard of known element/Ca ratios was used to correct for instrument bias and convert to molar elemental ratios. Solution-based standards were used in an attempt to match the high calcium matrix of the otoliths. The currently available solid standard materials, such as NIST 612 glass, are very different in composition from the otolith (Pearce et al. 1997). Solid calcium carbonate pellets have been prepared to solve this issue, but there are problems associated with heterogeneity.

Average detection limits for each element per sequence ($n = 20$ sequences), calculated as 3 × SD of the intensity of 1% HNO3 blanks and expressed as ratios of the isotope intensity and mean otolith Ca48 intensity, were 24 $\mu mol \cdot mol^{-1}$ for Mg/Ca, 0.95 $\mu mol \cdot mol^{-1}$ for Mn/Ca, 12.74 $\mu mol \cdot mol^{-1}$ for Fe/Ca, 6.77 $\mu mol \cdot mol^{-1}$ for Zn/Ca, 59 $\mu mol \cdot mol^{-1}$ for Sr/Ca, 0.06 $\mu mol \cdot mol^{-1}$ for Ba/Ca, 0.06 $\mu mol \cdot mol^{-1}$ for Pb/Ca.

We analyzed solid glass standard reference material (NIST 612) at the beginning and end of each workday to check overall analytical precision or repeatability for the entire study. Average percent relative standard deviations ($n = 10$ workdays) were 21% for Mg/Ca, 13% for Mn/Ca, 26% for Fe/Ca, 37% for Zn/Ca, 39% for Sr/Ca, 10% for Ba/Ca, 10% for Pb/Ca. The NIST 612 was ablated using the same method parameters as that used for the otolith samples. The constraint of using small 20 μm and 30 μm spot sizes resulted in low signal to noise for Sr/Ca estimates in the NIST 612 standard. An alternative reproducibility estimate for the study was determined for Sr/Ca by ablating material (three spots) along a randomly selected band of growth increments between the core and the edge of an otolith at the end of each sequence. The average percent relative standard deviation was 8% for Sr/Ca ($n = 19$ sequences).

We excluded negative blank-corrected intensity values from the sample spot dataset. We also excluded outliers from the sample spot dataset. Outliers were identified by examining frequency distribution and normal probability plots of the log-transformed element/Ca sample spot data. The outlier values were at least 1.5 times the interquartile range from the median of the log-transformed data (SPSS, Inc 2002). The exclusions reduced the original dataset of three samples of seven isotopes per otolith to 0–3 samples of each isotope per otolith. The abundance of each element from the otolith edge was estimated from either a single sample spot or the mean of two or three sample spots after the molar ratio dataset was log-transformed (i.e., log(μmol element/mol Ca)). A specimen lacked an abundance estimate for a given element if the readings from all three sample spots at the otolith edge were excluded.

## Relating otolith chemistry to in situ water masses

We examined the T-S distribution within the trawling depth ranges to resolve water mass groups. Principal component analysis (PCA) of T (15 1-m bin averaged variables, $T_{d = 0, 1, 2, ..., 14}$) and S (15 variables, $S_{d = 0, 1, 2, ..., 14}$) profile data over the depth interval of each haul was used to confirm the water mass groups. T and S values from each profile were normalized by subtracting the mean of all profiles and dividing by the standard deviation of all profiles.

Otoliths were assigned to the water mass groups by pairing hauls with vertical T-S profiles. All analyses were performed on log-transformed otolith element ratio data. Prior to the analyses, we assessed the normality and homoscedacity of the otolith data among water mass groups (SPSS, Inc 2002). We evaluated whether fish size was a confounding factor influencing the effect of water mass type on otolith element concentration. Rather than use log(element/Ca) in our analyses when the relationship between an element and fish size was significant (Pearson correlation coefficient r, $\alpha = 0.01$), we used the residual values from the least squares regression of log(element/Ca) as a function of log(SL) to remove the fish size effect on the concentration of an element in otoliths (Systat Software Inc 2006).

We performed one-way analysis of variance (ANOVA) to test for the effect of the water mass group on individual log-transformed element ratios in otoliths ($\alpha = 0.01$) (SPSS, Inc 2002). Although we would have preferred to use a nested ANOVA to account for within-group variability in individual elemental abundances due to station (i.e., water mass defined by the CTD cast data at an individual station) differences, the dataset was unsuitable because 10 of the 24 stations were represented by only one otolith, including all but one of the stations from central California. The Tukey's honestly significant difference (HSD) test was used to detect *a posteriori* differences among means ($\alpha = 0.01$).

We used PCA to examine whether variability among the 7-element signatures of otoliths showed natural groupings by water mass. Otoliths with excluded data from all three sample spots for at least one of the seven elements were not included in the PCA.

To test whether the elemental signature of otoliths identified water mass membership, we used canonical discriminant function analysis (DFA) (SPSS Inc 2002). The DFA included only the elements that showed significant differences among water masses as determined in the ANOVA. In comparison to the PCA, the DFA was run on an expanded subset of otoliths that had readings of the elements which showed a water mass grouping effect regardless of whether abundances of the other elements could be estimated.

The classification accuracy of each DFA was evaluated by leave-one-out cross-validation, also called jack-knife reclassification (SPSS Inc 2002; White and Ruttenberg 2007). The prior probabilities of group membership were assumed to be uniform. We used the randomization test of White and Ruttenberg (2007) to estimate the probability ($p$-value) that the observed jack-knife reclassification success rate was drawn from a null distribution of jack-knife values ($n = 5{,}000$ randomizations) given no difference among groups of samples.

We examined relationships between element abundances in the otolith and relationships between otolith elemental properties and T and S. We used the Pearson correlation coefficient, r, to determine whether otolith element abundances that were identified as the best discriminators of the water mass groupings co-varied (SPSS, Inc 2002). We estimated the correlation between the average otolith element abundance at a station and the average T or S. If the correlation was significant, we used least squares regression to model the relationship between T or S, the environmental predictor, and otolith log[element/Ca], the dependent variable, by fitting a linear equation to the observed data (Systat Software Inc 2006).

# RESULTS

## Water masses

Based on the T-S observations, the assayed fish resided in a range of water masses across the study area (Fig. 3). We resolved four water mass groups: central California (CC), southern California (SC), uplifted deep water in the eddy (UE), and stratified water in the eddy (SE) (Fig. 3 and Table 1). The CC samples (blue T-S curves) were cooler (T < 11.7°C) than the SC water masses which were sampled outside of the eddy (green T-S curves). Both CC and SC water mass groups were less saline (S < 33.70 psu) than the three water mass samples (station 7, black curve; stations 9 and 14, red curves) within the eddy in the Santa Barbara Channel defined by dynamic height ϕ < 0.37 (Fig. 1b; Nishimoto and Washburn 2002) and one station at the southwestern boundary of the eddy (station 8, black curve).

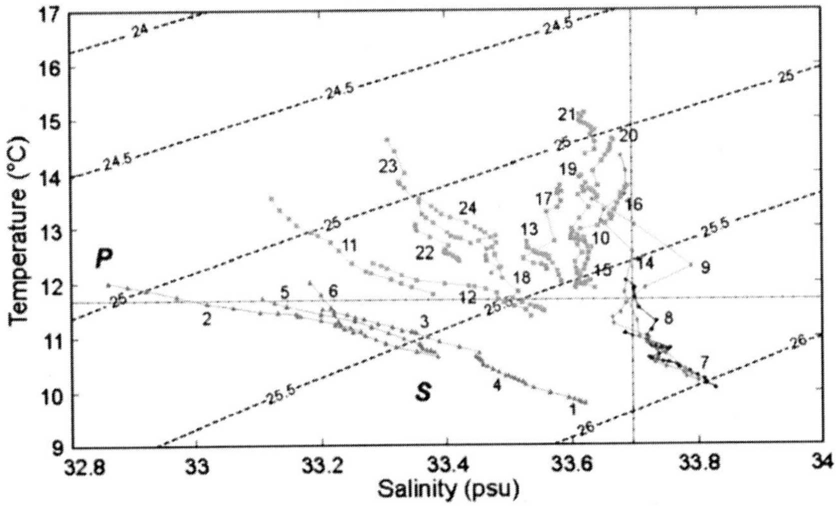


**Figure 3** Relationship of T-S profile data from the 14-m trawling stratum at stations where otoliths were sampled. The water mass samples were assigned to four groups: central California, CC, blue; uplifted deep water in the eddy, UE, black; stratified water mass in the eddy, SE, red; southern California, SC, green. The water mass groups are separated by grey horizontal and vertical reference lines at 11.7°C and 33.7 psu. Each T-S plot is labeled with a station number (see Fig. 1). Markers, *P* and *S*, in the plot are reference T-S of San Francisco Bay Plume water and subarctic water, respectively

The two water mass samples in the UE group (Fig. 3, black T-S curves, stations 7 and 8) have T and S properties within the trawling range of 19 m to 33 m that are consistent with deep water uplifted from 100 m within the eddy described by Nishimoto and Washburn 2002 (Fig. 2, $T < 11.7°C$ and $S > 33.7$ psu). The water mass samples in the SE group (Fig. 3, red T-S curves, stations 9 and 14), share T-S properties of uplifted deep water (UE group) and the most saline SC samples found outside of the eddy within the Santa Barbara Channel (Figs. 1b and 3, stations 10, 15, 19, 20).

The T-S properties of the central California samples are bounded by the characteristic T-S of two distinct end-members: San Francisco Bay Plume water (12.6, 32.85 psu, defined by Hurst and Bruland 2008) and subarctic water (mean temperature 10.06°C, mean salinity 33.36 psu, defined by Warn-Varnas et al. 2007) transported by the equatorward California Current (Fig. 3). The intermediate T-S properties of our central California samples from waters 50 m and shallower (stations 2, 3, 5 and 6; $d_0 \leq 36$ m) indicate advection and mixing of the two relatively fresh sources along the Central California coast. The two deeper CC samples (stations 1 and 4; $d_0 = 75$ and 77 m, respectively) and UE group are different "deep" water masses; although similar in temperature, the two CC samples are less saline than the UE group.

PCA of the T-S profile data was consistent with the distribution of water masses (i.e., T-S curves in Fig 3). The first two principal components, PC1 and PC2, explained 97% of the variance among the water mass samples (data not shown). PC1 separated CC samples from the southern California samples (SC, UE, and SE). PC2 separated the eddy samples, UE and SE, from SC. PC 3, which explained 2% of the variance, separated SE from UE, SC, and CC.

## Relationships between otolith signatures and water masses

Otolith concentrations of Mg, Ba, and Fe significantly varied among the four water masses (Fig. 4, Table 2). Mg/Ca was highest in otoliths from UE and significantly differed among the other three water masses (Tukey HSD test, $P < 0.001$). Ba concentration in the otoliths from the CC water mass was significantly higher than in otoliths from each of the water masses from the Santa Barbara Channel region (Tukey HSD test, $P < 0.001$). Fe concentration in otoliths from central California was lower than those from the UE (Tukey HSD test, $P = 0.009$), SE ($P = 0.024$, marginally significant), and SC water masses ($P < 0.001$). The estimated Fe concentrations of 30 otoliths from 18 stations, including all otoliths from central California, were below our calculated detection limit. Neither Ba nor Fe differed between water masses in the Santa Barbara Channel. The

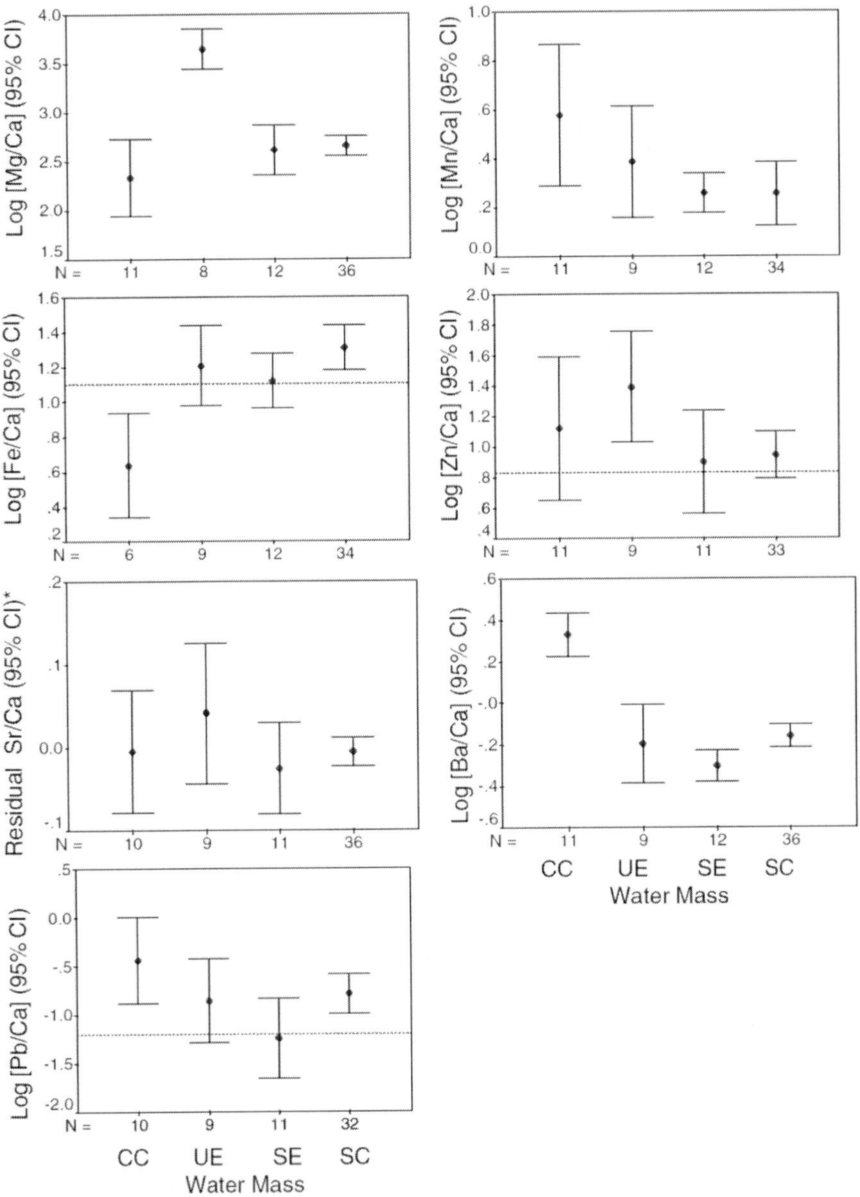


**Figure. 4** Average log concentration of Mg, Mn, Fe, Zn, Sr, Ba, and Pb (log[μmol element/mol Ca] ± 95% CI) in otoliths (N) from the four water masses: CC, UE, SE, and SC. The detection limit for an element/Ca is demarcated as a horizontal line if within the range of otolith element concentrations

**Table 2** ANOVA test of the effect of water mass group on element ratios in otoliths

| Element concentration | Source of variation | Sum of squares | DF | Mean square | F | P |
|---|---|---|---|---|---|---|
| log[Mg/Ca] | Groups | 8.803 | 3 | 2.934 | 21.59 | <0.001 |
| | Error | 8.562 | 63 | 0.136 | | |
| | Total | 17.365 | 66 | | | |
| log[Mn/Ca] | Groups | 0.948 | 3 | 0.316 | 2.67 | 0.06 |
| | Error | 7.328 | 62 | 0.118 | | |
| | Total | 8.276 | 65 | | | |
| log[Fe/Ca] | Groups | 2.389 | 3 | 0.796 | 7.35 | <0.001 |
| | Error | 6.177 | 57 | 0.108 | | |
| | Total | 8.565 | 60 | | | |
| log[Zn/Ca] | Groups | 1.704 | 3 | 0.568 | 2.22 | 0.10 |
| | Error | 15.350 | 60 | 0.256 | | |
| | Total | 17.054 | 63 | | | |
| log[Sr/Ca]* | Groups | 0.023 | 3 | 0.008 | 1.38 | 0.26 |
| | Error | 0.350 | 62 | 0.006 | | |
| | Total | 0.374 | 65 | | | |
| log[Ba/Ca] | Groups | 2.742 | 3 | 0.914 | 32.17 | <0.001 |
| | Error | 1.818 | 64 | 0.028 | | |
| | Total | 4.561 | 67 | | | |
| log[Pb/Ca] | Groups | 3.456 | 3 | 1.152 | 3.38 | 0.02 |
| | Error | 19.773 | 58 | 0.341 | | |
| | Total | 23.229 | 61 | | | |

Fish were significantly larger in UE than in CC (Tukey HSD test, $P = 0.047$) and SC ($P = 0.001$) (ANOVA, $n = 60$, $F = 4.665$, d. f. = 3, $P = 0.006$). Sr was the only element in fish otoliths that was related to fish size (Fig. 5; log[μmol Sr/mol Ca] = 3.63–0.25 log(mm SL), $n = 66$, $r^2 = -0.18$, $p = 0.0003$). We removed the confounding effect of fish size on otolith Sr by using the residual values from the regression of log(Sr/Ca) on log(SL) in the ANOVA and in the analyses reported herein.

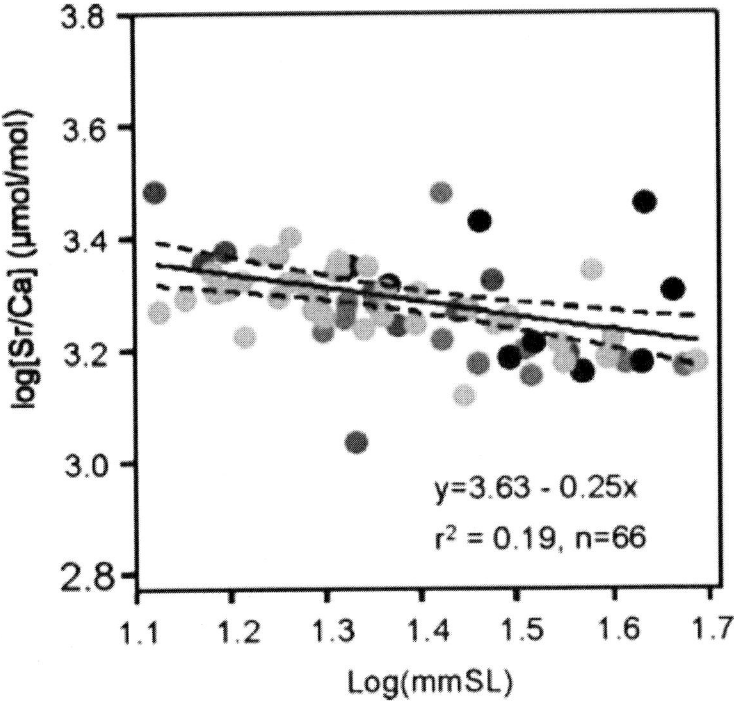


**Figure. 5** Otolith Sr v body length. Log(mm SL) was significantly but weakly correlated with Log [Sr/Ca]. Marker color indicates the water mass groups: CC, blue; UE, black; SE, red; SC, green

effect of water mass on Pb was marginally significant (ANOVA, $P = 0.024$); Pb in otoliths was higher in CC than in SE (Tukey HSD test, $p = 0.013$). Pb concentrations of 28 otoliths from 18 stations were below our calculated detection limit. The effect of water mass on the concentration of Mn, Zn, or Sr in otoliths was not significant.

Natural groupings of otolith signatures reflected the distinctiveness of the water masses where the fish resided. The first two principal components of PCA run on a subset of 49 otoliths with all seven elements explained 64% of the variance in elemental signatures among otoliths (Fig. 6). Inspection of the plot suggested that fish residing in central California had otolith signatures that differed from that of fish in southern California, and that fish residing in the UE water mass were distinct from fish from the SE and SC water masses. Separation of otolith signatures from CC and UE along PC1 are consistent with the lack of overlap in T and S properties of the water masses (Fig. 3). The lack of separation of otoliths from SE and SC (Fig. 4) is consistent with the intersection of the T-S properties of the two water masses for T > 11.6° C (Fig. 3). The separation of otoliths elemental signatures into water mass groupings in the PCA corresponds with the ANOVA findings of a water mass effect on the elemental concentrations of Mg, Ba, and Fe.

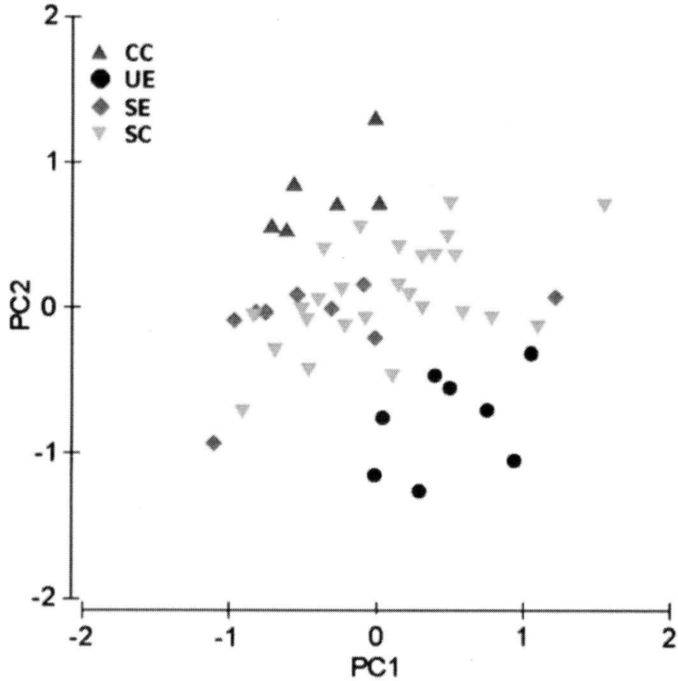


**Figure. 6** Principal components analysis for a subset of 49 otoliths using Mg, Mn, Fe, Zn, Sr, Ba, and Pb. Colored symbols indicate the water mass groups from where otoliths were sampled as in Fig. 3

We found that the otolith signature of three elements, Mg, Ba, and Fe was a good predictor of water mass residency. DFA was run on a subset of 60 otoliths that had readings of Mg, Ba, Fe, three elements that differed among water masses, regardless of whether the other four elements could be estimated (Fig. 7). Cross-validation showed that 70% of the otoliths correctly grouped into one of the four assigned water mass types (Table 3), significantly better than 25.3% (SD = 8.2% upper 95% quartile of the null distribution = 38.3% expected by chance ($p = 0.0002$)). The inclusion of Pb in the DFA (55 otoliths with Mg, Ba, Fe, and Pb) had an insignificant effect on the classification accuracy (plot of the DFA results not shown); cross-validation showed that 70.9% of the 4-element signatures correctly grouped into the assigned water masses.

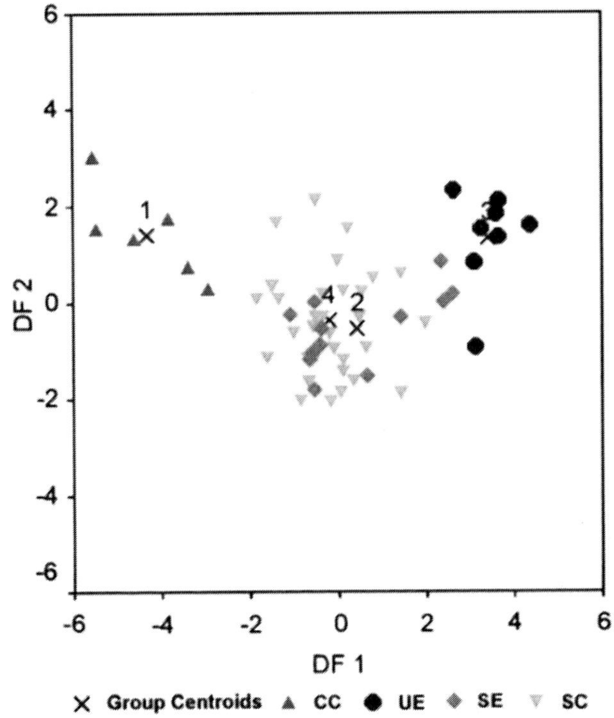


**Figure 7** Canonical Discriminant Analysis (CDA) of 60 otoliths using Mg, Ba, and Fe. Colored symbols indicate the water mass groups from where otoliths were sampled as in Fig. 6

**Table 3** Cross-validated classification of otoliths ($n = 60$) into the four water mass groups using Mg/Ca, Ba/Ca, and Fe/Ca in the canonical discriminant function

| **Predicted Group Membership** | | | | | |
|---|---|---|---|---|---|
| **Group** | **CC** | **UE** | **SE** | **SC** | **Total** |
| CC | 6 | 0 | 0 | 0 | 6 |
| UE | 0 | 8 | 0 | 0 | 8 |
| SE | 0 | 3 | 5 | 4 | 12 |
| SC | 0 | 0 | 11 | 23 | 34 |

The 3-element (Mg, Ba, and Fe) discriminant function correctly grouped 100% of the otoliths collected in CC, and 100% of the otoliths collected in UE corresponding with the distinct T-S properties of the two water masses

(Table 3). The signature poorly discriminated the residency of fish collected in SC and SE; however, none of the otoliths from these two groups were misclassified as CC otoliths. Fifty-eight percent of the otoliths from SE were incorrectly reassigned to UE (3 of 12 otoliths) and SC (4 of 12 otoliths). The lack of a distinctive SE otolith signature and the similarity of otolith signatures of fish collected from SE to otolith signatures from UE and SC (Figs. 4, 6 and 7) are consistent with the mixed water mass properties of SE (Fig. 3).

The strength in discriminating otoliths from the different water masses was associated with trends of decreasing otolith Ba and increasing otolith Mg with increasing S. Ba/Ca in the otoliths was negatively correlated with S and the regression explained 62% of the variance in Ba/Ca (Fig. 6d). Mg concentration was correlated with S although the regression explained only 25% of the variance in Mg/Ca (Fig. 8).

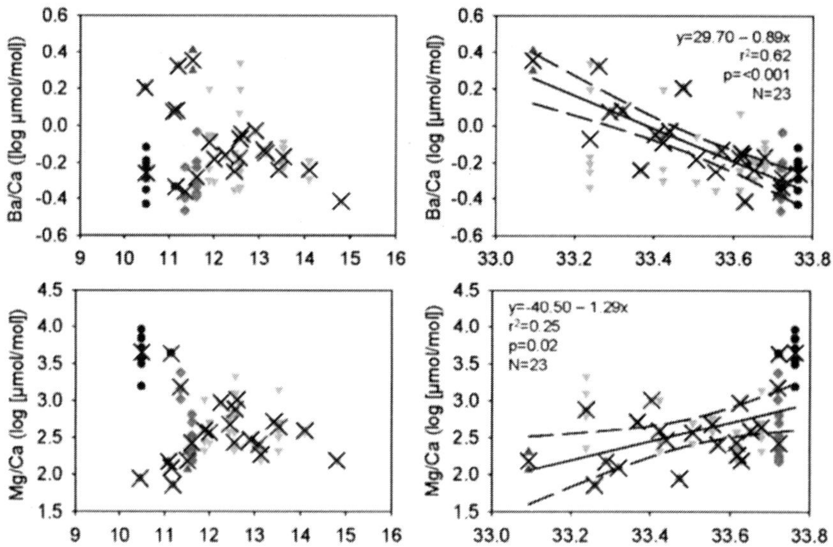

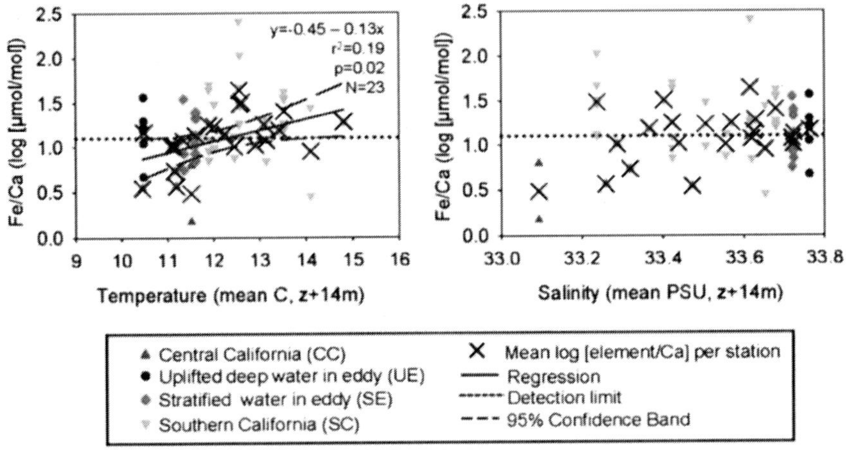


**Figure. 8** Least squares regression, y=(a±95%CI)x+(b±95%CI)y=(a±95%CI)x+ (b±95%CI), fit to the mean log otolith concentration of Ba, Mg, and Fe v mean T (black Xs in plots on left panel) and mean S (black Xs in plots on right panel) at each station ($n = 23$). Dashed lines are 95% confidence bands. Log[element/Ca] per otolith are plotted (colored symbols indicate the water masses from where otoliths were sampled as in Fig. 6). The detection limit for Fe/Ca, the dotted horizontal line in the plots, was within the range of otolith Fe concentrations

Ba and Mg concentrations in otoliths from UE and SE differed from those in otoliths from CC despite similar T between 10 and 12 C (Fig. 8). T was a weak predictor of Fe explaining only 19% of the variance (Fig. 8).

## DISCUSSION

Otolith element/Ca ratios varied among the sampled water masses in the California Current System. Examining the variability in otolith chemistry in relation to temperature and salinity allowed us to not only distinguish fish from geographically distant regions with contrasting oceanographic conditions but also to discriminate fish that resided in different water mass environments that were associated with a local, physical phenomenon —cyclonic eddy circulation in the Santa Barbara Channel. We found that Ba, Mg, and Fe varied significantly among the distinguishable water masses that we sampled. Fish from central California had elevated otolith Ba/Ca compared with fish from southern California. In contrast, otolith Fe/Ca was elevated in fish from southern California compared to fish from central California. Otolith Mg/Ca was highest in fish collected from deep water uplifted to shallow depths by cyclonic eddy circulation in the Santa Barbara Channel and lowest in fish from central California. The 3-element signature correctly predicted the residency of 100% of the fish collected in the cold core of the eddy in the Santa Barbara Channel and separated 100% of the fish collected in the Channel region from fish collected from Central California.

Two of the three elements, Ba/Ca and Mg/Ca, have been useful for distinguishing fish from different coastal and oceanic locations of other studies. Fe/Ca is typically excluded from otolith microchemistry analyses, because uncertainty in element abundance estimates is assumed to be high for signals below detection limits (Campana 1999). However, Ben-Tzvi et al. (2007) found that LA-ICPMS element/Ca measurements that would be discarded as noise using conventional DL standards (e.g., 3 × SD) often formed sequences that were symmetrical about the otolith core (palindrome). Considering their argument that these sub-DL signals can contain valuable information, we included sub-DL signals of Fe/Ca and other elements in our analyses.

We found that otolith Ba/Ca was negatively correlated with salinity of the water mass occupied by the wild-caught fish (Fig. 8). We did not measure elemental concentrations in water mass samples; however, we surmise that ambient Ba/Ca was negatively associated with salinity and that otolith Ba/Ca reflected the availability of Ba/Ca. Laboratory and field studies have determined that otolith Ba/Ca correlates with ambient Ba/Ca (Bath et al. 2000; Elsdon and Gillanders 2003, 2005b). Compared to the influence of ambient Ba/Ca, the effect of ambient temperature and salinity on otolith Ba/Ca has been shown to be minimal (Bath et al. 2000; Elsdon and Gillanders 2004, 2005b). Elsdon and Gillanders (2005a) attributed the inverse relationship between otolith Ba/Ca of black bream and ambient salinity in estuarine and coastal environments of southern Australia to the input of Ba from fluvial sediments in freshwater runoff. In the Southern California Bight, ambient Ba/Ca in estuarine waters (about 6 $\mu mol \cdot mol^{-1}$) were higher than coastal waters (about 3 to 5 $\mu mol \cdot mol^{-1}$) among several studies that were not conducted during our survey year (Forrester 2005; Warner et al. 2005; Lloyd et al. 2008). In a three-year study off central California, Brown (2006) found that whole otolith Ba/Ca of two flatfish species along central California was highest in 1998 (the year of our collections) and lowest in 1999, in correspondence with runoff levels.

River runoff is the major source of Ba to coastal waters (Chan et al. 1977) and may have contributed substantially to the Ba load off Central California during our study. Increased precipitation in California and coastal runoff are frequently associated with strong El Niño events, and the winter of 1997–1998, when an exceptionally strong El Niño was fully developed, was one of the wettest seasons on record. The impact of sediment and freshwater transport in nearshore waters from coastal watersheds was spatially extensive in satellite imagery during the winter and spring of 1998 (Mertes and Warrick 2001). Oceanographic surveying with the collection of the central California specimens indicate that outflow from the San Francisco Bay, draining the largest watershed of California, generated a markedly low salinity signal that extended 100 km down the central California coast to Monterey Bay (Sakuma et al. 2000). High turbidity and elevated nutrient levels were correlated with low salinity in Monterey Bay providing evidence that low salinity in the upper 15–20 m was the result of local and remote runoff and river outflow in 1998 (Friederich et al. 2002). The concentrations of some trace elements, Ba was not among the elements measured, have been found to be higher in the San Francisco Bay plume than in coastal and upwelled waters (Hurst and Bruland 2008).

Elevated Ba/Ca in otoliths and invertebrate hard parts has been associated with upwelling (Lea et al. 1989; Patterson et al. 1999; Zacherl 2005; Brown 2006) and deep water (Zumholz et al. 2007). Dissolved Ba in oceanic waters shows a nutrient-like distribution with very low concentrations in surface waters and higher values in deep water (Chan et al. 1977). In the Santa Barbara Channel, upwelling within the cyclonic eddy is a mechanism that can draw up nutrient-rich waters from below the thermocline into shallower depths where the juvenile fish were sampled; however, otolith Ba/Ca within the cyclonic eddy was not significantly higher than that outside of the eddy (Fig. 7). Bograd and Lynn (2001) regarded the depth of the 12°C isotherm a reliable proxy for the nutricline depth in the Southern California Bight and Santa Barbara Channel. Figure 3 suggests that the nutricline was uplifted within the core of the eddy; however, it is plausible that variation in ambient Ba did not significantly affect otolith Ba among the three water masses in southern California. Warner et al. (2005) found no correspondence between the concentrations of Ba in seawater and Ba in otoliths of adult or embryonic larval kelp rockfish (*Sebastes atrovirens*) from nearshore reefs within our southern California study region.

An alternate, plausible mechanism that would limit otolith Ba/Ca in the fish from the core of the eddy is localized Ba depletion from rapid plankton production in the euphotic zone where the fish were residing. Lochte and Pfannkuche (1987) observed nutrient depletion in the upper 50 m within the center of a cyclonic eddy. They proposed that the physical isolation of eddy surface waters due to the formation of a shallow thermocline lead to rapid utilization of nutrients. Off the Baja California coast, the waning stages of an upwelling-induced phytoplankton bloom were associated with Ba depletion in surface waters (Esser and Volpe 2002). The relationship between Ba removal from surface seawater, uptake by different organisms, and biological productivity are poorly understood and beyond the scope of this discussion.

In contrast to otolith Ba/Ca, Mg/Ca was relatively low in otoliths of fish from central California, where the waters were freshest in our study area, and highest in otoliths of fish residing in the core of the eddy, where waters were the most saline (Figs. 4 and 8). Unlike Ba which has a nutrient-like vertical distribution in the ocean, Mg, like Ca, has generally a conservative distribution and is present at relatively high concentration levels in constant proportion to salinity (Nozaki 1997). Ratios of conservative elements, Mg/Ca for example, are not expected to vary significantly in seawater (Lea et al. 1999). Rivers contribute Mg to coastal areas (Tipper et al. 2006); however, it is likely that the input of Mg from freshwater runoff would be swamped out by the concentration of Mg in seawater, which is about five times higher than Ca in seawater. Substantial differences in ambient Mg/Ca among the water masses we sampled seem unlikely, so variation in otolith Mg/Ca probably is due to other factors.

A large body of studies shows that Mg/Ca is useful for discriminating fish from different open coastal and oceanic locations exposed to water masses. For example, in nearshore coastal waters along Washington and Oregon, Miller and Shanks (2004) found that Mg/Ca was the most important discriminator among a suite of elements for assigning juvenile rockfish to their collection locations. Mg is one of the most important elements in otoliths separating fish from deepwater areas. Swan et al. (2006) found that Mg along with Pb were important for differentiating deep water fish stocks of the European hake, *Merluccius merluccius*, in the northeastern Atlantic. The otolith chemistry results were consistent with the treatment of a northern stock west of Norway and a southern stock off Portugal that were separated by an oceanographic and bathymetric feature, the Bay of Biscay. Ashford et al. (2005) used Mg along with Ba, Mn, and Sr in otoliths of deep-water Patagonian toothfish to successfully distinguish fish collected off South America from fish collected south of the subantarctic front in the Atlantic, Indian, and Pacific regions of the Southern Ocean.

The effects of ambient Mg/Ca on otolith Mg/Ca and the effects of salinity and temperature on Mg uptake are unclear. We found that otolith Mg/Ca was weakly correlated with S and not related to T. Most of the experimental studies have found no effect of salinity on otolith Mg/Ca (Fowler et al. 1995b; Hoff and Fuiman 1995; Martin and Thorrold 2005). Fowler et al. (1995a) found that Mg and Ca were more abundant in the whole otoliths of fish in low-temperature treatments. In contrast, in experimental manipulations of salinity and temperature, Martin and Thorrold (2005) observed that the uptake rate of Mg in otoliths increased with increasing temperature (although the relationship was not statistically significant) while ambient Mg/Ca did not vary among the treatments. With exception of Martin and Thorrold (2005), the limitation of most of the experimental studies was that dilution of a single natural water mass or artificial seawater was used to manipulate salinity, and neither ambient Mg/Ca nor the effect of dilution on ambient Mg/Ca were measured in the majority of studies. The effect of ambient Mg/Ca on otolith Mg/Ca and possible interactions with ambient temperature and salinity is yet to be rigorously tested.

It is arguable whether ambient Mg/Ca actually has an effect on otolith Mg/Ca, because Mg is under strict osmoregulatory control (Campana 1999). Elsdon and Gillanders (2002) found that Mg/Ca showed little or no effects due to temperature or salinity and varied considerably among individuals within temperature and salinity treatments. Endogenous controls would vary among individuals due to genetics, ontogeny, condition, metabolism, and somatic growth. Mg/Ca was highest in otoliths from UE where fish were larger than fish from CC or SC; however, Mg/Ca tended to be relatively low in otoliths from CC where fish were not the smallest among water masses. Ontogenetic differences may account for some of the variance in Mg/Ca observed among water masses. Fowler et al. (1995a, b) found that Mg in whole otoliths varied with ontogeny. Martin and Thorrold (2005) showed that uptake rate of Mg in otoliths and otolith Mg/Ca was related to otolith precipitation and somatic growth, which is often highly correlated with temperature in the field. If endogenous control of otolith Mg/Ca is not measurably mediated by ambient Mg, temperature, and/or salinity from the environment, then otolith Mg/Ca may not be a reliable indicator of water mass residency, but rather a useful discriminator for the separation of populations that have unique physiological traits meditated by genetics, growth, and condition.

It is tempting to speculate that otolith Mg/Ca was a good discriminator of the fish residing in the different water masses that we sampled, because the population of young-of-year fish within each water mass possessed unique intrinsic characteristics. Oceanographic processes that determined circulation patterns along with behaviors such as swimming capability, temperature or depth preferences, and schooling may have contributed to the isolation of these pelagic juvenile fish populations.

In conclusion, we found that otolith chemistry was a good predictor of water mass residency. Variability in otolith chemistry reflected the spatial patterns of both horizontal gradients and vertical gradients in water mass properties related to circulation. Neither ambient temperature nor salinity must necessarily relate to otolith chemistry in order to be useful for discriminating fish from different water masses; rather, we suggest that temperature and salinity are useful as identifiers of distinct water masses that are associated with unique otolith elemental signature in the residing fish. We found it sufficient to use only temperature and salinity to discriminate among water masses; however, utilizing other parameters such as $O_2$ and nutrients (often measured in oceanographic surveys) will likely improve our ability to identify water masses and refine future investigations of this kind. If shown to be reliable (i.e., temporally consistent), then the development of a reference or atlas of water mass-specific otolith elemental signatures is promising. Potentially, the use of an atlas of water mass-specific otolith signatures with water mass monitoring and current mapping could greatly reduce the uncertainty surrounding reconstructions of larval and adult movements and connectivity estimates among locations such as spawning and nursery areas and marine reserves.

## ACKNOWLEDGEMENTS

This work was funded by NOAA Grant # NA06RG0142, California Sea Grant College Program Project # R/CZ-178, through NOAA's National Sea Grant College Program, U.S. Dept. of Commerce. The microchemistry facility was supported in part by the Partnership for the Interdisciplinary Study of Coastal Oceans funded by the David and Lucile Packard Foundation. California Artificial Reef Enhancement Foundation provided additional matching funds. MMN received support from the U.S. Minerals Management Service. We thank D. Lea for providing ICPMS laboratory resources, and M. Sheehy for guidance on otolith microchemistry procedures. We thank K. Sakuma for otolith samples and the catch and oceanographic data from NOAA's central California surveys (NMFS/SWFSC, Santa Cruz, CA). Funding for the southern California trawling survey was provided by the Biological Resources Division of the U.S. Geological Survey under cooperative agreement 1445-CA09-95-0836. The California Department of Fish and Game provided the midwater trawl net. E. Cheng and A. Muto, high school interns in the UCSB Research Mentorship Summer Program, assisted in the laboratory. We thank S. Ralston, M. Carr, and anonymous reviewers for their useful comments on the manuscript. The statements, findings, conclusions and recommendations are those of the authors and do not necessarily reflect the views of California Sea Grant or the U.S. Dept. of Commerce.

## REFERENCES

1. Ashford JR, Jones CM, Hofmann E, Everson I, Moreno C, Duhamel G, Williams R (2005) Can otolith elemental signatures record the capture site of Patagonian toothfish (*Dissostichus eleginoides*), a fully marine fish in the Southern Ocean? Can J Fish Aquat Sci 62:2832–2840

2. Bath G, Thorrold SR, Jones CM, Campana SE, McLaren JW, Lam JWH (2000) Strontium and barium uptake in aragonitic otoliths of marine fish. Geochim Cosmochim Acta 64:1705–1714
3. Ben-Tzvi O, Abelson A, Gaines SD, Sheehy MS, Paradis GL, Kiflawi M (2007) The inclusion of sub-detection limit LAICPMS data, in the analysis of otolith microchemistry, by use of a palindrome sequence analysis (PaSA). Limnol Oceanogr Methods 5:97–105
4. Bograd SJ, Lynn RJ (2001) Physical-biological coupling in the California current during the 1997–99 El Niño-La Niña cycle. Geophys Res Lett 28:275–278
5. Brown JA (2006) Classification of juvenile flatfishes to estuarine and coastal habitats based on the elemental composition of otoliths. Estuar Coast Shelf Sci 66:594–611
6. Campana SE (1999) Chemistry and composition of fish otoliths: pathways, mechanisms and applications. Mar Ecol Prog Ser 188:263–297
7. Campana SE, Chouinard GA, Hanson JM, Frechet A, Brattey J (2000) Otolith elemental fingerprints as biological tracers of fish stocks. Fish Res 46:343–357
8. Chan LH, Drummond D, Edmond JM, Grant B (1977) On the barium data from the Atlantic GEOSECS expedition. Deep Sea Res 24(7):613–649
9. Dove SG, Kingsford MJ (1998) Use of otoliths and eye lenses for measuring trace-metal incorporation in fishes: a biogeographic study. Mar Biol 130:377–387
10. Elsdon TS, Gillanders BM (2002) Interactive effects of temperature and salinity on otolith chemistry: challenges for determining environmental histories of fish. Can J Fish Aquat Sci 59:1796–1808
11. Elsdon TS, Gillanders BM (2003) Relationship between water and otolith elemental concentrations in juvenile black bream *Acanthopagrus butcheri*. Mar Ecol Prog Ser 260:263–272
12. Elsdon TS, Gillanders BM (2004) Fish otolith chemistry influenced by exposure to multiple environmental variables. J Exp Mar Biol Ecol 313:269–284
13. Elsdon TS, Gillanders BM (2005a) Alternative life-history patterns of estuarine fish: barium in otoliths elucidates freshwater residency. Can J Fish Aquat Sci 62:1143–1152
14. Elsdon TS, Gillanders BM (2005b) Consistency of patterns between laboratory experiments and field collected fish in otolith chemistry: an example and applications for salinity reconstructions. Mar Freshw Res 56:609–617
15. Esser BK, Volpe AM (2002) At-sea high-resolution chemical mapping: extreme barium depletion in North Pacific surface water. Mar Chem 79:67–79
16. FitzGerald JL, Thorrold SR, Bailey KM, Brown AL, Severin KP (2004) Elemental signatures in otoliths of larval walleye pollock (*Theragra chalcogramma*) from the northeast Pacific Ocean. Fish Bull 102:604–616
17. Forrester GE (2005) A field experiment testing for correspondence between trace elements in otoliths and the environment and for evidence of adaptation to prior habitats. Estuaries 28:974–981
18. Fowler AJ, Campana SE, Jones CM, Thorrold SR (1995a) Experimental assessment of the effect of temperature and salinity on the elemental composition of otoliths using solution-based ICPMS. Can J Fish Aquat Sci 52:1421–1430
19. Fowler AJ, Campana SE, Jones CM, Thorrold SR (1995b) Experimental assessment of the effect of temperature and salinity on the elemental composition of fish otoliths using laser ablation ICPMS. Can J Fish Aquat Sci 52:1431–1441
20. Friederich G, Walz P, Burczynski M, Chavez FP (2002) Inorganic carbon in the central California upwelling system during the 1997–1999 El Nino-La Nina event. Prog Oceanogr 54:185–203
21. Gillanders BM (2002) Temporal and spatial variability in elemental composition of otoliths: implications for determining stock identity and connectivity of populations. Can J Fish Aquat Sci 59(4):669–679
22. Hamer PA, Jenkins GP, Gillanders BM (2003) Otolith chemistry of juvenile snapper *Pagrus auratus* in Victorian waters: natural chemical tags and their temporal variation. Mar Ecol Prog Ser 263:261–273

23. Hamilton SL, Regetz J, Warner RR (2008) Postsettlement survival linked to larval life in a marine fish. Proc Nat Acad Sci USA 105:1561–1566
24. Hoff GR, Fuiman LA (1995) Environmentally induced variation in elemental composition of red drum (*Scianops ocellatus*) otoliths. Bull Mar Sci 56:578–591
25. Hurst MP, Bruland KW (2008) The effects of the San Francisco Bay plume on trace metal and nutrient distributions in the Gulf of the Farallones. Geochim Cosmochim Acta 72:395–411
26. Lea DW, Shen GT, Boyle EA (1989) Coralline barium records temporal variability in equatorial Pacific upwelling. Nature 340:373–376
27. Lea DW, Mashiotta TA, Spero HJ (1999) Controls on magnesium and strontium uptake in planktonic foraminifera determined by live culturing. Geochim Cosmochim Acta 63:2369–2379
28. Lloyd DC, Zacherl DC, Walker S, Paradis G, Sheehy M, Warner RR (2008) Egg source, temperature and culture seawater affect elemental signatures in *Kelletia kelletii* larval statoliths. Mar Ecol Prog Ser 353:115–130
29. Lochte K, Pfannkuche O (1987) Cyclonic cold-core eddy in the eastern North Atlantic. II. Nutrients, phytoplankton and bacterioplankton. Mar Ecol Prog Ser 39:153–164
30. Love MS, Yoklavich M, Thorsteinson L (2002) The rockfishes of the northeast Pacific. University of California Press, Berkeley, p 404
31. Martin GB, Thorrold SR (2005) Temperature and salinity effects on magnesium, manganese, and barium incorporation in otoliths of larval and early juvenile spot *Leiostomus xanthurus*. Mar Ecol Prog Ser 293:223–232
32. Mertes LAK, Warrick JA (2001) Measuring flood output from 110 coastal watersheds in California with field measurements and SeaWiFS. Geology 29:659–662
33. Miller JA, Shanks AL (2004) Evidence for limited larval dispersal in black rockfish (*Sebastes melanops*): implications for population structure and marine reserve design. Can J Fish Aquat Sci 61:1723–1735
34. Milton DA, Chenery SR, Farmer MJ, Blaber SJM (1997) Identifying the spawning estuaries of the tropical shad, terubok *Tenualosa toli*, using otolith microchemistry. Mar Ecol Prog Ser 153:283–291
35. Nishimoto MM, Washburn L (2002) Patterns of coastal eddy circulation and abundance of pelagic juvenile fish in the Santa Barbara Channel, California, USA. Mar Ecol Prog Ser 241:183–199
36. Nozaki Y (1997) A fresh look at element distribution in the North Pacific Ocean. EOS Trans Am Geophys Union 78(21):221
37. Patterson HM, Thorrold SR, Shenker JM (1999) Analysis of otolith chemistry in Nassau grouper (*Epinephelus striatus*) from the Bahamas and Belize using solution-based ICP-MS. Coral Reefs 18:171–178
38. Pearce NJG, Perkins WT, Westgate JA, Gorton MJ, Jackson SE, Neal CR, Chenery SP (1997) A compilation of new and published major and trace element data for NIST SRM 610 and NIST SRM 612 glass reference materials. Geostandard Newslett 21:115–144
39. Ruttenberg BI, Hamilton SL, Hickford MJH, Paradis GL et al (2005) Elevated levels of trace elements in cores of otoliths and their potential for use as natural tags. Mar Ecol Prog Ser 297:273–281
40. Sakuma KM, Schwing FB, Pickett MH, Roberts D, Ralston S (2000) The physical oceanography off the central California coast during May–June 1998: a summary of CTD data from pelagic juvenile rockfish surveys. US Dep Commer NOAA Tech Memo, NOAA-TM-NMFS-SWFSC 297:1–77
41. Secor DH, Henderson-Arzapalo A, Piccoli PM (1995) Can otolith microchemistry chart patterns of migration and habitat utilization in anadromous fishes? J Exp Mar Biol Ecol 192:15–33
42. SPSS Inc (2002) SPSS for Windows. Release 11.5.0

43. Standish JD, Sheehy M, Warner RR (2008) Use of otolith natal elemental signatures as natural tags to evaluate connectivity among open-coast fish populations. Mar Ecol Prog Ser 356:259–268
44. Swan SC, Geffen AJ, Morales-Nin B, Gordon JDM, Shimmield T, Sawyer T, Massutı E (2006) Otolith chemistry: an aid to stock separation of *Helicolenus dactylopterus* (bluemouth) and *Merluccius merluccius* (European hake) in the Northeast Atlantic and Mediterranean. ICES J Mar Sci 63:504–513
45. Swearer SE, Caselle JE, Lea DW, Warner RR (1999) Larval retention and recruitment in an island population of a coral reef fish. Nature 402:799–802
46. Systat Software Inc (2006) SigmaPlot for Windows Version 10.0
47. Thorrold SR, Latkoczy C, Swart PK, Jones CM (2001) Natal homing in a marine fish metapopulation. Science 291:297–299
48. Tipper ET, Galy A, Gaillardet J, Bickle MJ, Elderfield H, Carder EA (2006) The magnesium isotope budget of the modern ocean: constraints from riverine magnesium isotope ratios. Earth Planet Sci Lett 250:241–253
49. Warner RR, Swearer SE, Caselle JE, Sheehy M, Paradis G (2005) Natal trace-elemental signatures in the otoliths of an open-coast fish. Limnol Oceanogr 50:1529–1542
50. Warn-Varnas A, Gangopadhyay A, Hawkins JA (2007) Water masses in the Monterey Bay during the summer of 2000. Cont Shelf Res 27:1379–1398
51. White JW, Ruttenberg BI (2007) Discriminant function analysis in marine ecology: some common oversights and their solutions. Mar Ecol Prog Ser 329:301–305
52. Wyllie Echeverria T, Lenarz WH, Reilly CA (1990) Survey of the abundance and distribution of pelagic young-of-the-year rockfishes, *Sebastes*, off central California. US Dep Commer NOAA Tech Memo, NOAA-TM-NMFS-SWFSC 147:1–125
53. Zacherl DC (2005) Spatial and temporal variation in statolith and protoconch trace elements as natural tags to track larval dispersal. Mar Ecol Prog Ser 290:145–163
54. Zacherl DC, Paradis G, Lea D (2003) Ba and Sr uptake into larval protoconchs and statoliths of the marine neogastropod *Kelletia kelletii*. Geochim Cosmochim Acta 67:4091–4099
55. Zumholz K, Klugel A, Hansteen T, Piatkowski U (2007) Statolith microchemistry traces the environmental history of the boreoatlantic armhook squid *Gonatus fabricii*. Mar Ecol Prog Ser 333:195–204

# Index